Lecture Notes in Artificial Intelligence   1701

Subseries of Lecture Notes in Computer Science
Edited by J. G. Carbonell and J. Siekmann

Lecture Notes in Computer Science

Edited by G. Goos, J. Hartmanis and J. van Leeuwen

Springer
Berlin
Heidelberg
New York
Barcelona
Hong Kong
London
Milan
Paris
Singapore
Tokyo

Wolfram Burgard   Thomas Christaller
Armin B. Cremers (Eds.)

KI-99: Advances in Artificial Intelligence

23rd Annual German Coference
on Artificial Intelligence
Bonn, Germany, September 13-15, 1999
Proceedings

 Springer

Series Editors

Jaime G. Carbonell, Carnegie Mellon University, Pittsburgh, PA, USA
Jörg Siekmann, University of Saarland, Saarbrücken, Germany

Volume Editors

Wolfram Burgard
Armin B. Cremers
Universität Bonn, Institut für Informatik III
Römerstraße 164, D-53117 Bonn, Germany
E-mail: {wolfram/abc}@cs.uni-bonn.de

Thomas Christaller
GMD - Forschungszentrum Informationstechnik GmbH
Schloß Birlinghoven, D-53754 Sankt Augustin, Germany
E-mail: thomas.christaller@gmd.de

Cataloging-in-Publication data applied for

Die Deutsche Bibliothek - CIP-Einheitsaufnahme

Advances in artificial intelligence : proceedings / KI-99, 23rd Annual German
Conference on Artificial Intelligence, Bonn, Germany, September 13 - 15, 1999.
Wolfram Burgard ... (ed.). - Berlin ; Heidelberg ; New York ; Barcelona ; Hong
Kong ; London ; Milan ; Paris ; Singapore ; Tokyo : Springer, 1999
(Lecture notes in computer science ; Vol. 1701 : Lecture notes in artificial
intelligence)
ISBN 3-540-66495-5

CR Subject Classification (1998): I.2, I.4, I.5

ISBN 3-540-66495-5 Springer-Verlag Berlin Heidelberg New York

© Springer-Verlag Berlin Heidelberg 1999
Printed in Germany

Typesetting: Camera-ready by author
SPIN 10705034   06/3142 – 5 4 3 2 1 0   Printed on acid-free paper

Preface

For many years, Artificial Intelligence technology has served in a great variety of successful applications. AI research and researchers have contributed much to the vision of the so-called Information Society. As early as the 1980s, some of us imagined distributed knowledge bases containing the explicable knowledge of a company or any other organization. Today, such systems are becoming reality. In the process, other technologies have had to be developed and AI-technology has blended with them, and companies are now sensitive to this topic.

The Internet and WWW have provided the global infrastructure, while at the same time companies have become global in nearly every aspect of enterprise. This process has just started, a little experience has been gained, and therefore it is tempting to reflect and try to forecast, what the next steps may be. This has given us one of the two main topics of the 23rd Annual German Conference on Artificial Intelligence (KI-99) held at the University of Bonn: The Knowledge Society. Two of our invited speakers, Helmut Willke, Bielefeld, and Hans-Peter Kriegel, Munich, dwell on different aspects with different perspectives. Helmut Willke deals with the concept of virtual organizations, while Hans-Peter Kriegel applies data mining concepts to pattern recognition tasks. The three application forums are also part of the Knowledge Society topic: "IT-based innovation for environment and development", "Knowledge management in enterprises", and "Knowledge management in village and city planning of the information society".

But what is going on in AI as a science? Good progress has been made in many established subfields such as Knowledge Representation, Learning, Logic, etc. The KI-99 technical program includes 15 full and 6 short papers out of 32 received; together with the workshop program and poster sessions, we feel the conference reflects the steady growth of the field and represents the forefront of AI research. It was a pleasure to work with the program committee, and especially to choose one of the accepted papers for the Best Paper Award donated by Springer-Verlag. This year, the prize goes to Bernhard Nebel, Freiburg, for his paper "Compilation Schemes: A Theoretical Tool for Assessing the Expressive Power of Planning Formalisms", which will also be published in the special issue of the "KI" journal for this conference.

But within the steady flow of research, new trends emerge from time to time, and an important one, which we as organizers felt should get special attention at KI-99, is the rediscovery of robotics or, more precisely, of robots for AI. We put this interest under the second main topic of the conference: Cognitive Robotics. Three invited speakers present their perspectives: Sebastian Thrun, Pittsburgh, on learning and probabilistic reasoning; Rolf Pfeifer, Zurich, on a general methodology for behavior-oriented AI; Hans-Hellmut Nagel, Karlsruhe, on image processing.

Another flavor of the KI-99 conference is the sharing of a part of the program with the pattern recognition conference DAGM-99 in the form of a joint invited talk by Takashi Matsuyama on cooperative distributed vision, as well as a common technical session, and a joint technical exhibition.

As applications and scientific questions evolve, so does this conference. We have introduced a new section into the program and into these proceedings, in which short papers are presented that deserve a technical presentation. This provides faster access to ideas and concepts than with the accepted long papers. Another reinvention is that of student posters, which give students the opportunity to present and discuss their work with the scientific community.

The annual AI conference in Germany now has a long tradition, evolving together with the maturing field of AI, reflecting every year a snapshot of the quality of research and development. Over the years, the KI conference has turned out to be one of the largest national Computer Science conferences in Germany. Yet, the trend to specialization and therefore to fragmentation and internationalization of AI raises the question: Do we need a regional event like this, adding another item to the market of proceedings? We as organizers have answered this question with a yes!, together with all who have contributed in one way or another to make this conference and this new volume of LNCS a success.

Many thanks to Christine Harms and Manfred Domke. They made sure that the details were taken care of and the deadlines were met. Many students from the Institute on Autonomous intelligent Systems at GMD and the Computer Science Department at the University of Bonn together with our secretaries Marie-Luise Liebegut and Myriam Jourdan helped us in preparing this conference. Special thanks to Sylvia Johnigk, who helped produce the complete proceedings in LaTeX.

July 1999

Wolfram Burgard
Thomas Christaller
Armin B. Cremers

23rd Annual German Conference on Artificial Intelligence (KI-99)

Conference Chairs

Wolfram Burgard, Bonn

Thomas Christaller, Sankt Augustin

Armin B. Cremers , Bonn

Program Committee Chair

Thomas Christaller, Sankt Augustin

Program Committee Members

Elisabeth André, Saarbrücken
Gerhard Brewka, Leipzig
Hans-Dieter Burkhard, Berlin
Armin B. Cremers, Bonn
Rüdiger Dillmann, Karlsruhe
Ulrich Furbach, Koblenz
Joachim Hertzberg, Sankt Augustin
Wolfgang Hoeppner, Duisburg
Alois C. Knoll, Bielefeld
Gisbert Lawitzky, München
Katharina Morik, Dortmund
Gunter Schlageter, Hagen

Clemens Beckstein, Jena
Wolfram Burgard, Bonn
Thomas Christaller, Sankt Augustin,
Kerstin Dautenhahn, Reading, UK
Wolfgang Förstner, Bonn
Horst-Michael Groß, Ilmenau
Otthein Herzog, Bremen
Knut Hinkelmann, Kaiserslautern
Jana Köhler, Freiburg
Hanspeter A. Mallot, Tübingen
Bernd Neumann, Hamburg
Stefan Wrobel, Magdeburg

Workshop Chair

Norbert Reithinger, Saarbrücken

Exhibition Chairs

Karlheinz Schunk, Sankt Augustin

Volker Steinhage, Bonn

List of Additional Referees

Baumgartner, Peter
Dahn, Ingo
Deister, Angelika
Güsgen, Hans
Hannebauer, Markus
Kühnel, Ralf
Lenz, Marie
Meyer auf'm Hofe, Harald
Müller, Jörg
Paaß, Gerhard
Plümer, Lutz
Stolzenburg, Frieder
Tobies, Stephan
Wendler, Jan
Wolter, Frank

Table of Contents

Section 3

Section 4

Section 5

Short Papers

From AI to Systemic Knowledge Management

Helmut Willke

Fakultät für Soziologie,
Universität Bielefeld, Postfach 100 131, D - 33501 Bielefeld
helmut.willke@post.uni-bielefeld.de

1 Introduction

AI has suffered severe drawbacks because it has been (mostly) unable as a concept and as an intellectual endeavor, to free itself from the narrow reigns of a rather restricted and psychologically and sociologically uneducated hard sciences" arrogance. AI has (mostly; for exceptions see Dreyfus/Dreyfus 1988) ignored the linguistic turn" in the hermeneutic sciences, and it has completely ignored the ongoing cognitivistic turn" in the sciences that are mostly interested in putting to use the vastly exaggerated promises of AI: management sciences, innovation management, business consulting, and knowledge management (see Ryan 1991).

So the time has come, it seems, to advance to a more integrated view on intelligence, artificial or natural, that is embedded and embodied in human and social systems whenever and before it is put to any use. At the same time there is an eminent and urgent need for innovative forms of embedded intelligence, of explicit knowledge, and of genuine organizational expertise. Approaching the new millenium, we are witnessing a tectonic shift from the traditional factors of production, that is land, capital, and labor, to a self-reinforcing predominance of a fourth factor: knowledge. We are just beginning to understand some parts of the script for the transition from industrial society via the information society to the knowledge society. We expect a painful transition of traditional industrial work to knowledge work, including major challenges for social policies; we expect equally demanding metamorphoses of tayloristic industrial organizations to intelligent firms" (Quinn 1992); and there are serious signs for an ubiquitious outphasing of dumb" products in favor of intelligent goods", products and services, that is goods whose value for the customer lies exclusively in their embedded intelligence - software, pharmaceuticals, communication satellites, health care goods, entertainment, traffic telematic systems, global communication services, etc. etc. The problem is that we still are almost completely ignorant of the complicated and complex interplay of all these factors, including the consequences of an explosive global communication infrastructure and equally unexpected changes in suprastructures (governance systems).

The hypocrisy of traditional hard science" AI has been brutally exposed by the fate of Long Term Hedge Fund". Founded by two extremely gifted model builders with an unbridled faith in the power of AI, and ennobled by Nobel prizes, the fund had attracted gigantic amounts of speculative money, before crashing in grand style. The economic and sociological naiveness of its managers had lead to a systemic risk", inducing an imminent breakdown of parts of the global financial system that was avoided only through costly interventions of the Federal Reserve and a number affiliated foreign banks. The net loss of the breakdown, Nobel prizes notwithstanding, has been around $ 5.5 billion: LTHF posed unacceptable risks to the American economy" (Wiliam McDonough, president of the New York Federal Reserve, Financial Times, October 17/18, 1998, p. 6).

Knowledge work is not a new type of work, but a form, which through changing organizational, economic and social contexts is turning from a special case into a common case. Almost all specialized work, and in particular the classic professional occupations (doctors, lawyers, teachers, academics) is knowledge work in the sense that it is based on the specialized expertise of individuals who have had to go through a laborious training process. Nevertheless, knowledge work is gaining a new relevance due to the interrelated dynamics of the Knowledge Society(Drucker 1994),the Intelligent Enterprises (Quinn 1992) and Symbolic Analysis (Reich 1991) which, at the end of this century, are causing a maelstrom in the developed industrial societies, changing the foundations of work itself.

Knowledge work has recently become a sociological topic because, within the context of a knowledge society, it has changed from a person-based occupation or from an activity of persons into an activity which touches upon an elaborate interplay of the personal and organizational elements of the knowledge base. It has permeated from professional businesses and laboratories, through to factories and offices. Studies on knowledge work don't ask (as do laboratory studies and knowledge engineering) about the construction of knowledge in the context of work, but about the reconstruction of work based on wide-spread and indispensable expertise. Rather like the workplace studies (Knoblauch 1996:352ff; Suchman 1993), studies on knowledge work question the forms of work within coordination settings, focusing less on the use of highly developed technology and more on the application and generation of knowledge within the context of intelligent organizations.

Theory and practice of knowledge work inexorably pose the problem of coping with complexity. The specific approach of modern sociological systems theory to researching and reconstructing knowledge management is to resist any reductionist simplification in confining knowledge work to the level of persons doing the work. Instead, a serious systems approach includes the organizational (systemic) side of knowledge work, too, and actually starts from the vantage point that the complexity and specifity of knowledge work only comes to the fore if both sides are taken into account.

In the following text I will briefly sketch a few of the characteristics of the emerging knowledge society, equally a few characteristics of intelligent organizations, in order to use this framework for a closer definition of knowledge work.

At the center of interest here is the newly constituted connection of organized work in types of businesses which, on the one hand, are dependent on efficient public (or quasi-public) intelligent infrastructures, such as data networks, telematic traffic systems and interoperational standards, and on the other hand, employees, who, as the cognitariat, have long outgrown the old stereotype of proletarian existence: The knowledge workers of the modern society, the cognitariat, are in charge of their own means of production: knowledge, information, evaluation. The cognitariat forms, by us,the majority of the working population (Toffler 1995: 60).

2 An Outline of the Knowledge Society

For a long time now the information society has been officially proclaimed: since the beginning of the 1960's in Japan, since the Clinton/Gore government in the USA and since the EU meeting of the Council of Ministers on Corfu in the summer of 1994. More fitting however is the term, knowledge society, since the envisaged qualitative changes touch not on information but on the new power, economic significance and political criticality of knowledge and expertise. Whereas information denotes relevant system specific differences, knowledge is created when such information is seen in connection with particular contexts of experience and communities of practice. One can talk of expertise when knowledge is related to definite or concrete decision making situations (distinguishing facts from data, information, knowledge, expertise and reflection s. Willke 1998). At the core of the formation of the knowledge society appears to be the quantity, quality and speed of innovations through new information, new knowledge and new expertise.

The concept of a knowledge society, or a knowledge based society, is only then a substantial one when the structures and processes of material and symbolic reproduction of a society are so permeated with knowledge dependent operations that information processing, symbolic analysis and expert systems take pride of place over other factors of reproduction. Although the fundamental studies of Fritz Machlup, on the production and economic value of knowledge, have been little appreciated, and although the definition of indicators and their measurement, in the area of information, knowledge and knowledge bases have to remain fuzzy, a clear trend can be seen from the observations available: in contrast to agriculture, industrial production and simple services, knowledge based occupations are on the increase: in contrast to products with a high proportion of their value in labor and material, products whose value lies primarily in embedded intelligence or embedded expertise, gain an advantage e.g. software, logic-chips, computers, color-displays, electronic games, films etc. At Siemens, in the meantime, over 50% of their net product comes from knowledge-intensive services (Pierer 1996). Whilst simple occupations and services are being taken over by robots, the need for professional expertise is increasing in all areas.

The knowledge society doesnt exist as yet, but it is already casting its shadow in advance. With the victory of the capitalist democratic form of society over socialism, the construction of more efficient global digital data networks and the intensification of global contexts for local trade, the modern nation-state is gradually losing elements of its significance. With the upgrading of products and services to knowledge based, professional goods, the conventional production factors (land, capital, labor) dramatically lose significance in relation to implicit or embedded expertise. Therefore the modern capitalist economy is gradually changing, step by step, to a post-capitalist, knowledge based form of production. These combined elements are changing the face of the modern labor and industrial society. For developed societies, the relevant form of work will become knowledge work, whilst the conventional forms of simple work will be taken over by machines or will move to the still remaining low-wage countries. The welfare-state is currently collapsing due to the high demands of the social services and subvention, which, in a globalizing economy can no longer be controlled or organized by the state. All these things are happening at the moment in front of our eyes. Not only politics but also many social scientists are hardly taking any notice of these developments because they are looking in another direction; the direction of the defense of the achievements of the 19th century, the defense of coal and steel, of capitalism and the nation-state.

The construction of a knowledge based infrastructure of the second generation (intelligent infrastructure) is becoming the driving force of the transformation of the industrial society into a post-capitalist society (Drucker 1994), in the context of a global, technological-economic competition of the nations. In the same way as the public infrastructures of energy supply grids, streets and railway lines brought about the full development of the first and second industrial revolutions, it appears that it will be the infrastructure of a highly efficient communications network which will realize the potential of the third industrial revolution. That the new society will be both a non-socialist and a post-capitalist society is practically certain. And it is certain also that the primary resource will be knowledge (Drucker 1994:4).

The idea of a knowledge society isn't new. Two of the best earlier texts on this subject come from Amitai Etzioni (1971) and Daniel Bell (1976). Etzioni wrote a very impressive chapter on the role of knowledge and the relationship between politics and knowledge in the context of his model of the active society (Etzioni 1971, chapters 6-9). Daniel Bell predicted, in his post-industrial society, the pervasive and dominant role of knowledge based services, symbol analysis and research (Bell 1976, chapters 2 and 3). Surprisingly, after these early drafts, the theme knowledge society remained at rest for two decades, whilst social scientists as well as AI-researchers turned their attention to the conditions and consequences of the triad competition of national economies and governance regimes (Feigenbaum/McCorduck 1984).

The present renaissance of the theme knowledge society comes remarkably, not from sociology or political science but from management theory. The main text is Peter Drucker's post-capitalist society, flanked by Robert Reich's (1991) concept of knowledge work through symbolic analysts and James Quinn's (1992) fundamental text on intelligent organization.

The idea of the knowledge society as a possible form of post-industrial and post-capitalistic society has nothing to do with the questionable model of a scientific society. At least in the case of a modern society a functionally specialized subsystem of society, whether it be politics, the economy or science, cannot stand for the whole without deforming the complete concept of a society. This means that one should only talk of knowledge society when a qualitative new form of knowledge base, digitalization and symbolic analysis permeates through all areas of a society and when the impact of special expertise is transforming all areas of society.

3 Characteristics of the Intelligent Organization

Management theory and practice have very much caught on to the idea and vision of the intelligent enterprise (Ghawla/Renesch 1995; Quinn 1992; Senge 1990; Senge et al. 1996), not only in the field of the professional service industries through consultation firms, legal practices, design and film studios, hospitals, research institutes, software firms, pharmaceutical businesses etc., but also through the whole spectrum of specialized industrial production, from special steels to bio-genetic products, through to car firms which realize that electronic parts make , in the meantime, up to 25% of the value of the car and forecast in the year 2000 up to 40% (Pountain 1995:19).

Whole sectors, such as the entertainment industry, commercial institutes for training and further education, commercial data bases and information services, the mass media and the publishing bodies are growing together due to the digitalization of the content and the transfer onto electronic media, and through this are opening up wide reaching global markets for knowledge content and expertise (Eisenhart 1994). Parallel to this, the digitalization of expertise is allowing the system specific construction of organizational intelligence in the form of proprietary data-bases, expert systems, rule systems and pattern recognition procedures for the available knowledge (e.g. data-mining), so that the knowledge from the members of the organizations - including the implicit and tacit knowledge - can be symbolically prepared, organized and step by step transformed into an independent knowledge of the organization (Huber 199; Willke 1995, chapter 7).

Scientific knowledge is only one form of knowledge among others. Next to it stand functional technologies, expertise, intelligence, implicit knowledge, organized symbolic-systems, organizational knowledge, knowledge-based operation forms, professional guidance-knowledge and many others. Gerhard Schulmeyer of Siemens-Nixdorf, spoke in the introduction to the business report 1996 about the change of the SNI enterprise from a bureaucratic into a knowledge-based enterprise (Computerwoche 1997). The academic/scientific system is no longer in the position to control the production and the use of specialized expertise which falls into foreign contexts. But above all because of this polycentric production of knowledge (Jasanoff 1990), the speed of the revision of knowledge has increased to such an extent, that the laborious detour through the academic system would be counterproductive.

It is striking, for example, that management expertise and new concepts of corporate guidance no longer stem from economic, organizational or management theory, but from either a reflective practice or from hybrid forms of knowledge creating organizations and consulting firms (Peters 1989; Peters 1992; Senge 1990; Womack et al. 1991). In the same way new derivatives of financial instruments aren't created within the financial sciences but in practice from combined expert teams (Kurtzman 1993; Sassen 1994: 82ff., 89ff.), who cannot and no longer want to afford the long-winded validation from the academic researchers.

These observations can be generalized with the innumerable case studies on especially innovative enterprises: Knowledge based, intelligent organizations develop a self-strengthening recursiveness of the use and the generation of knowledge. Above all Ikuyiro Nonaka (1992; 1994; 1995) extended this basic idea into a model of an organizational knowledge creation. He takes up and adapts the difference between tacit and explicit knowledge from Michael Polanyi. He is primarily interested, on the one hand in the intersection of these two types of knowledge, and on the other hand in the transitions between personal and organizational knowledge. In a table, four modes of the creation of knowledge are represented (see table 1) which in an optimum combination form a spiral of the organizational knowledge creation (Nonaka 1994: 20).

Transition from to	implicit knowledge	explicit knowledge
implicit knowledge	socialization	externalization
explicit knowledge	internalization	combination

Table 1 Modes of Knowledge Creation in Organizations

Nonaka's main statement is that a knowledge-based organization only becomes a generator of innovative knowledge when it has comprehended the precarious intersection points between explicit and implicit knowledge in routinized organizational processes, which demand that individual knowledge is articulated and spread through accessibility. The basis of such processes are recursive rounds of meaningful dialogue and the ability to use metaphors, which capture the implicit knowledge in comprehensible symbols (Nonaka 1994: 20).

Similarly elaborated, Dorothy Leonard-Barton (1995) also addresses the question of the generation of organizational knowledge. She distinguishes four main attributes of generating and diffusing knowledge, cross-referencing the dimensions of inside and outside on the one hand and present and future on the other: problem solving, implementation and integration, experimentation, and the importation of knowledge. She then diversifies this simple scheme with various short case studies in a detailed grid in order to search for sources and the mechanisms of knowledge creation. For example, she names as sources of external technological knowledge not only customers and universities but equally competing firms, non-competing firms, salespeople, major research institutes (in the USA national laboratories) and advisors (1995:15).

For many, it is difficult to even imagine organizational knowledge at all, that is knowledge which isn't stored in the minds of people, but which resides within the operational rule systems of an organization. Organizational or institutional knowledge is embedded in the person-independent anonymous rule systems which define the operational mode, that is the *modus operandi* of a social system. Above all these are standard operating procedures, guidelines, encodings, work processes and descriptions, established prescribed knowledge for particular situations, routines, traditions, specialized data bases, coded production and project knowledge, and the specific culture of an organization. Niklas Luhmann (1990:11) begins on the first page of his theory of knowledge with the preposition of freeing the attribution of knowledge from the individual consciousness.

In order to trace the knowledge-basis of an organization you have to ask for the available knowledge of the organization. How and in what form in this knowledge stored? How, from whom, and in which situations is it retrieved? How does the organization gain, store, administer and change this knowledge? Other questions are also the construction/development, the use and the management of the organizational knowledge (detailed conceptual considerations on this subject from Huber 1990). From the outset, it should be made clear that although the knowledge-base of an organization is separate from people, it cannot be activated independently of them. An important question therefore is the interplay between individual and organizational knowledge and the interplay of the appropriate learning processes.

The fundamental studies from James Quinn (1992) on intelligent organizations are seen today as the catalyst for the idea of corporate reengineering (Hammer/Champy 1994) and the restructuring of corporations, organizations and even administration along the lines of optimizing business processes. Like Drucker and others, Quinn stresses the new role of knowledge: With rare exceptions, the economic and production power of a modern corporation lies more in its intellectual and service capabilities than its hard assets - land, plant and equipment (Quinn 1992:241). But he goes further in developing the vision of an intelligent organization, in which work consists of remolding knowledge and expertise as raw materials: All the trends, new capabilities, and radical organizations analyzed in earlier chapters are converging to create a new enterprise, best described as: a highly disaggregated, knowledge and service based enterprise concentrated around a core of set knowledge and service skills. This is what we refer to more succinctly as the intelligent enterprise, in a firm that primarily manages and coordinates information and intellect to meet customer needs (Quinn 1992: 373).

4 Knowledge Work

The basic problems of knowledge work revolve around the question of how the interplay between personal and organizational knowledge can be understood and organized. In order to understand the new quality of today's knowledge-work, one must realize that it is not sufficient when either the people or the organization operate on a knowledge-base. Socrates, without a doubt, carried out knowledge work, but he didn't need an elaborate organization for his form of knowledge work. The big churches and the parliamentary systems of modern democracies are amazingly elaborate and intelligent organizations but knowledge work only happens by accident or sporadically in this context, because they are so arranged that also average people can use their operational procedures (Willke 1997).

The type of knowledge-work possible today only emerges when both sides, people and organizations, in a complementary way generate, use and in turn make their knowledge available to each other. This is exactly at which the four above mentioned modes of intermediation from Nonaka aim. James Quinn also aims at this when he says that an intelligent firm must learn to manage and coordinate differing knowledge elements: As a company focuses ever more on its own internal knowledge and service skills and those of its suppliers, it increasingly finds that managing shifts away from the overseeing and deployment of fiscal and physical assets and towards the management of human skills, knowledge bases, and intellect both within the company and in its suppliers. In fact, its raison d'on d'être becomes the systematic coordination of knowledge and intellect throughout its (often highly disaggregated) network to meet customer needs« (Quinn 1992: 72).

This seemingly straight forward, reasonable formula, that members and organizations should make their useful knowledge available to each other actually is extraordinarily difficult to put into practice. The main reason for this lies in the fact that we know quite a lot about the construction, development, and the application of knowledge, in other words the knowledge based operational forms of people, but almost nothing about organizational intelligence in the sense of a collective, a systematic or an emerging characteristic of an organized social system (Marshall et al. 1996). Early views on the problem of organizational intelligence talk of an intelligent organization when its members work with a knowledge-base, in other words when they are professionals or experts and when they don't hinder each others work to any great extent (Wilensky 1967).

In contrast to this it can be seen today that the actual difficulty in constructing organizational intelligence as a framework for knowledge-work lies in incorporating independent expertise in the anonymous trans-personal rule system of the organization. That is not to say that this organization specific knowledge-base is created or operates independently from people, but that it is independent of specific people, and therefore functions in the sense of a collective mind (Weick/Roberts 1993) or an institutionalized rule-structure, that instructs the actions (operational mode) of its members with a high degree of reliability and resilience.

In spite of its traditional fixation upon people and their activities, the idea of an emerging collective or a systemic quality of organized contexts is not entirely foreign to sociological thinking. Under the keywords autonomy, auto-dynamic or auto-logic which appear again and again in descriptions of complex social systems, is the suggestion of a reality beyond that of individual activity.

If one frees the idea of autopoiesis from its ballast of the natural sciences, the central components of self-reference and operational closure remain. Neither of which are new concepts. Self-reference has been a theme in communication theory since the development of the visions of meta-communication and the reflective mechanism of the stabilization of communication rules. Considering the extent to which the sociological systems theory makes communication central to the construction of social systems, it becomes clear that systems constitute and stabilize themselves in relation (reference) to themselves. Self-reference denotes the operational mode of a system in which the reproduction of the unity of the system concedes to the condition of the possibility of environmental contacts (external references). The system itself and the continuation of its operational closure in terms of its functions, becomes the measure of the suitability of the operations of the system. The environment offers possibilities and sets restrictions, which only can be recognized in reference to the features of the operational mode of the system. Therefore self-reference in non-trivial systems denotes a particular relationship to the outside world through the rules of the autonomous operational mode of the system. Self-reference therefore doesn't rule out contact with the outside world but selects the nature of the contacts by rules and criteria, which are laid down not from the outside but from the system itself. It is obvious that the organizational generation and processing of knowledge differ greatly from one another depending on whether self-referential processes are at work or not.

It certainly appears to be quite difficult to see and accept processes and effects of operative closure. Operational closure appears to contradict all our visions of the necessary openness of systems. How can systems enter into relationships amongst themselves, share and cooperate with each other if they are operationally closed? Doesn't the idea of operational closure have an isolationist and unworldliness about it?

Before one starts to have a heated discussion over these and further questions it would be quite useful to look in more detail at what is meant by the term operational closure. Operative closure is created when a system networks its specifically operative elements (e.g. its special communications, decisions or activities) in a circular form to one another, in other words self-referentially organized, and if in addition its specific processes are also concatenated in a circular form to one another so that an interlocking, self-strengthening and in this sense auto-catalyzing network of processes is formed (Varela 1981: 1435). If a self-strengthening circular networking materializes on both these levels it is possible to speak of the concept of a hypercycle (a concept formed by Manfred Eigen). Operational closure in this sense means the hypercyclic organizational mode of a system (Teubner 1990).

If one takes a sociological look at actual existing organizations it is not particularly difficult to recognize self-reference and operational closure when one knows what it is about. It is this inherent feature of organizations, for example enterprises, which brings about their own destruction, although this is contrary to the interests of their members (one can hardly fail to ask the question of how these actions relate to the idea of rational behavior), although warning signs are in abundance in the outside world, although other organizations behave differently in similar situations, although individual people or groups within the organization see the destruction coming and take measures against it (classical case studies on this subject by Allison 1971; Dörner 1989; Wilensky 1967; Wohlstetter 1962; 1966). What is in effect here is the operational logic of the system, deep frozen in its anonymous rule systems and its institutionalized collective identity. These rules frequently enough restrict the possibilities of action of the firm's members, even if the penalty is high losses or the destruction of the system. This shows that exactly because of established self-reference and »inner-directed« ways of operating, operational closure of organizations can also have a destructive effect when it promotes a »pathological« learning process within the system (for details on this see Argyris/Schön 1996).

On the other hand, there are, in the meantime, descriptions of organizations in which high quality knowledge-work only functions because personal and organizational knowledge-bases are understood as two separate entities, which have to relate to one another in a productive way but shouldn't disturb or hinder one another. This includes above all reports on the knowledge management of globally operating consulting firms (overview see Willke 1996), the descriptions of high-reliability organizations (LaPorte/Consolini 1991), and an abundance of case descriptions which come from applications of Peter Senge's model of the fifth discipline (Kim 1993; Senge et al. 1996). And there are some ventures in building a theory of knowledge management and knowledge society (Nonaka/Takeuchi 1995; Willke 1998).

5 Closing Remarks

The research into organizational knowledge work has only just begun. It is obviously not just restricted to classical professional work and the new" businesses of consulting firms and financial services. Organizational knowledge affects all organizations which use and produce knowledge-based goods. It is known that a whole range of global firms, from very differing sectors, are presently regrouping their businesses according to knowledge-work (Marshall et al 1996: 80).

The difficulties of researching and reconstructing knowledge work seem to contain a lesson for AI studies and knowledge studies in general. Most of these studies have been concerned with hard sciences and technology and have been subjected to a reductionist technological bias. Knowledge production and knowledge work in multiple centers of expertise (Jasanoff) raise the question, how the social and systemic embeddedness of knowledge work and of strategies of coping with the complexities of knowledge work will have repercussions on our understanding of any form of intelligence, including artificial intelligence. There seem to be some good reasons for a new beginning.

6 Literature

Allison, G. (1971). Essence of Decision. Explaining the Cuban Missile Crisis. Boston, Little, Brown & Co.

Argyris, C. and D. Schön (1996). Organizational learning II. Theory, method, and practice. Reading, Mass., Addison-Wesley.

Bell, D. (1976). The Coming of Post-Industrial Society. A Venture in Social Forecasting. 1. Auflg. 1973. New York, Basic Books.

Brentano, M. v. (1996). "Nur eine Handvoll Global Players. Investment-banken teilen den Weltmarkt neu auf." FAZ Nr. 31 v. 6.2.1996, S. B7.

Calverley, J. (1995). "The currency wars." Harvard Business Review(March-April 1995): 146.

Chawla, S. and J. Renesch, Eds. (1995). Learning Organizations. Developing cultures for tomorrows workplace. Portland, Oregon, Productivity Press.

Computerwoche (1997). "SNI-Chef betont Wandel zur wissensbasierten Firma." Computerwoche 12/97: 51.

Dörner, D. (1989). Die Logik des Mißlingens. Reinbek, Rowohlt.

Dreyfus, H. D. S. (1988). Making a Mind Versus Modeling the Brain: Artificial Intelligence Back at a Branchpoint. Daedalus 117, Winter 1988: 15-44.

Drucker, P. (1994). Post-Capitalist Society (first publ. in 1993). New York, HarperBusiness.

Eisenhart, D. (1994). Publishing in the information age. A new management framework for the digital age. Westport, London, Quorum Books.

Etzioni, A. (1971). The Active Society. Erstausgabe 1968. New York, Free Press.

Feigenbaum, E. and P. McCorduck (1984). The fifth generation. Artificial intelligence and Japans computer challenge to the world. New York, New American Library.

Hammer, M. and J. Champy (1994). Business Reengineering. Die Radikalkur für das Unternehmen. Englische Ausgabe 1993. Aus dem Englischen von Patricia Künzel. Frankfurt - New York, Campus.

Huber, G. (1990). "A theory of the effects of advanced information technologies on organizational design, intelligence, and decision making." Academy of Management Review 15(1): 47-71.

Jasanoff, S. (1990). "American Exceptionalism and the Political Acknowledgment of Risk." Daedalus. Proceedings of the American Academy of Arts and Sciences. Vol.119, Fall 1990, 61-82.

Kim, D. (1993). "The link between individual and organizational learning." Sloan Management Review(Fall 1993): 37-50.

Knoblauch, H. (1996). "Arbeit als Interaktion. Informationsgesellschaft, Post-Fordismus und Kommunikationsarbeit." Soziale Welt 47: 344-362.

Knorr-Cetina, K. (1984). Die Fabrikation von Erkenntnis: Zur Anthropologie der Naturwissenschaft. Frankfurt, Suhrkamp.

LaPorte, T. and P. Consolini (1991). "Working in practice but not in theory: theoretical challenges of "High-Reliability Organizations"." Journal of Public Administration Research and Theory 1(1): 19-47.

Leonard-Barton, D. (1995). Wellsprings of knowledge. Building and sustaining the sources of innovation. Boston, Mass, Harvard Business School Press.

Luhmann, N. (1990). Die Wissenschaft der Gesellschaft. Frankfurt, Suhrkamp.

Managermagazin (1996). "Tour de Suisse." Managermagazin(Juni 1996): 166-171.

Marshall, C., L. Prusak, et al. (1996). "Financial risk and the need for superior knowledge management." California Management Review 38(Srping 1996): 77-101.

Nonaka, I. (1992). "Wie japanische Konzerne Wissen erzeugen." Harvard manager(2/1992): 95-103.

Nonaka, I. (1994). "A dynamic theory of organizational knowledge creation." Organization Science 5(1, Feb. 1994): 14-37.

Nonaka, I. and H. Takeuchi (1995). The knowledge-creating company. How Janpanese companies create the dynamics of innovation. New Yor, Oxford, Oxford UP.

Pierer, H. v. (1996). "Statement in: Ist Wissen Geld?" managermagazin Mai 1996: 182-186.

Pountain, D. (1995). "Europe's chip challenge." Byte(March 1995): 19-30.

Quinn, J. (1992). Intelligent enterprise. A knowledge and service based paradigm for industry. Foreword by Tom Peters. New York, Free Press.

Rappaport, A. and S. Halevi (1991). "Chip- und Softwaredesign: Das Eldorado der Comoputerbauer." Harvard Business Manager: Informations- und Datentechnik, Band 2: 47-59.

Reich, R. (1991). The Work of Nations. Preparing ourselves for 21st Century Capitalism. New York, Knopf.

Ryan, B. (1991). AIs Identity Crisis. Byte, January 1991: 239-301

Sassen, S. (1994). Cities in a world economy. Thousand Oaks, London, New Delhi, Pine Forge Press.

Senge, P. (1990). The Fifth Discipline. New York, Doubleday.

Senge, P., A. Kleiner, et al. (1996). Das Fieldbook zur Fünften Disziplin. Stuttgart, Klett-Cotta.

Stehr, N. (1994). Knowledge Societies. London, Sage.

Stichweh, R. (1987). Die Autopoiesis der Wissenschaft. Dirk Baecker et al, Hg., Theorie als Passion. Niklas Luhmann zum 60. Geburtstag. Frankfurt, Suhrkamp.

Suchman, L. (1993). Technologies of accountability. Of lizards and aeroplanes. Technology in working order. Studies of work, interaction, and technology. G. Button. London, New York: 113-126.

Teubner, G. (1990). Die Episteme des Rechts. Zu erkenntnistheoretischen Grundlagen des reflexiven Rechts. Dieter Grimm, Hg., Wachsende Staatsaufgaben - sinkende Steuerungsfähigkeit des Rechts, 115-154. Baden-Baden, Nomos.

Toffler, A. (1995). "Das Ender der Romantik. Zukunftsforscher Alvin Toffler über das Überleben in der Informationsgesellschaft. Interview." Spiegel special 3/1995: 59-63.

Varela, F. (1981). Autonomy and autopoiesis. Self-organizing systems. G. Roth and H. Schwegler. Frankfurt , New York, Campus: 14-23.

Waldrop, M. (1994). Complexity. The emerging science at the edge of order and chaos. London, Penguin.

Weick, K. and K. Roberts (1993). "Collective mind in organizations: Heedful interrelating on flight decks." Administrative Quarterly **38**: 357-381.

Wilensky, H. (1967). Organizational Intelligence. Knowledge and POlicy in Government and Industry. New York, Basic Books.

Willke, H. (1996). Dimensionen des Wissensmanagements - Zum Zusammenhang von gesellschaftlicher und organisationaler Wissensbasierung. Jahrbuch für Managementforschung 6. Wissensmanagement. G. Schreyögg and P. Conrad. Berlin, de Gruyter: 263-304.

Willke, H. (1997). Dumme Universitäten, intelligente Parlamente. Wie wird Wissen wirksam? iff texte Band 1. R. Grossmann. Wien, New York, Springer: 107-110.

Willke, H. (1998). Systemtheorie III: Steuerungstheorie. 2. Auflg. Stuttgart, Fischer (UTB).

Willke, H., (1998. Systemisches Wissensmanagement. Stuttgart, Lucius&Lucius (UTB)

Wohlstetter, R. (1966 (zuerst 1962)). Pearl Harbor. Signale und Entscheidungen. Erlenbach-Zürich u. Stuttgart, E. Rentsch.

MINERVA: A Tour-Guide Robot that Learns

Sebastian Thrun[1], Maren Bennewitz[2], Wolfram Burgard[2], Armin B. Cremers[2], Frank Dellaert[1], Dieter Fox[1], Dirk Hähnel[2], Charles Rosenberg[1], Nicholas Roy[1], Jamieson Schulte[1], Dirk Schulz[2]

[1]School of Computer Science
Carnegie Mellon University
Pittsburgh, PA 15213

[2]Computer Science Department III
University of Bonn
53117 Bonn, Germany

Abstract. This paper describes an interactive tour-guide robot which was successfully exhibited in a Smithsonian museum. Minerva employed a collection of learning techniques, some of which were necessary to cope with the challenges arising from its extremely large and crowded environment, whereas others were used to aid the robot's interactive capabilities. During two weeks of highly successful operation, the robot interacted with thousands of people, traversing more than 44km at speeds of up to 163 cm/sec in the un-modified museum.

1 Introduction

This paper presents Minerva, the latest in a series of mobile tour-guide robots. The idea of a tour-guide robot goes back to Horswill [8], who implemented a vision-based tour-guide robot in the AI lab of MIT. More recently, Burgard and colleagues developed a more sophisticated tour-guide robot called *Rhino*, which that successfully installed in the Deutsches Museum Bonn [1]. Inspired by this, Norbakhsh and colleagues developed a similar robot called *Chips* for the Carnegie Museum of Natural History [14].

Building on these successes, the Minerva project pursued two primary goals:

1. **Scaling up:** Minerva's environment was an order of magnitude larger and more crowded than previous museum environments. The robot's top speed of 163 cm/sec was more than double than that of previous robots. An additional scaling goal was a reduction in lead-time for installing such robots (from several months to only three weeks). As in the Rhino project, the environment was not modified in any way to facilitate the robot's operation.

2. **Improved user interaction.** To aid the robot's interactive capabilities, Minerva was designed to exhibit a life-like character, strongly relying on common tokens in inter-human interaction. For example, Minerva possessed a moving head with motorized face and a voice. The robot's behavior was a function of its "emotional state," which adjusted in accordance to people's behavior. These new interactive means were essential for the robot's ability to attract, guide, and educate people—something which was recognized as a clear deficiency of previous tour-guide robots.

Fig. 1. Panoramic view of the Material World Exhibition, Minerva's major operation area which is located in the entrance area of the Smithsonian's National Museum of American History (NMAH).

These goals mandated the pervasive use of *learning*, at several levels of the software architecture. Minerva

- employs statistical techniques to acquire geometric maps of the environment's floor plan and its ceiling (other such robots did not learn their map), thereby facilitating the installation process,
- uses the ML estimator to acquire high-level models of time-to-travel, enabling the robot to follow much tighter schedules when giving tours, and
- employs reinforcement learning to acquire skills for attracting people, thereby increasing its overall effectiveness.

In addition, Minerva continuously estimates other important quantities, such as its own location, the location of people, and its battery charge.

During a two-week exhibition that took place in August/September 1998 in one of the world's largest museums, the Smithsonian's National Museum of American History (NMAH), Minerva successfully gave tours to tens of thousands of visitors. The "Material World" exhibition (see Figure 1), where most exhibits were located, was a huge and usually crowded area, which almost entirely lacked features necessary for localization (even the ceiling was uninformative, as the center area had a huge whole with a continuously moving pendulum inside). During the exhibition, more than 100,000 people visited the NMAH, and thousands of people interacted with the robot. The robot successfully traversed 44km, at an average speed of 38.8 cm/sec and a maximum speed of 163 cm/sec. The top speed, however was only attained after closing hours. During public hours, we kept the robot at walking speed, to avoid frightening people.

This article describes major components of Minerva's software. Since we adopted much of the software approach by Burgard and colleagues [1], the focus is here on research that goes beyond this work. In particular, we describe in detail Minerva's learning components, and only highlight other, previously published components .

2 Learning Maps

Previous tour-guide robots were unable to learn a map; instead they relied on humans to provide the necessary information. To facilitate the rapid installation in novel environments, Minerva learns maps of its environments. In particular, Minerva learns two types of maps, *Occupancy maps* (see also [3, 18]), and *ceiling*

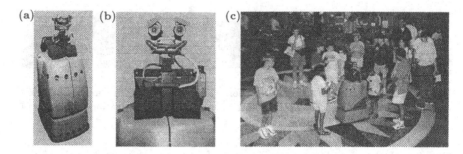

Fig. 2. (a) Minerva. (b) Minerva's motorized face (in happy mode). (c) Minerva gives a tour in the Smithsonian's National Museum of American History.

texture maps. The use of a dual map is unique; however, it is necessary in large, open and densely crowded environments such as the museum, where a single sensor modality is often insufficient to track the robot's position reliably.

2.1 Statistical Mapping with EM

Both mapping algorithms are variants of a single, overarching approach originally presented in [19]. The general problem is typically referred to as *concurrent
mapping and localization*, which indicates its chicken-and-egg nature: Building
maps when the positions at which sensor readings were taken are *known* is relatively simple, and so is determining the robot's positions when the map is *known*;
however, if neither the map nor the robot's positions are known, the problem is
hard.

The key idea is to treat the concurrent mapping and localization problem as
a *maximum likelihood estimation problem*, in which one seeks to determine the
the most likely map given the data.

$$Pr(m|d) \tag{1}$$

Here m denotes the map, and d the data (odometry and range data/camera im-
ages). The likelihood $Pr(m|d)$ takes into account the consistency of the odometry
(small odometry errors are more likely than large ones), and it also considers the
perceptual consistency (inconsistencies in perception are penalized). As shown
in [19], the likelihood function can be re-expressed as

$$Pr(m|d) = \alpha \int \cdots \int \prod_{t=0}^{T} Pr(s^{(t)}|m, x^{(t)}) \tag{2}$$

$$\prod_{t=0}^{T-1} Pr(x^{(t+1)}|u^{(t)}, x^{(t)}) \, dx^{(0)}, \ldots, dx^{(T)}.$$

where $x^{(t)}$ denotes the robot's pose at time t, $s^{(t)}$ denotes an observation (laser,
camera) and $u^{(t)}$ an odometry reading.

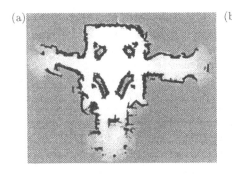 

Fig. 3. (a) Occupancy map and (b) ceiling mosaic of oddly-shaped ceiling (center area has a hole and a moving pendulum).

Finding the *global* maximum of $Pr(m|d)$ is computationally infeasible. However, local maxima can be found efficiently using the EM algorithm, a well-known statistical approach to likelihood maximization. This approach interleaves phases of localization (in the E-step) with mapping phases (the M-step). More specifically, in the E-step our approach estimates the various past robot locations denoted x

$$Pr(x|m,d) \qquad (3)$$

assuming knowledge of the map m (initially, there is no map and x is exclusively estimated from odometry). The M-step computes the most likely map based on the previously computed x:

$$\operatorname*{argmax}_{m} Pr(m|x,d) \qquad (4)$$

As argued in [19], iteration of both steps leads to a local maximum in likelihood space. In practice, this approach has been shown to generate maps of unprecedented complexity and size. However, the approach in [19] required that a person put tape on the floor and pushed a button every time the robot traversed a piece of tape—something that we want to avoid since it is an obstacle in rapid, robust installation.

2.2 Occupancy Maps

Our approach deviates from [19] in two aspects: First, we omit the "backward phase" in the E-step (localization), and second, we replace the density $Pr(x|m,d)$ by its maximum likelihood estimate $\operatorname{argmax}_x Pr(x|m,d)$. We found that both simplifications have only a marginal effect on the result; however, they greatly reduce the computational and memory requirements. This, in turn, makes it possible to use *raw laser data* for localization and mapping, instead of the "button pushes" that were required in the original approach to reduce the computation.

Simplified speaking, the resulting mapping algorithm performs Markov localization [2, 10, 15] (E-step), interleaved with occupancy grid mapping [3, 18]

(M-step). Both are easily done in real-time. In practice, we found that ≤ 5 iterations are sufficient, so that the resulting mapping algorithm is fast. A map of the museum is shown in Figure 3a. This map, which was used during the entire exhibition, is approximately 67 by 53 meter in size.

2.3 Texture Maps of the Ceiling

The sheer size, openness, and density of people in the present museum made it necessary to learn a map of the museum's ceiling, using a (mono) B/W camera pointed up. The ceiling map is a large-scale *mosaic* of a ceiling's texture (c.f., Figure 3b). Such ceiling mosaics are difficult to generate, since the *height* of the ceiling is unknown, which makes it difficult to map the image plane to world coordinates.

Just like the occupancy grid algorithm, our approach omits the backwards phase. In addition, we make the restrictive assumption that all location distributions are normal distributed (Kalman filters), which makes the computation extremely fast. Unfortunately, this assumption requires that all uncertainty in the robot's pose is *unimodal*—an assumption that is typically only true when the position error is small at all times (c.f., [12]).

The plain EM approach, as described above, does *not* produce small error. However, a modified version does, for environments of the size of the museum and for sensors as rich in information as cameras. The idea is to interleaves the E-step and the M-step *for each sensor item*, which is much finer grained a level than the approach described above. Whenever a sensor item is processed (new and past data alike), it is first localized and then the map is modified accordingly. A mosaic of the museum's ceiling is shown in Figure 3b. One can clearly see the ceiling lights and other structures of the museum's ceiling (whose height varied drastically).

3 Localization

In everyday operation, Minerva continuously tracks its position using its maps. Position estimates are necessary for the robot to know where to move when navigating to a specific exhibit, and to ensure the robot does not accidentally leave its operational area. Here we adopt Markov localization, as previously described in [2, 10, 15]. Markov localization is a special case of the E-step above.

Figure 4 illustrates how Minerva localizes itself from scratch (global localization). Initially, the robot does now know its pose; thus, $Pr(x|m, d)$ is distributed *uniformly*. After incorporating one sensor reading (laser and camera), $Pr(x|m, d)$ is distributed as shown in Figure 4a. While this distribution is multi-modal, high likelihood is already placed near the correct pose. After moving forward and subsequently incorporating another sensor reading, the final distribution $Pr(x|d, m)$ is centered around the correct pose, as shown in Figure 4b. We also employed a filtering technique described in depth in [7], to accommodate the crowds that frequently blocked the robot's sensors (and thereby violated the Markov assumption that underlies Markov localization).

(a) 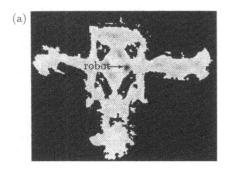   (b)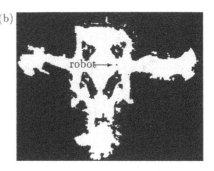

Fig. 4. Global localization: (a) Pose probability $Pr(x)$ distribution after integrating a first laser scan (projected into 2D). The darker a pose, the more likely it is. (b) $Pr(x)$ after integrating a second sensor scan. Now the robot knows its pose with high certainty/accuracy.

4 Collision Avoidance

Minerva's collision avoidance module controls the momentary motion direction and velocity of the robot so as to avoid collisions with obstacles—people and exhibits. At velocities of up to 163 cm/sec, which was Minerva's maximum speed when under exclusive Web control, inertia and torque limits impose severe constraints on robot motion which may *not* be neglected. Hence, we adopted a collision avoidance method called μDWA, originally developed by Fox and colleagues [5, 6]. This approach considers torque limits in collision avoidance, thereby providing safety even at high speeds.

The μDWA algorithm was originally designed for circular robots with synchro-drive. Minerva, however, possesses a non-holonomic differential drive, and the basic shape resembles that of a rectangle. Collision avoidance with rectangular robots is generally regarded more difficult. However, μDWA could easily be extended to robots of this shape by adapting the basic geometric model. The approach was able to safely steer the robot at speeds of 1.63 cm/sec, which is more than twice as high as that of any autonomous robot previously used in similar applications. This suggests that the μDWA approach applies to a much broader class of robots than previously reported.

5 "Coastal" Navigation

Minerva's *path planner* computes paths from one exhibit to another. The problem of path planning for mobile robots has been solved using a variety of different methods [11]. Most mobile robot path planners, however, do not take into account the danger of getting lost; instead, they minimize path length (see [17] for an exception). In wide, open environments, the choice of the path influences the robot's ability to track its position. To minimize the chances of getting lost, it is therefore important to take uncertainty into account when planning paths.

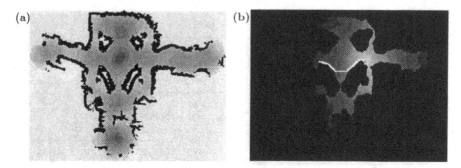

Fig. 5. Coastal navigation: The entropy map, shown in (a), characterizes the information loss across different locations in the unoccupied space. The darker an area, the less informative it is. (b) Path generated by the planner, taking both information loss and distance into account. Minerva avoids the center area of the museum.

Our idea is simple but effective: In analogy to ships, which typically stay close to coasts to avoid getting lost (unless they are equipped with a global positioning system), Minerva's path planner is called a *coastal planner*[1]. In essence, paths are generated according to a mixture of two criteria: path length and *information content*. The latter measure, information content, reflects the amount of information a robot is expected to receive at different locations in the environment. A typical map of information content is shown in Figure 5a. Here the grey scale indicates information content: the darker a location, the less informative it is. This figure illustrates that the information content is smallest in the center area of the museum.

Formally, information content is defined as the *expected reduction in entropy upon sensing*, i.e.,

$$H[Pr(x)] \ - \ \int Pr(x) \ E[s|x] \ H[Pr(x'|s)] \ dx. \tag{5}$$

Here $E[s|x]$ denotes the expected sensor reading at pose x. When constructing the map shown in Figure 5a, this expression is computed off-line for every location, making the assumption that the robot knows its position within a small, Gaussian-distributed uncertainty margin. Our approach also exploits the fact that the robot's sensors cover a 360° field of view, which allows us to ignore the orientation θ when considering information content. When computing (5), the presence of people is taken into account by modeling noise in sensing (assuming 500 randomly positioned people in the museum).

As described above, paths are generated by simultaneously minimizing path length and maximizing information content, using dynamic programming [9]. A typical result is shown in Figure 5b. Here the path (in white) avoids the center region of the museum, even though the shortest path would lead straight through this area. Instead, it generates a path that makes the robot stay close to

---

[1] The name "coastal navigation" was suggested by Thomas Christaller in a private communication

1	2	3	4	5	6	7	8	9	10	11	12	13	14	15	16	17	18	19	20	21
1		26	68	14	28															
2					23	38	13													
3				81	66	51		66						60						
4											62					76	22			
5											62								49	
6		41						44												
7			44	1	55			42						51						
8									44	63										
9																				
10																				
11				34								16	69							
12				61	53	69		72	32				87	55						
13													28							
14	33		39																	
15									60											
16																		46		68
17								59						13		57				
18					46	42		31	36					31						12
19			1		25			58			69		12							
20		57	62															37		
21				55	24	20		15						74						
22		208		66	46	38		38	23					56	39					
23				113	76	59		24	46					59						

Table 1. Time (in sec) it takes to move from one exhibit to another, estimated from 1,016 examples collected in the museum. These times, plus the (known) time used for explaining an exhibit, form the basis for the decision-theoretic planner.

obstacles, where chances of getting lost are much smaller. In comparative tests, we found that this planner improved the certainty in localization by a factor of three, when compared to the shortest-path approach.

6 Learning to Compose Tours

It was generally desirable for tours to last approximately six minutes—which was determined to be the duration the average visitor would like to follow the robot. Unfortunately, in practice the rate of progress depends crucially on the number and the behavior of the surrounding people. Thus, the duration of tours can vary widely if the the exhibits are pre-selected. To meet the target duration as closely as possible, tours are composed dynamically, based on the crowdedness in the museum.

To address this problem, Minerva uses a flexible high-level control module, capable of composing tours on-the-fly. This module *learns* the time required for moving between pairs of exhibits, based on data recorded in the past (using the maximum likelihood estimator). After an exhibit is explained, the interface chooses the next exhibit based on the remaining time. If the remaining time is below a threshold, the tour is terminated and Minerva instead returns to the center portion of the museum. Otherwise, it selects exhibits whose sequence best fit the desired time constraint. The learning algorithm (maximum likelihood estimator) and the decision algorithm were implemented in RPL, a language for reactive planning [13].

Table 2 illustrates the effect of dynamic tour decomposition on the duration of tours. Minerva's environment contained 23 designated exhibits, and there were 77

	average	min	max
static	398 ± 204 sec	121 sec	925 sec
with learning	384 ± 38 sec	321 sec	462 sec

Table 2. This table summarizes the time spent on individual tours. In the first row, tours were pre-composed by static sequences of exhibits; in the second row, tours were composed on-the-fly, based on a learned model of travel time, successfully reducing the variance by a factor of 5.

sensible pairwise combinations between them (certain combinations were invalid since they did not fit together thematically). In the first days of the exhibition, all tours were static. The first row in Table 2 illustrates that the timing of those tours varies significantly (by an average of 204 seconds). The average travel time, shown in Table 1, was estimated using 1,016 examples, collected during the first days of the project. The second row in Table 2 shows the results when tours were composed dynamically. Here the variance of the duration of a tour is only 38 seconds. Minerva's high-level interface also made the robot return to its charger periodically, so that we could hot-swap its batteries.

7 Spontaneous Short-Term Interaction

Interaction with people was Minerva's primary purpose—it is therefor surprising that previous tour-guide robots' interactive capabilities were rather poor. The type of interaction was *spontaneous* and *short-term*: Visitors of the museum typically had no prior exposure to robotics technology, and they could not be instructed beforehand as to how to "operate" the robot. The robot often had to interact with crowds of people, not just with single individuals. Most people spent less then 15 minutes (even though some spend hours, or even days). This type of interaction is characteristic for robots that operate in public places (such as information kiosks, receptionists). It differs significantly from the majority of interactive modes studied in the field, which typically assumes long-term interaction with people one-on-one.

To maximize Minerva's effectiveness, we decided to give Minerva "human-like" features, such as a motorized face, a neck, and an extremely simple finite state machine emulating "emotions," and use reinforcement learning to shape her interactive skills.

7.1 Emotional States

When giving tours, Minerva used its face, its head direction, and its voice to communicate with people, so as to maximize its progress. A stochastic finite state machine shown in Figure 6 is employed to model simple "emotional states" (moods), which allowed the robot to communicate its intent to visitors in a social context familiar to people from human-human interaction [4]. Moods ranged from happy to angry, depending on the persistence of the people who blocked its

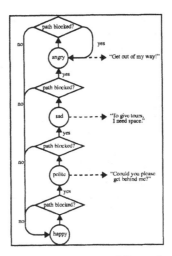

Fig. 6. The finite state automaton that governs Minerva's "emotional" states.

path. When happy, Minerva smiled and politely asked for people to step out of the way; when angry, its face frowned and her voice sounded angry. Most museum visitors had no difficulty understanding the robot's intention and "emotional state." In fact, the ability to exhibit such extremely simplified "emotions" proved one of the most appreciated aspect of the entire project.

The effect of this approach is best evaluated by comparing it with Rhino [1], which uses a similar approach for navigation but mostly lacks these interactive capabilities. We consistently observed that people cleared the robot's path much faster than reported by the Rhino team. We found that both robots maintained about the same average speed (Minerva: 38.8 cm/sec, Rhino: 33.8 cm/sec), despite the fact that Minerva's environment was an order of magnitude more crowded. These numbers shed some light on the effectiveness of Minerva's interactive approach. People clearly "understood' the robot's intentions, and usually cooperated when they observed the robot's "mood" changed.

7.2 Learning to Attract People

How can a robot attract attention? Since there is no obvious answer, we applied an on-line learning algorithm. More specifically, Minerva used a memory-based reinforcement learning approach [16] (*no* delayed award). Reinforcement was received in proportion of the proximity of people; coming too close, however, led to a penalty (violating Minerva's space). Minerva's behavior was conditioned on the current density of people. Possible actions included different strategies for head motion (e.g., looking at nearest person), different facial expressions (e.g., happy, sad, angry), and different speech acts (e.g., "Come over," "do you like robots?"). Learning occurred during one-minute-long "mingling phases" that took place between tours. During learning, the robot chose with high probability the best known action (so that it attracted as many people as possible); however,

24

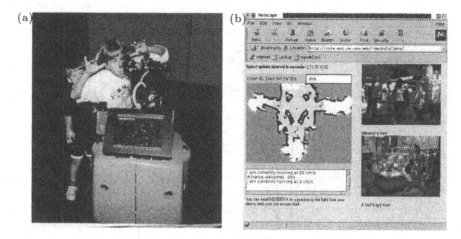

Fig. 7. (a) User interaction; (b) The Web interface for monitoring and control.

with small probability the robot chose a random action, to explore new ways of interaction.

During the two weeks, Minerva performed 201 attraction interaction experiments, each of which lasted approximately 1 minute. Over time, Minerva developed a "positive" attitude (saying friendly things, looking at people, smiling). As shown in Figure 8, acts best associated with a "positive" attitude attracted the most people. For example, when grouping speech acts and facial expressions into two categories, friendly and unfriendly, we found that the former type of interaction performed significantly better than the first (with 95% confidence). However, people's response was highly stochastic and the amount of data we were able to collect during the exhibition is insufficient to yield statistical significance in most cases; hence, we are unable to comment on the effectiveness of individual actions.

8 Conclusion

This article described the software architecture of a mobile tour-guide robot, which was successfully exhibited for a limited time period at the Smithsonian's National Museum of American History. During more than 94 hours of operation (31.5 hours of motion), Minerva gave 620 tours and visited 2,668 exhibits. The robot interacted with thousands of people, and traversed more than 44km. Its average speed was 38.8 cm/sec, and its maximum speed was 163 cm/sec. The map learning techniques enabled us to develop the robot in 3 weeks (from the arrival of the base platform to the opening of the exhibition). A much improved Web interface (Figure 7b) gave people direct control of the robot when the museum was closed to the public.

Our approach contains a collection of new ideas, addressing challenges arising from the size and dynamics of the environment, and from the need to interact

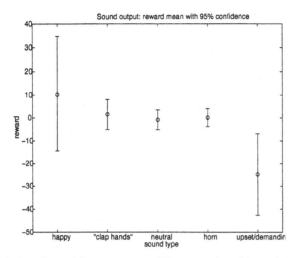

Fig. 8. Statistics of people's response to different styles of interaction (from friendly on the left to upset/demanding on the right). The data were input to a reinforcement learning algorithm, which learned on-line optimal interaction patterns.

with crowds of people. Special emphasis has been placed on learning, to accommodate the challenges that arose in this unprecedently large, open and crowded environment. In particular, Minerva differed from previous tour-guide robots in its ability to learn maps (of the floor-plan and the ceiling), models of typical navigation times for scheduling tours, and patterns of interaction when attracting people. In addition, Minerva differed from previous tour-guide robots in that she exhibited a "personality," adopting various cues for interaction that people are already familiar with. The empirical results of the exhibition indicate a high level of robustness and effectiveness. Future research issues include the integration of speech recognition, to further develop the robot's interactive capabilities.

Acknowledgments

We are deeply indebted to the Lemelson Center of the National Museum of American History, for making this project possible and their superb support. We also thank Anne Watzman for managing our relation to the media, and Greg Armstrong for his excellent hardware support. Special thanks also to IS Robotics and its Real World Interface Division for excellent hardware support, without which this project would not have been possible.

This research is sponsored in part by DARPA via AFMSC (contract number F04701-97-C-0022), TACOM (contract number DAAE07-98-C-L032), and Rome Labs (contract number F30602-98-2-0137). Additional financial support was received from Daimler Benz Research and Andy Rubin, all of which is gratefully acknowledged.

References

1. W. Burgard, A.B., Cremers, D. Fox, D. Hähnel, G. Lakemeyer, D. Schulz, W. Steiner, and S. Thrun. The interactive museum tour-guide robot. AAAI-98.
2. W. Burgard, D. Fox, D. Hennig, and T. Schmidt. Estimating the absolute position of a mobile robot using position probability grids. AAAI-98.
3. A. Elfes. *Occupancy Grids: A Probabilistic Framework for Robot Perception and Navigation*. PhD thesis, CMU, 1989.
4. C. Breazeal (Ferrell). A motivational system for regulating human-robot interaction. AAAI-98.
5. D. Fox, W. Burgard, and S. Thrun. The dynamic window approach to collision avoidance. *IEEE Robotics and Automation*, 4(1), 1997.
6. D. Fox, W. Burgard, and S. Thrun. A hybrid collision avoidance method for mobile robots. ICRA-98.
7. D. Fox, W. Burgard, S. Thrun, and A.B. Cremers. Position estimation for mobile robots in dynamic environments. AAAI-98.
8. I. Horswill. Specialization of perceptual processes. TR 1511, MIT, AI Lab.
9. R. A. Howard. *Dynamic Programming and Markov Processes*. MIT Press, 1960.
10. L.P. Kaelbling, A.R. Cassandra, and J.A. Kurien. Acting under uncertainty: Discrete bayesian models for mobile-robot navigation. IROS-96.
11. J.-C. Latombe. *Robot Motion Planning*. Kluwer, 1991.
12. F. Lu and E. Milios. Globally consistent range scan alignment for environment mapping. *Autonomous Robots*, 4, 1997.
13. D. McDermott. The RPL manual, 1993. Can be obtained from http://www.cs.yale.edu/HTML/YALE/CS/HyPlans/mcdermott.html
14. I.R. Nourbakhsh. The failures of a self-reliant tour robot with no planner. Can be obtained at http://www.cs.cmu.edu/~illah/SAGE/index.html, 1998.
15. R. Simmons and S. Koenig. Probabilistic robot navigation in partially observable environments. IJCAI-95.
16. R.S. Sutton and A.G. Barto. *Reinforcement Learning: An Introduction*. MIT Press, 1998.
17. H. Takeda, C. Facchinetti, and J.-C. Latombe. Planning the motions of mobile robot in a sensory uncertainty field. *IEEE Trans. on Pattern Analysis and Machine Intelligence*, 16(10):1002–1017, 1994.
18. S. Thrun. Learning metric-topological maps for indoor mobile robot navigation. *Artificial Intelligence*, 99(1), 1998.
19. S. Thrun, D. Fox, and W. Burgard. A probabilistic approach to concurrent mapping and localization for mobile robots. *Machine Learning*, 31, 1998.

Dynamics, Morphology, and Materials in the Emergence of Cognition

Rolf Pfeifer

Artificial Intelligence Laboratory, Department of Information Technology, University of Zurich, Winterthurerstrasse 190, CH-8057 Zurich, Switzerland pfeifer@ifi.unizh.ch, phone: +41 1 63 5 4320/4331, fax: +41 1 635 68 09, www.ifi.unizh.ch/~pfeifer

Abstract. Early approaches to understanding intelligence have assumed that intelligence can be studied at the level of algorithms which is why for many years the major tool of artificial intelligence (AI) researchers has been the computer (this has become known as classical AI). As researchers started to build robots they realized that the hardest issues in the study of intelligence involve perception and action in the real world. An entire new research field called "embodied intelligence" (or "New AI", "embodied cognitive science") emerged and "embodimen t" became the new buzzword. In the meantime there has been a lot of research employing robots for the study of intelligence. However, embodiment has not been taken really seriously. Hardly any studies deal with morphology (i.e. shape), material properties, and their relation to sensory- motor processing. The goal of this paper is to investigate – or rather to raise – some of the issues involved and discuss the far-reaching implications of embodiment which leads to a new perspective on intelligence. This new perspective requires considerations of "ecological balance" and sensory-motor coordination, rather than algorithms and computation exclusively. Using a series of case studies, it will be illustrated how these considerations can lead to a new understanding of intelligence.

Key phrases: embodiment; relation between materials, morphology and control; cheap design; ecological balance; sensory-motor coordination

1 Introduction

Early approaches to understanding intelligence have abstracted from physical properties of individual organisms. The generally accepted assumption was that behavior can be studied at the level of algorithms which is why for many years the major tool of AI researchers has been the computer (this has become known as classical AI). As researchers started to build robots they realized that the hardest issues in the study of intelligence involve perception and action in the real world. There is a growing consensus that if we are ever to understand human intelligence we must first understand the basics of how individuals interact with their environment: human intelligence must be grounded in sensory-motor behavior.

Rodney Brooks of the MIT AI Lab who was among the first to recognize the importance of system-environment interaction for intelligence, started a new research field called "behavior-based robotics" (or "embodied AI", or "New AI") (e.g. [1]). The concept of embodiment became the main focus of research in AI, psychology, and what is normally subsumed under the label "cognitive science". By embodiment we mean that agents in the real world have a body, i.e. they are realized as physical systems and can thus exhibit behavior that can be observed in the real world. The physical characteristics of sensory and motor systems as well as their morphology, i.e. their form and structure, are important aspects of embodiment. In biology morphology has always been a central research topic. In the cognitive sciences, especially in AI, cognitive psychology, and in neurobiology it has been largely neglected which implies that an essential explanatory component is missing. For example, if we want to understand the function of the neural substrate, the brain, it is not sufficient to look at the neural substrate itself but it has to be known how the neural network is embedded in the physical agent and what the properties, i.e. the morphology and the physical characteristics of the sensory and the motor systems are to which the neural network is connected. Through embodiment it is determined what signals the neural system has to process in the first place. During the course of evolution, neural systems have started to exploit embodiment in ingenious ways. Problems that seem intractable at the purely computational level are often dramatically simplified if embodiment is taken advantage of.

It is obvious and undisputed that embodiment is required for behavior, for the interaction with the environment. But many researchers in cognitive science doubt that this has anything to do with higher cognitive processes. It has been argued elsewhere that, in essence, so-called high-level cognition is emergent from sensory-motor processes, i.e. it is grounded through embodiment [6, 11, 25, 29, 31]. For example, the embodied origins of categorization, i.e. the ability to make distinctions in the real world and to form abstract categories, and memory processes have been discussed by these authors. It turns out that if we look at the mechanisms underlying the so-called high-level processes like categorization and memory the distinction between high-level processes and sensory-motor processes starts to blur. In this paper the focus is on fundamental implications of embodiment for intelligence. In this context, a number of ideas and principles are introduced, in particular sensory-motor coordination, cheap design, and ecological balance.

We proceed as follows. First we elaborate the notion of embodiment and its main implications. We then illustrate first the physical implications with several examples, the passive dynamic walker, the Lee-Shimoyama artificial hand [17], the Face Robot, and the doll "Bit". We will see that these physical aspects have direct consequences for control. We then discuss a number of case studies illustrating the interdependence of morphology, sensory, and sensory-motor processes. Then in order to illustrate potential ways in which ecological balance can be explored, we discuss the evolution of the morphology of an "insect eye" on a real robot, the use of humanoid robots, and morpho-functional machines. Finally, we try to bring all of this together and show how we can bring these diverse topics and ideas to bear on the issue of emergence of cognition.

2 Implications of embodiment

Two kinds of closely related types of implications can be distinguished, physical and information theoretic ones. The physical ones relate to the physical setup of the agent, the body, the limbs, its mass, physical forces acting on the agent, gravity, friction, inertia, and energy dissipation. The neural systems of animals and humans have evolved to cope with these effects and are therefore tuned to the particular embodiment. This implies that the signals the neural system has to process depend on the physical characteristics of the agent which is an aspect of the information theoretic implications of embodiment: depending on the body, the neural system's tasks will be different (We use the term information theoretic in its commonsense meaning and not in the strict technical sense). The point becomes even more obvious if we look at the sensory side. Information about the environment can only be acquired through an agent's sensory system. Depending on the physical characteristics of the sensors, their morphology, and their position on the agent, the sensory signals will be different. The basic issues involved are best explained by a number of examples.

2.1 Control

As pointed out in the introduction there is a close interdependence of embodiment and neural processing. We first discuss control and then sensory processing.

The passive dynamic walker

The passive dynamic walker was originally suggested by Tad McGeer [20, 21]. A passive dynamic walker is a robot capable of walking down an incline without control, i.e. without actuation (Figure 1). The requires that the dynamics (the physical characteristics) be carefully taken into account. The strategy behind the passive dynamic walker is that once the robot can walk down an incline, little control is required to make it walk on a flat surface and once this ability is in place it is again easy to add control to make it go over obstacles. A picture of a passive dynamic walker is shown in figure 1. The passive dynamic walker is an example of "cheap design", meaning that it exploits the physics and the system-environment interaction which makes it cheap and parsimonious (see also [23, 25]).

This approach is quite in contrast to the one taken by the Honda design team where the goal was to design a humanoid robot (i.e. a robot that looks like a human) that could move its limbs into every possible position and thus perform a large number of possible movements. By contrast, all the passive dynamic walker

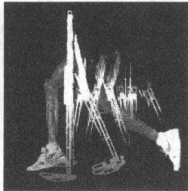

Fig. 1. The passive dynamic walker built at Cornell University, Dep. of Theoretical and Applied Mechanics. It walks down a shallow slope, driven solely by gravity. There is no electronic control whatsoever on the robot: it has no "brain". Still, the movements of the robot look very natural and human-like. This is a beautiful example of exploitation of dynamics (http://www.msc.cornell.edu/~ruinalab/pwd.html#overview).

can do is walk, and it can only walk down an incline; but within that narrow ecological niche it can walk very well. The Honda robot is much more flexible, it can walk uphill, downhill, even up and down the stairs, but it always requires a lot of control effort and looks a bit unnatural. The Honda approach reflects the currently dominating philosophy of top down design whereas the passive dynamic walker is an epitome of the bottom-up approach. It seems that there is a kind of "ecological balance" between task-environment, dynamics and control: by exploiting the dynamics of a particular task environment less control effort may be needed.

If we take these ideas one step further, we realize that in natural systems the joints are actuated by muscles and tendons and that they have their own dynamic properties which, in turn, can be utilized by the control system. The next case study, artificial muscles, illustrates the point.

Artificial muscles

The traditional approach to robot control is illustrated by the Puma arm. The Puma arm is a very popular beautifully designed robot arm that is used in many research laboratories and industrial applications. It is, like most robots today, made up mostly of rigid materials and actuation is achieved by electrical motors. If the arm-hand system has many degrees of freedom, we are faced with a hard control problem. One of the reasons is that all the degrees of freedom are "full" or unconstrained degrees of freedom, i.e. the joint can, within its boundary condition (maximum and minimum angle), be moved freely. This is true because of the way the arm was engineered by its developers and it is also what users who develop applications actually want: They want to be able to freely move the arm into every desired position. By contrast, natural arms – of humans, for example – typically have constrained degrees of freedom: Because the joints, the tendons and the muscles have intrinsic material properties like stiffness and elasticity that provide the joint with a preferred (minimal energy) position, some positions are more difficult to achieve than others. If the joint is moved away from this equilibrium point, there is a natural tendency for the joint to move back to its natural position without the need for neural control. In some sense the materials have their own control properties. And these are very nice properties because they are local (i.e. there is no need to communicate with a central processor) and do not have to be controlled by the nervous system (i.e. it requires no central resources). Of course, because the human arm is so awesomely complex, there will always be neural control involved at some level. What we are trying to do here is make an in principle point. While at first sight accounting for material properties of muscles and tendons seems to complicate matters, it leads in fact to a significant simplification because much of the control task is taken over by material properties.

Another point is worth mentioning. The Puma arm has been designed such that a user can simply specify the target position and the arm will move there. He does not have to worry about physics, about forces acting on the arm, inertia, etc. In a sense, the physical problem has been turned into a purely geometric one. As a consequence, the dynamics, the physical forces, do not have to be taken into account by the applications programmer. However, in natural systems, the muscles and the system as a whole have their own intrinsic dynamics which, again, can be exploited by the neural system (but such natural properties may be undesirable in an industrial environment).

Because natural systems by exploiting these properties can deal with the complexity of control of systems with many degrees of freedom, it seems like a good idea to draw inspiration from biological designs, e.g. natural muscle-skeletal systems. One idea is to use (artifical) muscles or muscle-like devices as actuators. There is already a considerable literature on artificial muscles: springs, rubber bands, and springs simulated with electrical motors are the most simple examples. The series elastic actuators (e.g. [26]) fall into this category. The main idea is to attach a spring between the motor and the actuator. This spring has some muscle-like properties like elasticity and damping and thus provides its own (local) dynamics. The spring can be simulated by electrical motors which has the advantage that the particular dynamics, i.e. the spring constant and the damping properties can be changed. Again, this control can be done locally and no central resources are required. A relatively popular artificial muscle is the McKibben pneumatic actuator [3] (figure 2a) An illustration of the application of pneumatic actuators is the artificial hand by Lee and Shimoyama [17] (figure 2b). Pneumatic actuators have many advantages: they are fast, they have a good contraction ratio, they produce considerable force, and they can be made small. The important point to be made here is that using pneumatic actuators instead of electrical motors completely changes the control problem. If we are using neural networks, they will look completely different from controllers for hands based on electrical motors. Once again, we see the interdependence of materials, or more generally embodiment, and neural control. Many other technologies have been suggested for artificial muscles (for a partial review, see, for example, Kornbluh et al. [14]). Because of their potentially favorable properties, in particular the potential for exploitation of intrinsic material dynamics, we suspect that advances in artificial muscle technology will lead to a quantum leap in robot technology in general.

a. b.

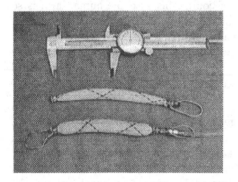 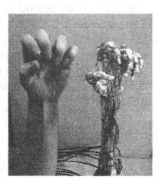

Fig. 2. Pneumatic actuator and artificial hand. (a) A pneumatic actuator contracts if air pressure is applied. (b) Artificial and natural hand.

Facial expression

An example that illustrates the potential ways in which properties of muscles and characteristics of tissue in general can be exploited concerns facial expression. Hiroshi Kobayashi and Fumio Hara of the Science University of Tokyo have developed the world-famous "Face Robot" [13]. The goal of the project is the investigation of human-robot communication and more generally, human-machine communication. We use the case study here for a different purpose, namely to illustrate the implications of materials for control. The face robot is based on 18 actuators driven by a hydraulic technology and 6 driven by electrical motors, yielding a total of 24 degrees of freedom. Figure 3a shows the frame which is then covered by a silicon rubber skin. Figure 3b shows some of the facial expressions the face robot can display: the neutral face, and some basic emotions, i.e., surprise, fear, anger, and happiness. The robot's ability for facial expression is indeed impressive; it is also capable of mimicking the facial expressions of a person looking at it (there are cameras inside its eyes). However, one of the problems is that this setup requires a lot of resources.

a.   b.

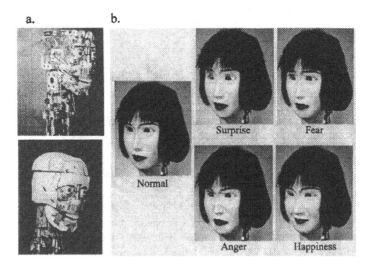

Fig. 3. The "Face Robot". (a) Frame. (b) Some facial expressions. The Face Robot consists of 18 hydraulic actuators and 6 electrical motors to produce a large variety of facial expression based on FACS, the facial action coding system.

By contrast, the robot doll "Bit" is equipped with much simpler mechanisms than the Face Robot to produce facial expression. Because the doll should eventually be sold in toy stores it was clear that a much cheaper solution for the facial expression had to be found. The solution was to find the right materials for the face. As a result, "Bit" can do with five actuators only and displays a considerable number of distinct facial expressions with a surprising feel of reality. For example, if two actuators are pulling the material in between apart, the facial "tissue", given the right materials, will exhibit the appropriate shape for a particular facial expression such as happiness. Figure 4 shows a few facial expressions of "Bit." Clearly, the flexibility of "Bit" is lower than the one of the Face Robot, but the design is much simpler than the one of the Face Robot and its performance is still impressive, another beautiful example of cheap design. What we see nicely in this illustration is that, given the right materials, the control problem to achieve certain desired behaviors, in this case displaying facial expressions, becomes much simpler.

So far we have seen how material properties, physical characteristics, and morphology influence control, i.e. the motor system. But the import of embodiment is even more far-reaching. Let us now look at the consequences of embodiment for sensory processes. Later we will discuss how the two, control and sensory processes, are in fact intimately coupled.

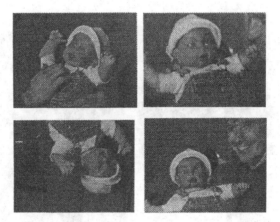

Fig. 4. Some facial expressions of the robot doll "Bit", developed by IS Robotics, a spin-off company of the MIT Artificial Intelligence Laboratory.

2.2 Sensor signal processing: Sensor morphology and sensor positioning

There is a strong dependence of behavior on sensor morphology and sensor position which can best be illustrated by a set of examples. The first is about a number of robots that, given the right sensor positioning, are cleaning an arena, the second concerns categorization and sensory-motor coordination, and the third one shows how motion parallax can be exploited.

The "Swiss Robots"

The "Swiss Robots" are simple robots equipped with infrared sensors, as shown in figure 5. They are controlled by a very simple neural network that implements the following rule: if there is sensory stimulation on the left turn right (and vice versa), a rule intended for obstacle avoidance. If put into an arena with Styrofoam cubes, they move the cubes into clusters, and some cubes end up along the wall (figure 5b). The reason is given in figure 5c. Normally the robots simply avoid obstacles. If they happen to encounter a cube head on, they push the cube. However, they are not searching for cubes and then pushing them: because of the particular geometry and the arrangement of the sensors, they push the cubes if they happen to encounter them head on. They will push the cube until there is another cube on the side that will provide sufficient sensory stimulation for the robot to turn away. But now there are already two cubes together and the probability that an additional cube will be deposited near them is higher. If now the position of the infrared sensors is changed as shown in figure 5d, the "Swiss Robots" will no longer move the cubes into clusters. For a more complete discussion of these experiments, see [19].

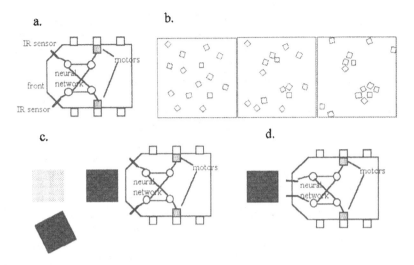

Fig. 5. Cluster formation with robots. The figure illustrates how simple robots, the "Swiss Robots", form clusters. (a) The robots with two sensors and a simple neural network that implements an avoidance reflex. (b) The clustering process (overall duration about 20 minutes). (c) Explanation of cluster formation. (d) Didabot with changed positions of the sensors. In this case its behavior will be entirely different even though the neural network is identical: the robots no longer clean the arena!

This example illustrates the strong interdependence of behavior, morphology, and environment (e.g. block size). Although obvious in this example, it is surprising that this idea is normally not explicitly taken into account in agent design. Of course, clustering can be achieved in many ways. For example, one might put a CCD camera on the robots which can be used to recognize the cubes, to move towards them. The robots could then look for a cluster and move the cube towards the cluster where it is to be deposited. This seems like a very natural solution, but it requires that the robots have highly sophisticated visual abilities, it is an expensive solution. The solution of the "Swiss Robots", by contrast, is cheap because it exploits the specific properties of the ecological niche and the morphology of the robots. However, if the geometry changes, e.g. the cubes get bigger, this solution no longer works.

2.3 Categorization as sensory-motor coordination

While the examples shown demonstrate some basic abilities of behaving systems, it can be argued that the kinds of tasks the agents had to perform do not require cognitive skills. If the nature of the task is changed and made more complex, so one could continue to reason, additional mechanisms, cognitive mechanisms, will be required that are independent, or at least much less dependent, on embodiment. Let us pursue this point a little further.

One of the most fundamental abilities of agents – animals, humans, and robots – in the real world, is the capacity to make distinctions: food has to be distinguished from non-food, predators from con-specifics, the nest from the rest of the environment, and so forth. Many models of categorization have been proposed in the past, the most prominent one presumably the ALCOVE model [15]. Indeed, ALCOVE is an excellent model: It can predict a large part of the experimental data published in the psychological categorization literature. In essence, ALCOVE is a connectionist model in which certain nodes, the category nodes, are activated whenever an instance of a particular category is encountered. In other words, these category nodes are representations of the categories. The task of ALCOVE can then be seen as one of mapping the input feature vector onto an internal representation of the category.

The main problem with ALCOVE, as is the problem with most models in classical cognitive psychology, is that it is not connected to the outside world: its inputs are feature vectors, its output activation levels of nodes in a connection-ist network. In the real world, agents are exposed to a stream of continuously changing sensory stimulation, not to feature vectors, and they require a conti-nous stream of motor control. Moreover, there is the problem of object constancy, i.e. the phenomenon that the sensory stimulation from one and the same object varies enormously depending, for example, on distance, orientation, and lighting conditions. It turns out – and it has been discussed extensively in the literature – that categorization in the real world requires a completely different approach. The attempt to map sensory stimulation onto an internal representation has not been met with much success, at least not in a general sense, i.e. in a world in which the agent only has incomplete knowledge, as the history of computer vision teaches.

The insight that categorization in the real world is not a computational problem, or at least not an exclusively computational one and requires that embodiment be taken into account is gaining increasing acceptance: It has been demonstrated that categorization is best viewed as a process of sensory-motor coordination [6, 23]. The term sensory-motor coordination which goes back to John Dewey 1896 [5] designates processes where there is a coupling of sensory and motor processes with respect to a particular purpose. For example, a robot which is turning about its own axis is not involved in a sensory-motor coordination whereas a robot that is moving up to and grasping a particular object is (with respect to this object). By definition, sensory-motor coordination involves both the sensory and motor systems, in other words, it involves the agent's body. Moreover, the sensory stimulation that the neural system has to process depends on the physical characteristics and on the positioning of the sensors on the agent. But not only that, it also crucially depends on the agent's behavior. It has been demonstrated that through its own actions, the agent can generate stable sensory patterns in different sensory channels that can be exploited to form cross-modal associations (e.g. [23–25]). The creation of cross-modal associations seems to be at the heart of the process of concept formation (e.g. [29]) and concepts are fundamental to cognition. In turn, these cross-modal associations crucially depend on the agent's morphology. For example, touching a bottle with a stiff hand yields entirely different sensory stimulation than fully grasping the bottle with the entire hand bent around the bottle. Note that this is a change in the morphology of the hand which leads to a change in the sensory stimulation. So, there are two closely related factors influencing the sensory stimulation, morphology, and sensory-motor coordination.

If we want to construct robots capable of acquiring their own concepts in the interaction with the real world, they have to be designed such that they are exposed to, or can generate, appropriate sensory stimulation, e.g. in order to form cross-modal associations. As of now, there is no general solution to the problem of how to optimally design sensory systems with different channels and where to position the sensors. This is because the optimal design strongly depends on the task environment. Thus, a good bet, is once again to draw inspiration from natural systems, hoping that evolution did in fact do a good designer's job. The essential point here is not so much finding optimal solutions to particular problems but to understand the ecological balance between task environment, sensory positioning – or more generally sensor morphology – , neural processing and behavior. In the following section we will explore aspects of this ecological balance.

3 Exploring ecological balance

In a number of places in the paper we mentioned the idea of ecological balance, the interdependence of task environment, morphology, and neural processing. Given that there is currently no theory about these interdependencies, we can try to gain some insights through empirical exploration. There are essentially two ways in which this can be done. First, we can study existing systems, natural systems, where we know that they are ecologically balanced because they are the result of natural evolution. Second, we can ask ourselves if the distribution of "labor" between morphology and neural substrate of a system, natural or artificial, is really optimal, given a certain task environment. Of course, the task environment of humans is almost impossible to describe comprehensively, and thus this is really a mute point. But we can define simple task environments and explore different ecological balances in these environments. Once the task environments are defined, we can use artificial evolution, as we will illustrate below.

The idea of exploring ecological balance can be nicely illustrated in the context of motion parallax which we present next. Thereafter we discuss the use of humanoids and morpho-functional machines to study ecological balance.

3.1 Motion parallax and the evolution of an insect eye

The basic phenomenon of motion parallax is very simple and everyone is familiar with it. Very briefly, when sitting in a train, a light spot, say from a tree or a house, travels very slowly across the visual field if it is towards the front and far away. The closer it gets and the more the spot moves to the side, the faster it moves across the visual field. This phenomenon has been studied in the context of insects. For example, Franceschini et al. [9] showed that in the eye of the fly there is a non-uniform layout of the visual axes such that sampling of the visual space is finer towards the front than laterally. If the fly is moving at constant speed the time for a light spot to move from one facet to the next remains essentially the same which implies that at the neural level a uniform array of elementary motion detectors can be used, i.e. neural processing for motion detection becomes very simple.

Let us follow up on this idea a little further and put artificial evolution to work on this problem of exploring ecological balance [18]. In contrast to standard evolutionary robotics practice the authors did not prescribe the morphology and then evolve the neural control. Rather, they started with a fixed neural substrate, i.e. an array of simple, uniform motion detectors. In this experiment the neural network is kept constant and the morphology of the "insect eye" is changed instead.

a.   b.

Fig. 6. Evolving eye morphology. (a) Top view of the setup of the experiments. The robot is equipped with a number of tubes that can be moved under program control. Light sensitive diodes have been placed at the inner end of the tubes (which mimic the facets) to ensure that a particular aperture is maintained. The vertical light tube represents the obstacle. (b) One distribution of the facets. The non-uniform spacing is clearly visible.

Figure 6a shows the experimental setup and the robot used for the experiments. The facets of the robot can be adjusted individually under program control. In the experiment the facets are adjusted by means of an evolution strategy [27]. The environment consists of vertical light tubes. The task of the robot is to maintain a critical lateral distance (not too far away, not too close) to the obstacle which is given by the vertical light tubes. Its fitness is increased if it manages to maintain the critical distance. The robot experiment takes about

5 to 6 hours until a stable distribution of the facets is achieved. Figure 6b shows the results of some runs with different initial conditions. The result is what one would theoretically expect: a non-uniform distribution which is more densely spaced towards the front. In other words, evolution has found a solution that enables the agent to perform its task with the neural substrate with which it was endowed. This is an example of how sensory morphology was not predesigned but emergent from an evolutionary process in which task environment and neural substrate was given. In this case evolution has exploited the potential for adaptation in the morphology of the sensory system.

To conclude this case study it is worth mentioning that in the field of computer and robot vision, aspects of this trade-off have been taken into account and exploited for various robot tasks. The approach is called space-variant sensing or space-variant vision (e.g. [8, 30]). We now turn to the next method for investigating ecological balance, humanoid robots

3.2 Humanoid robots

In recent years many researchers and research labs have taken to building humanoid robots, or briefly humanoids [2, 16, 33]. Humanoids are robots that superficially resemble humans. In the context of exploring ecological balance, they are of interest mainly for the following reasons. They have complex morphologies that can change dynamically (e.g. the hands, the flexible body) which is in contrast to the insect examples we just looked at. Complex morphologies are important if we want to investigate sensory-motor coordination. In contrast to humans, where little is known about the sensory data the neural system has to process, we can simply record sensory data on humanoids (and, of course, on all robots) and we can calculate various statistical measures. In this way we can acquire a good understanding of the sensory stimulation in relation to the morphology. And the sensory stimulation constitutes the "raw material" that the neural system has to process and on top of which, for example, cross-modal associations and later more abstract concepts can be formed. Let us now turn to the last approach for exploring ecological balance that we discuss in this paper, the morpho-functional machines.

3.3 Morpho-functional machines

Very briefly, morpho-functional machines which were originally proposed by Fumio Hara and his collaborators (e.g. [12]) are machines that change their morphology based on a particular situation, which in turn implies that they can perform different functions. Changing morphologies are very common in nature, but his idea has not been explored to date for engineering purposes. With only few exceptions, machines don't change their shape significantly during their lifetimes. As an example, Kawai and Hara [12] suggested a linear-cluster robotic system that has to solve a "baggage carrying problem". A linear-cluster robotic system consists of linearly connected autonomous mobile modules, each equipped with sensory and motor components. The "baggage carrying problem" consists of three subtasks: passing through a narrow and winding corridor, approaching and enclosing the "baggage", and carrying the "baggage" to a goal location.

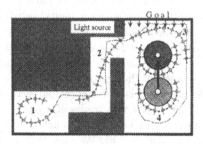

Fig. 7. Change of morphology as the linear-cluster robot is solving the "baggage carrying problem."

The neural system for controlling the individual modules has been hand-designed and is not changed during the experiments: what changes is only the morphology. The way different morphologies are achieved is by means of a head module that can pull in a particular direction, by friction, and by the movement of the individual modules. Luggage and goal are marked with different light sources that the modules can detect. Figure 7 shows the result of a simulation. (1) is the starting position. Then there are three types of morphologies or shapes: S-shaped as the robot is passing through the corridor (2), a kind of "arch" shape as the robot passes the corner in the upper right (3), and the semi-circle shape as the robot is enclosing the "baggage" and moving it towards the goal (top of figure). The head and tail modules send signals to the other modules as they are touching something. If they simultaneously touch a "baggage", the individual modules start moving towards the light source. Because they cooperate in this way they can move large pieces of "luggage." This case study illustrates that by applying the idea of morpho-functional machines, the space of potential designs can be enlarged by orders of magniute so that original solutions can be found by using flexible hardware components, i.e. flexible morphology, that would not be possible if the flexibility were restricted to software.

4 Discussion – moving towards cognition

We have looked at a large number of different case studies illustrating various ways in which embodiment – morphology, material properties – are related to neural processing. The question we have to ask now is how this all connects to the study of cognition. Before we discuss this point, a terminological comment is in order. Cognition is a descriptive term used by observers to describe certain kinds of activities of individuals. Clearly, performing a logical deduction is very different from riding a bicycle. The former would be termed cognitive, the latter sensory-motor. Ultimately, we would like to understand not only sensory-motor processes, but also cognitive ones. Now, the goal of AI and cognitive science has always been to discover the mechanisms underlying intelligent behavior. It turns out that as we try to model the mechanisms underlying cognitive processes, we soon discover that the distinction between cognitive and sensory-motor starts to blur, as mentioned earlier. Categorization, perception, but even memory processes turn out to be directly coupled to sensory-motor processes and thus to embodiment (e.g. [6, 10, 25]. Moreover, it has been demonstrated that performance on mental rotation tasks which are just about as "high-level" and cognitive as one can get, depends not only on the size of the angle through which the figure has to be rotated, but also on embodiment, i.e. on the particular sensory-motor task the subject is engaged in [32]. It has also been suggested that abstract relationships like transitivity can be explained as emergent from embodied interactions with the environment (Linda Smith, pers. comm.).

Now how does what we call cognition come about? What we have shown is the basic ways in which neural processing, morphology, and environment are interconnected. It is also clear that the neural substrate subtends all cognitive processes and so an explanation of cognition will have to resort to neural processes. A number of people have argued that we have to understand ontogenetic development if we are to explain cognition and we fully agree with their rationale (e.g. [4, 6, 7, 29]). From a developmental perspective the question is what the mechanisms are through which, over time, those processes which are initially highly sensory motor and are directly coupled to the system-environment interaction, become more and more "decoupled" from the environment, to our knowledge an unresolved research issue.

From developmental studies we also learn that exploration strategies play an essential role. For example, through object exploration, i.e. a particular kind of sensory-motor coordination, correlated sensory stimulation is generated in various sensory channels. From the perspective of embodiment, the resulting sensory stimulation depends on the position of the sensors, on the types of manipulations that are possible in the first place, i.e. on the morphology of the arm-hand system, and on how this arm-hand system can be controlled. The kind of control depends on the materials, which leads to particular movements, to particular sensory-motor coordinations, which in turn generate sensory stimulation. In some sense "everything is connected," which makes it hard to investigate these issues and which may, at least partially, explain why many researchers in AI prefer to stay with the computational paradigm even tough it is clear that it falls short of explaining intelligent behavior. By trying to come up with a set of principles (sensory-motor coordination, cheap design[1], ecological balance) and by the series of case studies presented we tried to demonstrate how, in spite of the "connectedness" of all aspects of agents, relevant issues in the study of intel-

---

[1] We should perhaps mention that in this paper we focused on cheap design. There is a complementary principle about redundancy that we did not discuss here for reasons of space limitations. However, redundancy is the prerequisite for adaptive behavior. For more details, see [25].

ligence can be productively investigated. We have been focusing on fairly basic processes. But if we want to explore the neural mechanisms by which abstract thinking eventually comes about we must understand the "raw material" the neural system has to process, and this raw material must not be seen, for example, exclusively in sensory stimulation but in the ecological balance of sensory stimulation, task environment, dyanmics, morphology, materials, and control.

One of the problems with the examples and ideas presented in this paper is that they are mostly qualitative. For sensory- motor coordination of a humanoid robot, we can calculate the correlations, but it is as yet unclear how we can quantitatively capture something like "ecological balance". Clearly, more quantitative statements will be required to make the story more compelling. But we hope that researchers will take up the challenges posed by embodiment.

The frequently asked question that must be asked at this point is whether this bottom-up approach, starting from the "raw material" will indeed scale up to high-level cognitive processes. Above we suggested that this is indeed the case. The jury is still out on whether this is a sound intuition or will turn out to be flawed. All we can do at the moment is outline a research program. But because embodiment provides a new perspective and many ideas for empirical studies on natural and artificial systems, as well as for new kinds of agents, we are optimistic that we can achieve a better understanding of cognition in the future.

Acknowledgments

I would like to thank Fumiya Iida, Lukas Lichtensteiger and Ralf Salomon for many valuable discussions and for their comments and their critical review of this paper.

References

1. Brooks, R.A. (1991). Intelligence without reason. Proc. IJCAI-91, 569-595.
2. Brooks, R.A., Breazeal, C., Marjanovic, M., Scassellati, B., Willimason, M.M. (1999). The Cog project: Building a humanoid robot. http://www.ai.mit.edu/projects/cog.
3. Chou, C.P., and Hannaford, B. (1997). Study of human forearm posture maintenance with a physiologically based robotic arm and apinal level neural controller. Biological Cybernetics, 76, 285-298.
4. Clark, A. (1997). Being there. Putting brain, body, and world together again. Cambridge, Mass.: MIT Press.
5. Dewey, J. (1896). The reflex arc in psychology. Psychol. Review, 3, 1981, 357-370. Reprinted in J.J. McDermott (eds.): The Philosophy of John Dewey. Chicago, IL: The University of Chicago Press, 136-148.
6. Edelman, G.E. (1987). Neural Darwinism. The theory of neuronal group selection. New York: Basic Books.
7. Elman, J.L, Bates, E.A., Johnson, H.A., Karmiloff-Smith, A., Parisi, D., Plunkett, K. (1996). Rethinking innateness. A connectionist perspective on development. Cambridge, Mass.: A Bradford Book, MIT Press, 421-430.

8. Ferrari, F., Nielsen, P.Q.J., and Sandini, G. (1995). Space variant imagin. Sensor Review, 15, 17-20.

9. Franceschini, N., Pichon, J.M., and Blanes, C. (1992). From insect vision to robot vision. Phil. Trans. R. Soc. Lond. B, 337, 283-294.

10. Glenberg, A. M. (1997). What memory is for? Behavioral and Brain Sciences. 20, 1-56.

11. Johnson, M. (1987). The body in the mind: The bodily basis of meaning, imagination, and reason. Chicago: Chicago University Press.

12. Kawai, N., and Hara, F. (1998). Formation of morphology and morpho-function in a linear-cluster robotic system. In R. Pfeifer, B. Blumberg, J.-A. Meyer, and S.W. Wilson (eds.). From Animals to Animats. Proc. of the 5th Int. Conference on the Simulation of Adaptive Behavior, SAB'98, 459-464.

13. Kobayashi, H., and Hara, F. (1995). A basic study on dynamic control of facial expressions for face robot. Proc. IEEE Int. Workshop on Robot and Human Communication, 275-280.

14. Kornbluh, R., Pelrine, R., Eckerle, J., Joseph, J. (1998). Electrostrictive polymer artificial muscle actuators. In Proceedings of the 1998 IEEE International Conference on Robotics and Automation. IEEE, New York, NY, USA, 2147-2154.

15. Kruschke, J.K. (1992). ALCOVE: An exemplar-based connectionist model of category learning. Psychological Review, 99, 22-44.

16. Kuniyoshi, Y., and Nagakubo, A. (1997). Humanoid as a research vehicle into flexible complex interaction. In Proc. IEEE/RSJ Int. Conf. Intelligent Robots and Systems (IROS). Grenoble, France, 811-819.

17. Lee, Y.K., and Shimoyama, I. (1999). A skeletal framework artificial hand actuated by pneumatic artificial muscles, IEEE International Conference on Robotics and Automation, 926-931.

18. Lichtensteiger, L., and Eggenberger, P. (1999). Evolving the morphology of a compound eye on a robot. To appear in Proc. of Eurobot'99.

19. Maris, M., and te Boekhorst, R. (1996). Exploiting physical constraints: heap formation through behavioral error in a group of robots. In Proc. IROS'96, IEEE/RSJ International Conference on Intelligent Robots and Systems, 1655–1660.

20. McGeer, T. (1990a). Passive dynamic walking. Int. Journal of Robotics Research, 9, 62-82.

21. McGeer, T. (1990b). Passive walking with knees. Proc. of the IEEE Conference on Robotics and Automation, 2, 1640-1645.

22. Pfeifer, R. (1996). Building "Fungus Eaters": Design principles of autonomous agents. In P. Maes, M. Mataric, J.-A. Meyer, J. Pollack, and S.W. Wilson (eds.): From Animals to Animats. Proc. of the 4th Int. Conf. on Simulation of Adaptive Behavior. Cambridge, Mass.: A Bradford Book, MIT Press, 3-12.

23. Pfeifer, R., and Scheier, C. (1997). Sensory-motor coordination: The metaphor andbeyond. Practice and future of autonomous agents [Special issue, R. Pfeifer and R. Brooks (Eds.)]. Robotics and Autonomous Systems, 20, 157–178.

24. Pfeifer, R., and Scheier, C. (1998). Representation in natural and artificial agents: an embodied cognitive science perspective. Zeitschrift für Naturforschung, 53c, 480-503.

25. Pfeifer, R., and Scheier, C. (1999). Understanding intelligence. Cambridge, Mass.: MIT Press.

26. Pratt, G.A., Williamson, M.M. (1995). Series Elastic Actuators. Proc. of IEEE/RSJ International Conference on Intelligent Robots and Systems (IROS), Pittsburgh, PA. 1, 399-406.

44

27. Rechenberg, I. (1973). Evolutionsstrategie: Optimierung Technischer Systeme nach Prinzipien der Beiologischen Evolution. Stuttgart: Frommann-Holzboog.
28. Scheier, C., Pfeifer, R., and Kuniyoshi, Y. (1998). Embedded neural networks: exploiting constraints. Neural Networks, 11, 1551-1569.
29. Thelen, E., and Smith, L. (1994). A dynamic systems approach to the development of cognition and action. Cambridge, Mass.: MIT Press, Bradford Books.
30. Toepfer, C., Wende, M., Baratoff, G., and Neumann, H. (1998). Robot navigation by combining central and peripheral optical flow detection on a space-variant map. Proc. Fourteenth Int. Conf. on Pattern Recognition. Los Alamitos, CA: IEEE Computer Society, 1804-1807.
31. Varela, F. J., Thompson, E., and Rosch, E. (1991). The embodied mind: Cognitive science and human experience. Cambridge, MA: MIT Press.
32. Wexler, M. (1997). Is rotation of visual mental images a motor act? In K. Donner (Ed.). Proceedings of the 20th European Conference on Visual Perception ECUP'97. London, UK: Pion Ltd.
33. Yamaguchi, J., Soga, E., Inoue, S., and Takanishi, A. (1999). Development of a bipedal humanoid robot: Control method of whole body cooperative dynamic biped walking. Proc. of the 1999 IEEE Int. Conference on Robotics and Automation, 368-374.

Natural Language Description of Image Sequences as a Form of Knowledge Representation

H.-H. Nagel[1,2]

[1] Institut für Algorithmen und Kognitive Systeme
Fakultät für Informatik der Universität Karlsruhe (TH)
D-76128 Karlsruhe, Germany
email nagel@ira.uka.de
[2] Fraunhofer-Institut für Informations- und Datenverarbeitung (IITB)
Fraunhoferstr. 1, D-76131 Karlsruhe, Germany

Abstract. An image sequence *evaluation* process combines information from different information sources. One of these sources is a *camera* which records a scene and provides the acquired information as a digitized image sequence. A different source provides *knowledge regarding signal processing and geometry*, exploited in order to map the image sequence signal to a system-internal representation of visible bodies and their movement in the depicted scene. Still another type of source provides *abstract conceptual knowledge* linking the system-internal geometric representation to tasks and goals of agents which act within the depicted scene or may influence it from the outside.
Rather than providing this third type of information for inference engines by 'handcrafted' rules or sets of axioms, it is postulated that this type of knowledge should be derived by algorithmic analysis of a suitably formulated natural language text: natural language text is considered as a genuine represention of abstract knowledge for an image sequence evaluation process. This hypothesis is studied for the example of a system which transforms *video sequences* of road scenes into *natural language text* describing the recorded actual traffic.
Keywords: Machine Vision, Image Sequence Evaluation, Situation, Fuzzy Metric-Temporal Logic, Knowledge Representation, Natural Language Text Understanding, Road Traffic, Vehicle Behavior.

1 Introduction

A natural language text constitutes a genuine represention of conceptual knowledge which can be exploited by an algorithmic system. Although such a statement may sound preposterous to many, I shall defend it in the remainder of this contribution. Habitual rules of scholarship recommend at least to qualify such a general statement which I shall do – eventually. First, however, I want to discuss why this subject is brought up *now*: it is considered neither an idiosyncrasy of the author nor an accidental event in the development of image (sequence) evaluation in general. In Section 3, I then outline an approach towards the evaluation of video sequences recording road traffic at innercity intersections. This should provide a framework for the introduction of a representation – called 'Situation Graphs (SG)' – of conceptual knowledge about the development of traffic situations at road intersections in Section 4. Section 5 illustrates the exploitation of SGs in order to track vehicles which *change their behavior while they are (more or less) occluded.*

A closer analysis of SGs in Section 6 will show that, not surprisingly, they exhibit a strong affinity to a logic representation of simple natural language statements describing what may happen at a road intersection. At this point, it is suggested to substitute the *explicit* construction of SGs by a process which algorithmically 'understands' a corresponding natural language text and transforms such a text into the required system-internal knowledge representation. Some of the difficulties to be expected for such an approach will be discussed, followed by suggestions how to overcome these problems. At this point of the exposition, the overall approach hopefully should be understood well enough so that a cautious qualification of the introductory statement may ease its acceptance as a serious contender for further investigations.

2 Extracting Complex Information from a Large Volume of Signal Data: Precursors to Current Developments

In the early seventies, it required a major effort to digitize videos of road traffic scenes and to tranfer the digitized image sequence into a digital computer. 25 years later, the 'Hamburg Taxi Sequence' – one of the first results of such an effort – still serves as a test sequence, documented by recent publications in leading international journals. Nowadays, using a PC with fast memory and bus transfer rates exceeding 30 MByte per second, digitization and storage of road traffic image sequences are no longer a real problem, not even for a small laboratory. Whereas in the seventies even simple change detection experiments required hours on the small to medium size computers available then for such experiments, a modern workstation can track a vehicle, depicted in an image region comprising up to several thousand pixels, from frame to frame in about a second. Simple (and thus fast, but usually less robust) approaches may even run in real-time.

As a consequence, experiments are no longer restricted to the detection of moving vehicles and to attempts to track them for short distances. The challenge currently consists in tracking vehicles under difficult boundary conditions (small vehicle images, highly varying contrasts, shadows, partial occlusions, etc.) for thousands of frames. Consider a vehicle which approaches an intersection, becomes partially occluded by a preceding vehicle or by others on neighboring lanes, slows down in front of a traffic light and progresses in stop-and-go cycles until it can finally cross the intersection, see Figure 1. More than mere computing power (and patience) is needed to cope with such challenges, in particular a *very robust tracking* approach. Otherwise, the vehicle might be lost – with the consequence that its behavior can not be determined reliably, at least not

Fig. 1. An image from the sequence 'stau02' recorded at the Durlacher-Tor-Platz Karlsruhe in order to study problems associated with the build-up and dissolution of vehicle queues in front of traffic lights.

without interactive help. Since such interactions are time consuming, they do not constitute a way out if a sample of interest for traffic engineers needs to be accumulated on short notice at reasonable cost.

The *progress* in image sequence evaluation indicated in the preceding paragraphs thus *dramatically increases the spatio-temporal scale of problems which become amenable to a systematic investigation*: rather than merely counting vehicles which pass a particular sampling area on a selected lane, one may now begin to study the behavior of specific vehicle configurations given particular traffic conditions at road intersections. The boundary conditions and results of such investigations will most likely be described at a conceptual rather than at a numerical level. The *quantitative* improvement in speed and reliability of image sequence evaluation thus induces the desire for a *qualitative* improvement in handling the results, namely an abstraction from the geometry of individual vehicle trajectories to the conceptual description of vehicle behavior in a particular traffic situation. *Technological* improvements in the design and production of computers – combined with the *methodological* improvements in image sequence evaluation – increasingly cause research groups, therefore, to study links between image sequences and their conceptual descriptions.

A simple, but not particularly powerful possibility consists in the provision of *sentence templates* into which selected items of information have to be inserted – see, e. g., [18]. In order to provide a *versatile* Human-Computer-Interface (HCI), the transformation of *geometric* descriptions resulting from an evaluation of image sequences into *conceptual* descriptions in the form of natural language texts is best decomposed into two component steps: on the one hand, based on formal logic, the selection and combination of instances of individual concepts and – on the other hand – the transformation of these selected pieces of information into a natural language text. Such an approach exploits the increasingly closer interaction between formal logic and natural language processing. Both of these disciplines have advanced considerably during the past three decades, independently from – but in parallel with – image sequence evaluation. A recent contribution to *AI magazine* documents the high esteem held by the international community regarding research into automatic deduction in Germany [15]. For a non-specialist, advances in natural language text understanding – in particular its close connection to formal logic – are treated in an exemplary manner by Kamp and Reyle in the book 'From Discourse to Logic' [13].

These advances for themselves, however, would not suffice to facilitate the design and implementation of an algorithmic link between image sequence evaluation and natural language processing. Image sequences are inherently related to time-varying scenes. In order to accomodate this aspect, formal logic had to take time into account – a significant extension, increasingly investigated during the past decades. Similarly, logic has been extended in order to cope with *spatial* in addition to *temporal* reasoning, see [2]. There are still two other aspects for which solution approaches had to become available in order to link image sequence evaluation and logic-based natural language text processing: a further extension to a *metric*-temporal logic and the accommodation to various types of uncertainty regarding the transition from video signals to natural language terms.

Consumer video recording in Europe samples the incoming spatio-temporal light intensity distribution at 50 (interlaced) half-frames per second, thus providing a 'natural' unit of time, namely 20 msec: according to Shannon's sampling theorem, any change occurring within a time interval substantially shorter than 40 msec can not be properly evaluated from a digitized video sequence. The discretization of time as a concatenation of basic (20 msec) intervals allows to treat *metric* temporal aspects relevant for the evaluation of digitized videos.

It is advantageous, moreover, to distinguish three types of uncertainty since each type is associated with different difficulties. Starting from digitized video signals, image sequence evaluation has to take into account the *stochastic nature* of the basic data to be evaluated. A second aspect concerns the *evaluation artefacts* due to unavoidable simplifications designed into the algorithms and – unfortunately, but realistically – due to undetected errors or inappropriate parameterisations of the evaluation program. This second type of uncertainty can result in a much more erratic and brittle performance than properly treated stochastic errors. Usually, the more prominent evaluation artefacts have been detected and removed in the course of extended tests so that the remaining artefacts can be difficult to distinguish from stochastic effects. The third type of uncertainty is related to the *vagueness* of concepts encountered at the logical and natural language processing level. In order to accommodate such uncertainties, a metric-temporal logic has been extended even further by the incorporation of *fuzzy reasoning*, see [20].

All of the methodological developments mentioned in the preceding paragraphs *have been necessary* in order to link the evaluation of digitized video sequences to a level of conceptual processing, facilitating an appropriate Human-Computer-Interface. It thus should not come as a surprise that *only now* serious efforts are started to investigate natural language text as a potential form of knowledge representation in the context of image sequence evaluation systems.

3   Outline of an Overall System Structure

The importance of *explicitly* represented conceptual knowledge about an application domain will be illustrated by an example which in turn can be best understood by first sketching the overall structure of an advanced image sequence evaluation system. A more detailed recent discussion of this approach can be found in [16]. This system has been implemented, but is not considered to be finished! Salient aspects of this structure are presented in the form of a hierarchy of layers which emphasizes the forward/upward flow of information, suppressing feedback links and layer-jumping connections. In addition, I do not distinguish between *processing* (or transformation) layers and layers *representing intermediate results*.

1. The signal or *source* layer provides digitized image sequences, either directly input from a video camera or read-in from background storage.
2. The *Picture Domain (PD)* processing layer transforms digitized video signals into image regions which tentatively correspond to stationary scene components or to images of moving vehicles. In addition, the 2D frame-to-frame motions of such regions in the image plane are estimated, see [17].
3. The *Scene Domain (SD)* processing layer introduces *explicit* knowledge about the three-dimensional structure of a depicted scene in the form of body models – representing either passive 'objects' or active 'agents' in the scene – and their admissible 3D trajectories. Based upon explicit body models (usually approximated by polyhedra) and information about light sources, processing takes into account shadow casting and occlusion relations, see [6].
4. The *Geometric/Conceptual-Interface (G/C-I)* layer links the geometric representation resulting from the processing at the preceding SD layer to a conceptual representation. This representational transformation applies both to bodies and to their motion. Based on 'fuzzy logic' conversion rules, stochastic uncertainties as well as processing artefacts originating in the preceding layers will thereby be merged with the vagueness of conceptual terms. Although these transformations did not yet generate occasions for concern so far, no claim is made that they are free from distortion or bias.
5. The *Conceptual Processing (CP)* layer (based upon a fuzzy metric-temporal logic as mentioned above [20]) allows to select and combine elementary concepts provided by the G/C-I layer in order to generate a conceptual representation of the output according to what the system's user specified.

6. The *Natural Language Processing (NLP)* layer accepts the conceptual representation provided by the CP layer and converts it into a natural language text, see [4].

7. The *Human-Computer-Interface (HCI)* layer combines the natural language text generated by the NLP layer with (possibly magnified) image regions cropped from original image frames in areas specified implicitly by intermediate results or explicitly by the user – potentially supplemented by overlays illustrating particular processing aspects – and makes these available to the user. Any user commands will be accepted at this layer, too, and will be decoded and forwarded to the required layer for further processing.

4 Representing Conceptual Knowledge about Vehicle Behavior: the Situation Graph (SG)

Exploitation of a-priori knowledge about a scene – in particular about its temporal variation while it is recorded by a video camera – can substantially improve the extraction of information about temporal developments in the depicted scene. This has been shown repeatedly for the extraction of geometric information regarding vehicle trajectories, i. e. for SD processing. Although it may appear plausible that such a statement applies to conceptual processing, too, a note of caution could be advisable in the case of image sequence evaluation.

As outlined above, information about vehicle behavior is extracted from geometric information provided by SD processing and forwarded via the G/C-I layer. In such a feed-forward structure, a-priori conceptual knowledge about vehicle behavior *represented at the CP layer* can not influence the geometric processing in the SD layer. Knowledge at the CP layer about behavior could thus only facilitate the recognition of inconsistencies, allowing to reject inconsistent intermediate results fed forward from the preceding G/C-I layer. Apart from the potential for such consistency checks, however, a-priori knowledge about vehicle behavior needs to be available in order to combine elementary conceptual representations into more complex ones for the subsequent preparation of appropriate natural language text.

Vehicle behavior is formalized here as a concatenation of *generically describable situations (gd_situations)*: a gd_situation schema comprises a state description schema and an associated action schema. The state description schema is formulated as a conjunction of predicates relating to particular (conceptually formulated) properties of an agent and his environment. It is understood that an agent will execute the action associated with a gd_situation if his current state can be instantiated from the state-schema component of the selected gd_situation. Instantiation is performed by a process operating on the information provided by the G/C-I layer. The action schema can be parameterized by functions of variables which appear in the argument expressions of predicates in the state schema. A successful instantiation of a state schema will thus provide parameter values which completely specify the action to be performed by an agent in the respective gd_situation.

Representations for different gd_situations can be considered as nodes forming a graph, the so-called 'Situation Graph (SG)'. A path through such a SG corresponds to a particular concatenation of gd_situation nodes. Each node implies the execution of an action by an agent traversing the SG along such a path. In the case of a SG comprising vehicle states at road traffic intersections, the concatenation of actions along a path through the SG corresponds to a sequence of maneuvers to be performed by a vehicle.

In addition to the a-priori knowledge encoded into state and action schemata and their association, a-priori knowledge about vehicle behavior is coded, too, into the selection of edges admitted between nodes of the SG as well as into edge priorities: edges linking a node to different successor nodes are prioritized according to the design preferences with which a process traversing the SG should attempt to instantiate successor nodes. An edge leading back to the node from which it originates provides the opportunity to have a vehicle being observed in a particular state for several consecutive time units.

Since situations at innercity road intersections can be quite complex, related SGs thus quickly become difficult to analyze and to modify. It has been decided, therefore, to represent SGs in the form of *hypergraphs* where a node may be decomposed hierarchically into subgraphs which 'inherit' attributes of their ancestor nodes. The traversal rule has been modified so that the process corresponding to a vehicle will search the SG in depth-first manner, always attempting to instantiate the most specialized node compatible with the elementary conceptual data about a vehicle as forwarded from the SD layer via the G/C-I layer.

Figure 2 shows a five-level hierarchy representing in schematic form the behavior of a vehicle which crosses an intersection. In order to fit this schematic presentation within the space available, some of the predicates qualifying a node and the action part of each node have been omitted in this Figure. The root node at Level I corresponds to the notion 'cross an intersection'. The hierarchical decomposition introduces at the next more specific Level II a concatenation of gd_situations for 'drive_to_intersection', 'drive_on_intersection', and 'leave_intersection'. Level III introduces further specializations – which address cycles of stopping, waiting, restarting, and proceeding – either during the approach phase or during the phase where the vehicle actually is on the intersection proper. Level IV essentially refines the admissible situations further by taking into account whether or not a vehicle in question follows another one. Level V eventually differentiates between the situations according to whether a vehicle turns off or proceeds straight ahead across the intersection.

5 Feedback from the Conceptual Processing to the Scene Domain Layer: an Example

The fastback enclosed by a box in Figure 1 is about to be occluded by a large traffic direction indicator board – see, too, Figure 3. In order to cope with such a configuration, one could assume that the fastback continues at about the same speed and direction which have been estimated prior to the beginning occlu-

52

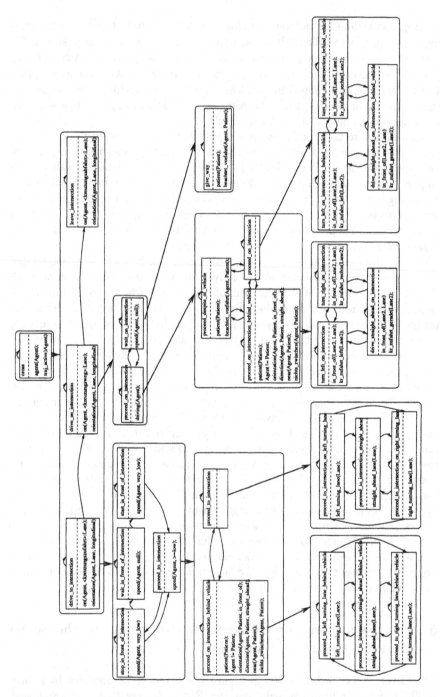

Fig. 2. 'Situation Graph' representing a-priori knowledge about the behavior of a vehicle which crosses a road intersection (see text for further explanations).

sion: based on this assumption, the trajectory could be extrapolated during the occlusion period.

In this particular case, however, such an approach will fail. Since a van in the same lane as the fastback has come to a stop in front of a traffic light, the fastback has to slow down and stop, too, to avoid crashing into the van. The 'behavior' of the fastback thus changes significantly during the occlusion period which invalidates a simple extrapolation on the basis of prior estimates. Such a situation, therefore, requires to extend the spatio-temporal scale of considerations which have to be taken into account.

Since the model-based tracking approach operating in the SD 'knows' about the spatial arrangement of the recording camera, the traffic direction indicator board, and the fastback in the scene, it can predict – based on the trajectory of the fastback estimated *prior* to the impending occlusion – that an occlusion is imminent. Although the van has been tracked, too, and thus its position in the scene is known to the system, the implicit relation between the stopped van and the approaching fastback, however, is not taken into account *by SD tracking*. One could debate whether this should be done at the SD level, but this alternative has been ruled out since a purely kinematic approach does not appear to constitute the appropriate level of abstraction to handle such a situation: the number of potential relations to be taken into account and the complex numerical quantification of potential reactions by the 'agents' involved (in this case the fastback and, as will be seen in a later stage, also the van) tipped the balance against such an attempt.

It is at this point where feedback from the CP layer appears to be the better choice. The leftmost bottom box from Figure 2 comprises the most specialized maneuver 'proceed_to_intersection_straight_ahead_behing_vehicle'. Its 'grandparent' node comprises – at a less specific level – a subgraph representing a stop-and-go cycle. Since this state schema is inherited, the leftmost bottom node from Figure 2 represents in schematic form the 'behavior' expected from the fastback shown in Figure 1. The system thus has the capability to instantiate a case where the fastback switches from a situation of unrestricted approach towards the intersection into one where 'it follows the preceding van'.

As is illustrated in Figure 3, the fastback has to *approach* the van which has stopped on the same lane in front of the fastback. Since the fastback becomes occluded by the direction indicator board, this expected slowdown *can not be observed directly*. As a consequence, the state vector describing the motion of the fastback has to be subjected to an additional input, counteracting a mere prediction of the current velocity vector during the next time unit. 'Distance control' – which prevents a crash into the preceding vehicle by assuring that a minimum distance between vehicles is preserved – replaces 'velocity control', i. e. the tendency to retain the same velocity as at previous time intervals.

The G/C-I layer constitutes the appropriate level where to realize this switch between control regimes. The CP layer can infer qualitatively that the lane ahead of the fastback is blocked by the van. As a consequence, the situation schema 'proceed_to_intersection_straight_ahead_behing_vehicle' can be instantiated for

54

State schema:

modus(Agens ,driving)
modus(Patiens,standing)

State instantiation:

modus(obj_7,driving)
modus(obj_9,standing)

Action schema:

approach_preceding_
vehicle (Agens,Patiens)

Action instantiation:

approach_preceding_
vehicle (obj_7,obj_9)

State schema:

modus(Agens,standing)
modus(Patiens,standing)

State instantiation:

modus(obj_7,standing)
modus(obj_9,standing)

Action schema:

wait_behind_preceding_
vehicle (Agens,Patiens)

Action instantiation:

wait_behind_preceding_
vehicle (obj_7,obj_9)

State schema:

modus(Patiens,driving)

State instantiation:

modus(obj_9,driving)

Action schema:

start_up_behind_preced-
ing_vehicle(Agens,Patiens)

Action instantiation:

start_up_behind_preced-
ing_vehicle(obj_7,obj_9)

Fig. 3. Supplementary predicates for the SG of Figure 2, corresponding to a specialization of 'drive_to_intersection' via the set of alternatives 'stop_in_front_of_intersection', 'wait_in_front_of_intersection', 'start_in_front_of_intersection', 'proceed_to_intersection' to the corresponding further specializations obtained by adding to each alternative a new condition 'behind_preceding_vehicle' in the action schema. The left column shows the window selected in Figure 1 for half-frame 1080 (top left panel), 1200 (center), and 1450 (bottom) with the corresponding estimated projected model positions for two vehicles superimposed ('obj_7' in blue and 'obj_9' in green). Note the superposition (in violet) of the projected model for the occluding direction indicator board and the mast carrying this board, as well as the shadows cast by modeled objects onto the road.

the 'fastback agent'. The resulting action consists in 'fastback approaching van', implying that the fastback has to slow down. This qualitative conclusion is fed back to the G/C-I layer where the quantitative data about both agents are available in order to determine (so far based on default assumptions about normal deceleration in such a case) the numerical value for deceleration of the fastback. This additional control input is fed into the prediction phase of the Kalman-Filter which tracks the fastback at the SD layer. In due course, the fastback model will come to a complete stop although the fastback itself is invisible to the recording camera – see the panel in the center row of Figure 3.

As soon as the traffic light has switched to green and the van begins to accelerate, the wait-state for the fastback will be assumed to be terminated, too, since it is still in the situation 'proceed_to_intersection_behind_vehicle'. The distance control regime at the G/C-I layer obtains the conclusion from inferences at the CP layer that the wait for the fastback can be replaced again by a motion. The G/C-I layer begins to feed a (default) value for an acceleration back to SD tracking whereupon the model for the fastback begins to move again. After several time units, the state-update phase of the Kalman-Filter which tracks the fastback will determine that the model for the fastback is no longer fully occluded by the direction indicator board: edge elements around the visible segments of the model are matched again to the latter. Provided all these steps are parameterized and executed properly, the projection of the fastback model catches onto the reappearing fastback image and tracking can continue in the normal manner, illustrated by the bottom panel in Figure 3.

This is but one example for feedback from the CP layer down to the SD layer studied by [6]. It demonstrates the interaction between qualitative inferences at the CP layer and quantitative tracking operations in the SD layer. Each layer performs the tasks best suited to it, thereby considerably simplifying the design, implementation, debugging, and maintainance of the overall system.

6 A Critical Look at Situation Graphs

The preceding section illustrated how a-priori knowledge about vehicle behavior is represented by a SG and exploited by a fuzzy metric-temporal inference engine. The interaction between qualitative reasoning at the CP layer and quantitative processing at the SD layer demonstrated by this example supports the hypothesis that SGs constitute a reasonable representation for (vehicle) behavior – to be investigated further.

As explained in Section 4, the SG in Figure 2 is structured into five levels, each level consisting of one row comprising one or more *outer* boxes. Each outer box contains either a single situation node (in the top row) or a sub-SG. A *specialization link* (downward pointing arrow) connects a parent node to the corresponding subgraph which is enclosed by an outer box, too. At any given time, an agent can activate only one node at each level, but an entire branch in the tree-like structure from the root to a leaf node. The packaging of knowledge as a (situation) *graph* should not distract from the fact that this knowledge is eventually represented as logic formulas. A given SG is (hand-)translated into a formal langual SIT++ (see [20]) from where it is automatically compiled into the logic representation (see [7]) evaluated by a fuzzy metric-temporal inference engine [20].

Upon closer inspection of Figure 2, each node identifier (given at the top of an inner box representing a SG-node, above the dotted line) can be looked at as a predicate. This becomes true if the conjunction of constituent predicates given below the dotted line are satisfied by an appropriate mapping of their arguments to individuals extracted by image sequence evaluation, i. e. by an instantiation of this SG node schema. An outer box encloses a subgraph which corresponds to a disjunction of the predicates represented by node identifiers. Since instantiation of a child node in a SG implies instantiation of its parent node, a particular situation is characterized by the conjunction – one node at each level in a sub-SG-hierarchy – of a disjunction of situation nodes.

Predicate identifiers can be read immediately as verbphrases. With some exceptions – due to historical traces of the development – such a verbphrase comprises more prepositional noun phrases the closer the corresponding node is to a leaf node, i. e. the more specifications have been added in the course of conceptual differentiation. Usually, the specification added upon further differentiation of a node corresponds to a(n) additional predicate(s) which turn(s) up below the dotted line of the child-node. Not surprisingly, a node is constituted by the conjunction of predicates: one corresponds to a verb and the others either to a prepositional noun phrase, an adverbial phrase, or an attribute. The subject noun phrase is missing since it is by definition represented by the agent on behalf of which a process traverses the SG.

During the gradual development of this SG, pushed by the necessity to accomodate a growing and increasingly complex set of situations encountered during the evaluation of ever more image sequences, knowledge about vehicle behavior at intersections has been encoded in conceptual terms which come rather close to a set of natural language sentences. This observation more or less reflects the inductive manner in which conceptual knowledge about traffic situations has been gradually encoded into a SG. Since – as has been shown in the preceding section – this knowledge representation can be exploited algorithmically, let us take a next step: can we replace the hand-construction of a SG by natural language text analysis, for example according to the Discourse Representation Theory (DRT)? Kamp and Reyle [13] describe in detail how entire discourses are to be converted by an algorithm into Discourse Representation Structures (DRS) and how a set of DRSs can be turned systematically into logic formulas, see [13, Chapter 1.5]. Rather than handcrafting a SG which is then translated by a human specialist into a SIT++ representation, it appears more attractive to *formulate knowledge* about the behavior of vehicles at intersections *as a text*. This text can be translated automatically via DRSs into a set of logic formulas which then have to be instantiated by suitable mapping to the results of image sequence evaluation – with other words, the textual description of expected vehicle behavior at intersections directly specifies a (logic) knowledge representation schema.

It appears worthwhile at this point to study some of the implications of this suggestion. There is no question that the SG as shown in Figure 2 – or, better, the knowledge to be supplied to an evaluation system for image sequences of road intersection scenes – has to be supplemented by additional variants of vehicle behavior. Sticking to a tree-like structure has the consequence that certain predicates, for example the one corresponding to the prepositional phrase 'behind_vehicle', will show up repeatedly in different branches. This observation brings up another question, namely in which order predicates should be appended during repeated specialization of a concept. Subgraphs in the lower right quadrant of Figure 2 illustrate – albeit not in a fully expanded form, due to space limitations – that in principle one has to incorporate both permutations of two differentiating attributes, 'behind_vehicle' and the disjunction of 'turn_left', 'drive_straight_ahead', or 'turn_right'. A tree-like specialization structure lets the SG expand quickly, potentially impeding clarity of understanding. In a similar vein, the prioritized links between nodes of the same subgraph can create difficult questions during the design phase and might necessitate experiments in order to determine an appropriate collection of situation nodes for a subgraph.

On the positive side, the prioritized links guide the instantiation process and thus enable a more efficient graph traversal. The tree-like specialization structure of the SG and the depth-first search during traversal never leaves in doubt to which parent node the search has to return if a predicate can no longer be instantiated. Since each node is associated with an action, the next action to be performed thus is always well defined.

It is expected to overly complicate knowledge input by natural language text if *prioritized* links between nodes have to be specified. Should graph traversal eventually become fast enough, this aspect could simply be dropped. An alternative would consist in implementing an analogy to the 'working set' approach in managing virtual memory spaces, i. e. one would have to monitor the (time-varying) frequency with which subsets of nodes will be visited successfully and prioritize the associated links dynamically during subsequent graph traversals.

Much more difficult appears the following problem: the history of a graph traversal – corresponding to the sequence of maneuvers executed by a vehicle agent – can easily be lost if the next situation node (and thereby the next action to be executed) is selected merely according to the satisfiability of a set of logic formulas at a particular instant in time. As has been observed, even minor changes in intermediate results – due to, for example, sensor noise – can severely influence selections unless these are stabilized by a kind of low-pass temporal filtering based on (at least short-term) history. We have experimented with beam-search approaches in order to be able to retain some situation selection even if it is not the most favored one at a particular instant, for example if (presumably by accident) some more special node can be successfully instantiated for a few half-frames. The SIT++ language provides means to specify minimal and maximal durations for certain states, thus allowing to accomodate some of the more qualitative criteria into the logic formalism.

The graphical representation of a SG in the form outlined above thus comprises some means which simplify the design of experiments. No experience is available yet regarding alternative means if knowledge about vehicle behavior at intersections will be provided in the form of a natural language text: although it will always be possible to introduce a natural language equivalent for direct specification of particular paths in a SG, the substitution of conventions by automatic in-depth analysis of alternative path configurations within a tentative SG could be more attractive in the long run.

7 Discussion and Outlook

The effort to extend image sequence evaluation into an area where it has to be performed efficiently with large volumes of data necessitates to integrate methods from several disciplines: signal processing, geometry, control, spatial and temporal reasoning, natural language text processing, and systems engineering certainly have to be taken into account. The abstraction process – necessitated by the large volume of data to be handled – enforces an incorporation of formal logic into the structure of an image sequence evaluation system. The representation and use of knowledge about behavior increasingly illustrates this need.

Although road traffic constitutes an application domain which comprises numerous challenges for image (sequence) evaluation as well as for spatial and metric-temporal reasoning, it appears too early to rank published approaches. More experience should be collected with various alternatives regarding knowledge representation for (vehicle) behavior, see [16] for pointers to earlier literature. Additional contenders for knowledge representation regarding behavior are *Bayesian Networks* – e. g., the survey in [1], the study of [8], and experiments reported in [18] – and *Semantic Networks*, see for example the monograph by Sagerer and Niemann [19]. Research by Intille and Bobick [10], although not refering to the road traffic domain, exemplifies still another approach towards the representation of behavior. These authors emphasize the recognition of activities performed by an entire group of interacting agents, instantiating their models on data from (interactively) evaluated video recordings of American Football.

Links between behavioral modeling and control engineering clearly begin to emerge – albeit in a rather early stage, see [3]. In a similar vein, the connection between image (sequence) understanding and logic becomes closer as exemplified by the study of [9]. Whereas the approaches mentioned so far explored either the use of particular representation formalisms (Bayesian Networks, Semantic Networks, Formal Logic) or were primarily oriented towards control ([3]), a number of recent investigations study predominantly data-driven approaches towards image sequence evaluation in order to cluster road traffic developments for subsequent conceptual categorization as exemplified by [5, 14, 11], with [12] as precursor.

Discussion of an encompassing, already fairly complex system for the evaluation of digitized video sequences recorded at road intersections has shown that knowledge represented at the conceptual level can contribute to system robustness. Upon closer inspection, it turned out that the knowledge about the behavior of vehicles crossing road intersections has been represented more or less unnoticed in a form which exhibits close affinity to a natural language text. This led to the idea to use established means of natural language text analysis in order to transform knowledge formulated as a text into the logic structure required by the Conceptual Processing layer of an image sequence evaluation system. The realization of such an approach would obviate the ever more complicated task of handcrafting 'Situation Graphs' for subsequent translation into the required logic structures. Knowledge represented in the form of natural language text could be converted algorithmically into logic formulas which would allow to access the a-priori knowledge at any desirable level of abstraction. The increasingly closer interaction between computational linguistics and logic lets it appear plausible that such tools will become available in the near future. It will have the advantage that standard tools for extension, adaptation, and maintainance of texts may be used – avoiding the necessity to construct the required tools ad-hoc in the context of designing image (sequence) evaluation systems.

The qualification promised at the beginning: I consider the initial statement as a hypothesis, a challenging one, as this contribution attempted to show. It will certainly require considerable efforts to validate it – even if it is restricted to road traffic scenes.

Acknowledgments

My thanks go to Th. Christaller whose invitation to discuss links between image evaluation and logic at KI '99 stimulated this exposition of current research, to R. Gerber and M. Haag for fruitful discussions, and to Th. Müller for comments on a draft version. Figure 3 has been adapted from a German version designed by M. Haag. I thank the Deutsche Forschungsgemeinschaft for encouraging support extended over a long period.

References

1. H. Buxton and S. Gong: Visual Surveillance in a Dynamic and Uncertain World. *Artificial Intelligence* **78** (1995) 431–459.
2. A.G. Cohn: Qualitative Spatial Representation and Reasoning Techniques. In G. Brewka, Ch. Habel, and B. Nebel (Eds.): *KI-97: Advances in Artificial Intelligence*, Proc. 21st Annual German Conference on Artificial Intelligence, 9-12 September 1997, Freiburg, Germany. LNAI 1303, Springer-Verlag Berlin Heidelberg New York/NY 1997, pp. 1-30.
3. E.D. Dickmanns: Vehicles Capable of Dynamic Vision: A New Breed of Technical Beings? *Artificial Intelligence* **103** (1998) 49–76.
4. R. Gerber and H.-H. Nagel: (Mis-?)Using DRT for Generation of Natural Language Text from Image Sequences. In *Proc. Fifth European Conference on Computer Vision (ECCV'98)*, 2-6 June 1998, Freiburg/Germany; H. Burkhardt and B. Neumann (Eds.), Lecture Notes in Computer Science LNCS 1407 (Vol. II), Springer-Verlag Berlin Heidelberg New York/NY 1998, pp. 255-270.

5. J. Fernyhough, A.G. Cohn, and D.C. Hogg: Building Qualitative Event Models Automatically from Visual Input. In *Proc. Sixth International Conference on Computer Vision (ICCV'98)*, 4–7 January 1998, Bombay/India, pp. 350–355

6. M. Haag: *Bildfolgenauswertung zur Erkennung der Absichten von Straßenverkehrsteilnehmern*. Dissertation, Fakultät für Informatik der Universität Karlsruhe (TH), Juli 1998. Erschienen in 'Dissertationen zur Künstlichen Intelligenz' DISKI **193**, infix–Verlag St. Augustin 1998 (in German).

7. M. Haag and H.-H. Nagel: Incremental Recognition of Traffic Situations from Video Image Sequences. In *Proc. ICCV-98 Workshop on Conceptual Descriptions of Images (CDI-98)*, H. Buxton and A. Mukerjee (Eds.), 2 January 1998, Bombay/India; Indian Institute of Technology, Kanpur/India 1998, pp. 1–20.

8. R.J. Howarth: Interpreting a Dynamic and Uncertain World: Taks-Based Control. *Artificial Intelligence* **100** (1998) 5–85.

9. T. Huang and S. Russell: Object Identification: A Bayesian Analysis with Application to Traffic Surveillance. *Artificial Intelligence* **103** (1998) 77–93.

10. St. Intille and A. Bobick: Visual Recognition of Multi-Agent Action Using Binary Temporal Relations. In *Proc. IEEE Conference on Computer Vision and Pattern Recognition (CVPR'99)*, 23–25 June 1999, Fort Collins, Colorado, Vol. 1, pp. 56–62.

11. Y. Ivanov, Ch. Stauffer, A. Bobick, and W.E.L. Grimson: Visual Surveillance of Interactions. In *Proc. Second IEEE International Workshop on Visual Surveillance (VS'99)*, 26 June 1999, Fort Collins, Colorado, pp. 82–89.

12. N. Johnson and D. Hogg: Learning the Distribution of Object Trajectories for Event Recognition. *Image and Vision Computing* **14**:8 (1996) 609–615.

13. H. Kamp and U. Reyle: *From Discourse to Logic*. Kluwer Academic Publishers Dordrecht Boston London 1993.

14. V. Kettnaker and M. Brand: Minimum-Entropy Models of Scene Activity. In *Proc. IEEE Conference on Computer Vision and Pattern Recognition (CVPR'99)*, 23–25 June 1999, Fort Collins, Colorado, Vol. 1, pp. 281–286.

15. D.W. Loveland: Automated Deduction: Looking Ahead. *AI magazine* **20**:1 (Spring 1999) 77-98.

16. H.-H. Nagel: Kann ein Rechner schon wahrnehmen oder sieht ein Informatiker die Dinge nur anders? In *Dynamische Perzeption*, S. Posch und H. Ritter (Hrsg.), Proceedings in Artificial Intelligence Vol. 8, infix-Verlag Sankt Augustin 1998, pp. 192-215 (in German).

17. H.-H. Nagel and A. Gehrke: Bildbereichsbasierte Verfolgung von Straßenfahrzeugen durch adaptive Schätzung und Segmentierung von Optischen-Fluß-Feldern. In *20. DAGM-Symposium 'Mustererkennung 1998'*, 29. September–1. Oktober 1998, Stuttgart/Germany; P. Levi, R.-J. Ahlers, F. May und M. Schanz (Hrsg.), Informatik aktuell, Springer-Verlag Berlin Heidelberg New York/NY 1998, pp. 314–321 (in German).

18. P. Remagnino, T. Tan, and K. Baker: Multi-Agent Visual Surveillance of Dynamic Scenes. *Image and Vision Computing* **16**:8 (1998) 529–532.

19. G. Sagerer and H. Niemann: *Semantic Networks for Understanding Scenes*. Plenum Press New York/NY 1997.

20. K. Schäfer: *Unscharfe zeitlogische Modellierung von Situationen und Handlungen in Bildfolgenauswertung und Robotik*. Dissertation, Fakultät für Informatik der Universität Karlsruhe (TH), Juli 1996. Erschienen in 'Dissertationen zur Künstlichen Intelligenz (DISKI)' **135**, infix-Verlag Sankt Augustin 1996 (in German).

Knowledge Discovery in Spatial Databases

Martin Ester, Hans-Peter Kriegel, Jörg Sander

Institute for Computer Science, University of Munich
Oettingenstr. 67, D-80538 Muenchen, Germany
{ester I kriegel I sander}@dbs.informatik.uni-muenchen.de
http://www.dbs.informatik.uni-muenchen.de

Abstract. Both, the number and the size of spatial databases, such as geographic or medical databases, are rapidly growing because of the large amount of data obtained from satellite images, computer tomography or other scientific equipment. Knowledge discovery in databases (KDD) is the process of discovering valid, novel and potentially useful patterns from large databases. Typical tasks for knowledge discovery in spatial databases include clustering, characterization and trend detection. The major difference between knowledge discovery in relational databases and in spatial databases is that attributes of the neighbors of some object of interest may have an influence on the object itself. Therefore, spatial knowledge discovery algorithms heavily depend on the efficient processing of neighborhood relations since the neighbors of many objects have to be investigated in a single run of a typical algorithm. Thus, providing general concepts for neighborhood relations as well as an efficient implementation of these concepts will allow a tight integeration of spatial knowledge discovery algorithms with a spatial database management system. This will speed-up both, the development and the execution of spatial KDD algorithms. For this purpose, we define a small set of database primitives, and we demonstrate that typical spatial KDD algorithms are well supported by the proposed database primitives. By implementing the database primitives on top of a commercial database management system, we show the effectiveness and efficiency of our approach, experimentally as well as analytically. The paper concludes by outlining some interesting issues for future research in the emerging field of knowledge discovery in spatial databases.

1 Introduction

Knowledge discovery in databases (*KDD*) has been defined as the process of discovering valid, novel, and potentially useful patterns from data [9]. *Spatial Database Systems* (*SDBS*) (see [10] for an overview) are database systems for the management of spatial data. To find implicit regularities, rules or patterns hidden in large spatial databases, e.g. for geo-marketing, traffic control or environmental studies, spatial data mining algorithms are very important (see [12] for an overview).

Most existing data mining algorithms run on separate and specially prepared files, but integrating them with a *database management system* (*DBMS*) has the following advantages. Redundant storage and potential inconsistencies can be avoided. Furthermore, commercial database systems offer various index structures to support different types of database queries. This functionality can be used without extra implementation effort to speed-up the execution of data mining algorithms. Similar to the relational standard query language SQL, the use of standard primitives will speed-up the development of new data mining algorithms and will also make them more portable.

In this paper, we introduce a set of database primitives for mining in spatial databases. [1] follows a similar approach for mining in relational databases. Our database primitives (section 2) are based on the concept of neighborhood relations. The proposed primitives are sufficient to express most of the algorithms for spatial data mining from the literature (section 3). We present techniques for efficiently supporting these primitives by a DBMS (section 4). Section 5 summarizes the contributions and discusses several issues for future research.

2 Database Primitives for Spatial Data Mining

The major difference between mining in relational databases and mining in spatial databases is that attributes of the neighbors of some object of interest may have an influence on the object itself. Therefore, our database primitives (see [7] for a first sketch) are based on the concept of spatial neighborhood relations.

2.1 Neighborhood Relations

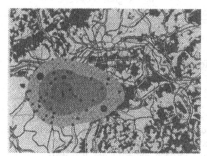

Fig. 1. Regions of pollution around a planned industrial plant [3]

The mutual influence between two objects depends on factors such as the topology, the distance or the direction between the objects. For instance, a new industrial plant may pollute its neighborhood depending on the distance and on the major direction of the wind. Figure 1 depicts a map used in the assessment of a possible location for a new industrial plant. The map shows three regions with different degrees of pollution (indicated by the different colors) caused by the planned industrial plant. Furthermore, the influenced objects such as communities and forests are depicted.

We introduce three basic types of binary spatial relations: topological, distance and direction relations. Spatial objects may be either points or spatially extended objects such as lines, polygons or polyhedrons. Spatially extended objects may be represented by a set of points at its surface, e.g. by the edges of a polygon (vector representation) or by the points contained in the object, e.g. the pixels of an object in a raster image (raster representation). Therefore, we use *sets of points* as a generic representation of spatial objects. In general, the points $p = (p_1, p_2, \ldots, p_d)$ are elements of a d-dimensional Euclidean vector space called *Points*. In the following, however, we restrict the presentation to the 2-dimensional case, although, all of the introduced notions can easily be applied to higher dimensions d. Spatial objects O are represented by a set of points, i.e. $O \in 2^{Points}$. For a point $p = (p_x, p_y)$, p_x and p_y denote the coordinates of p in the x- and the y-dimension.

Topological relations are relations which are invariant under topological transformations, i.e. they are preserved if both objects are rotated, translated or scaled simultaneously.

Definition 1: (*topological relations*) The topological relations between two objects A and B are derived from the nine intersections of the interiors, the boundaries and the complements of A and B with each other. The relations are: A *disjoint* B, A *meets* B, A *overlaps* B, A *equals* B, A *covers* B, A *covered-by* B, A *contains* B, A *inside* B. A formal definintion can be found in [5].

Distance relations are those relations comparing the distance of two objects with a given constant using one of the arithmetic operators.

Definition 2: (*distance relations*) Let *dist* be a distance function, let σ be one of the arithmetic predicates $<, >$ or $=$, let c be a real number and let A and B be spatial objects, i.e. $A, B \in 2^{Points}$. Then a distance relation A *distance*$_{\sigma\ c}$ B holds iff $dist(A, B)$ σ c.

In the following, we define 2-dimensional direction relations and we will use their geographic names. We define the direction relation of two spatially extended objects using one representative point $rep(A)$ of the *source* object A and all points of the *destination* object B. The representative point of a source object is used as the origin of a virtual coordinate system and its quadrants define the directions.

Definition 3: (*direction relations*) Let $rep(A)$ be the representative of a source object A.
- B *northeast* A holds, iff $\forall b \in B: b_x \geq rep(A)_x \wedge b_y \geq rep(A)_y$
    southeast, *southwest* and *northwest* are defined analogously.
- B *north* A holds, iff $\forall b \in B: b_y \geq rep(A)_y$. *south*, *west*, *east* are defined analogously.
- B *any_direction* A is defined to be TRUE for all A, B.

Obviously, for each pair of spatial objects at least one of the direction relations holds but the direction relation between two objects may not be unique. Only the special relations *northwest*, *northeast*, *southwest* and *southeast* are mutually exclusive. However, if considering only these special directions there may be pairs of objects for which none of these direction relations hold, e.g. if some points of B are northeast of A and some points of B are northwest of A. On the other hand, all the direction relations are partially ordered by a specialization relation (simply given by set inclusion) such that the smallest direction relation for two objects A and B is uniquely determined. We call this smallest direction relation for two objects A and B the *exact direction relation* of A and B.

Topological, distance and direction relations may be combined by the logical operators $\wedge$ (and) as well as $\vee$ (or) to express a *complex neighborhood relation*.

Definition 4: (complex neighborhood relations) If r_1 and r_2 are neighborhood relations, then $r_1 \wedge r_2$ and $r_1 \vee r_2$ are also (*complex*) neighborhood relations.

2.2 Neighborhood Graphs and Their Operations

Based on the neighborhood relations, we introduce the concepts of neighborhood graphs and neighborhood paths and some basic operations for their manipulation.

Definition 5: (*neighborhood graphs and paths*) Let *neighbor* be a neighborhood relation and $DB \subseteq 2^{Points}$ be a database of spatial objects.

a) A *neighborhood graph* $G_{neighbor}^{DB} = (N, E)$ is a graph where the set of nodes N corresponds to the set of objects $o \in DB$. The set of edges $E \subseteq N \times N$ contains the pair of nodes (n_1, n_2) iff *neighbor*(n_1, n_2) holds. Let n denote the cardinality of N and let e denote the cardinality of E. Then, $f := e / n$ denotes the average number of edges of a node, i.e. f is called the "fan out" of the graph.

b) A *neighborhood path* is a sequence of nodes $[n_1, n_2, \ldots, n_k]$, where $neighbor(n_i, n_{i+1})$ holds for all $n_i \in N$, $1 \le i < k$. The number k of nodes is called the *length* of the neighborhood path.

Lemma 1: The expected number of neighborhood paths of length k starting from a given node is f^{k-1} and the expected number of all neighborhood paths of length k is then $n * f^{k-1}$.

Obviously, the number of neighborhood paths may become very large. For the purpose of KDD, however, we are mostly interested in a certain class of paths, i.e. paths which are "leading *away*" from the starting object in a straightforward sense. Therefore, the operations on neighborhood paths will provide parameters (filters) to further reduce the number of paths actually created.

We assume the standard operations from relational algebra such as *selection, union, intersection* and *difference* to be available for sets of objects and for sets of paths. Furthermore, we define a small set of basic operations on neighborhood graphs and paths as database primitives for spatial data mining. In this paper, we introduce only the two most important of these operations:

```
neighbors: NGraphs x Objects x Predicates --> 2^Objects
extensions: NGraphs x 2^NPaths x Integer x Predicates -> 2^NPaths
```

The operation `neighbors(graph,object,pred)` returns the set of all objects connected to `object` via some edge of `graph` satisfying the conditions expressed by the predicate `pred`. The additional selection condition `pred` is used if we want to restrict the investigation explicitly to certain types of neighbors. The definition of the predicate `pred` may use spatial as well as non-spatial attributes of the objects.

The operation `extensions(graph,paths,max,pred)` returns the set of all paths extending one of the elements of `paths` by at most `max` nodes of `graph`. All the extended paths must satisfy the predicate `pred`. Therefore, the predicate `pred` in the operation `extensions` acts as a filter to restrict the number of paths created using domain knowledge about the relevant paths.

2.3 Filter Predicates for Neighborhood Paths

Neighborhood graphs will in general contain many paths which are irrelevant if not "misleading" for spatial data mining algorithms. The task of spatial trend analysis, i.e. finding patterns of systematic change of some non-spatial attributes in the neighborhood of certain database objects, can be considered as a typical example. Detecting such trends would be impossible if we do not restrict the pattern space in a way that paths changing direction in arbitrary ways or containing cycles are eliminated. In the following, we discuss one possible filter predicate, i.e. *starlike*. Other filters may be useful depending on the application.

Definition 6: (*filter starlike*) Let $p = [n_1, n_2, \ldots, n_k]$ be a neighborhood path and let rel_i be the exact direction for n_i and n_{i+1}, i.e. $n_{i+1} \, rel_i \, n_i$ holds. The predicates *starlike* and *variable-starlike* for paths p are defined as follows:

$starlike(p) :\Leftrightarrow (\exists j < k: \forall i > j: n_{i+1} \, rel_i \, n_i \Leftrightarrow rel_i \subseteq rel_j)$, if $k > 1$; TRUE, if $k=1$.

 *starlike*

Fig. 2. Illustration of two different filter predicates

The filter *starlike* requires that, when extending a path p, the exact "final" direction rel_j of p cannot be generalized. For instance, a path with "final" direction *northeast* can only be extended by a node of an edge with exact direction *northeast* but not by an edge with exact direction *north*.

Under the following assumptions, we can calculate the number of all *starlike* neighborhood paths of a certain length l for a given fanout f of the neighborhood graph.

Lemma 2: Let A be a spatial object and let l be an integer. Let *intersects* be chosen as the neighborhood relation. If the representative points of all spatial objects are uniformly distributed and if they have the same extension in both x and y direction, then the number of all *starlike* neighborhood paths with source A having a length of at most l is $O(2^l)$ for $f = 12$ and $O(l)$ for $f = 6$. (see [6] for a proof)

The assumptions of this lemma may seem to be too restrictive for real applications. Note, however, that *intersects* is a very natural neighborhood relation for spatially extended objects. To evaluate the assumptions of uniform distribution of the representative points of the spatial objects and of the same size of these objects, we conducted a set of experiments to compare the expected numbers of neighborhood paths with the actual number of paths created from a real geographic database on Bavaria. The database contains the ATKIS 500 data [2] and the Bavarian part of the statistical data obtained by the German census of 1987.

We find that for $f = 6$ the number of *all* neighborhood paths (starting from the same source) with a length of at most *max-length* is $O(6^{max\text{-}length})$ and the number of the *starlike* neighborhood paths only grows approximately linear with increasing *max-length* - as stated by lemma 2. For $f = 12$ the number of *all* neighborhood paths with a length of at most *max-length* is $O(12^{max\text{-}length})$ as we can expect from the lemma. However, the number of the *starlike* neighborhood paths is significantly less than $O(2^{max\text{-}length})$. This effect can be explained as follows. The lemma assumes equal size of the spatial objects. However, small destination objects are more likely to fulfil the filter starlike than large destination objects implying that the size of objects on starlike neighborhood paths tends to decrease. Note that lemma 2 nevertheless yields an upper bound for the number of starlike neighborhood paths created.

3 Algorithms for Spatial Data Mining

To support our claim that the expressivity of our spatial data mining primitives is adequate, we demonstrate how typical spatial data mining algorithms can be expressed by the database primitives introduced in section 2.

3.1 Spatial Clustering

Clustering is the task of grouping the objects of a database into meaningful subclasses (that is, clusters) so that the members of a cluster are as similar as possible whereas the members of different clusters differ as much as possible from each other. Applications of clustering in spatial databases are, e.g., the detection of seismic faults by grouping the entries of an earthquake catalog or the creation of thematic maps in geographic information systems by clustering feature spaces.

Different types of spatial clustering algorithms have been proposed. The basic idea of a *single scan algorithm* is to group neighboring objects of the database into clusters based on a *local* cluster condition performing only one scan through the database. Single scan clustering algorithms are efficient if the retrieval of the neighborhood of an object can be efficiently performed by the SDBS. Note that local cluster conditions are well supported by the `neighbors` operation on an appropriate neighborhood graph. The algorithmic schema of single scan clustering is depicted in figure 3.

```
SingleScanClustering(Database db; NRelation rel)
    set Graph to create_NGraph(db,rel);
    initialize a set CurrentObjects as empty;
    for each node O in Graph do
        if O is not yet member of some cluster then
            create a new cluster C;
            insert O into CurrentObjects;
            while CurrentObjects not empty do
                remove the first element of CurrentObjects as O;
                set Neighbors to neighbors(Graph, O, TRUE);
                if Neighbors satisfy the cluster condition do
                    add O to cluster C;
                    add Neighbors to CurrentObjects;
end SingleScanClustering;
```

Fig. 3. Schema of single scan clustering algorithms

Different cluster conditions yield different notions of a cluster and different clustering algorithms. For example, *GDBSCAN* [16] relies on a density-based notion of clusters. The key idea of a density-based cluster is that for each point of a cluster its ε-neighborhood has to contain at least a minimum number of points. This idea of "density-based clusters" can be generalized in two important ways. First, any notion of a neighborhood can be used instead of an ε-neighborhood if the definition of the neighborhood is based on a binary predicate which is symmetric and reflexive. Second, instead of simply counting the objects in a neighborhood of an object other measures to define the "cardinality" of that neighborhood can be used as well. Whereas a distance-based neighborhood is a natural notion of a neighborhood for point objects, it may be more appropriate to use topological relations such as *intersects* or *meets* to cluster spatially extended objects such as a set of polygons of largely differing sizes.

3.2 Spatial Characterization

The task of *characterization* is to find a compact description for a selected subset (the *target set*) of the database. A *spatial characterization* [8] is a description of the spatial and non-spatial properties which are typical for the target objects but not for the whole database. The relative frequencies of the non-spatial attribute values and the relative frequencies of the different object types are used as the interesting properties. For instance, different object types in a geographic database are communities, mountains, lakes, highways, railroads etc. To obtain a *spatial* characterization, not only the properties of the target objects, but also the properties of their neighbors (up to a given maximum number of edges in the relevant neighborhood graph) are considered.

A spatial characterization rule of the form $target \Rightarrow p_1 \ (n_1, freq\text{-}fac_1) \wedge ... \wedge p_k \ (n_k, freq\text{-}fac_k)$ means that for the set of all targets extended by n_i neighbors, the property p_i is $freq\text{-}fac_i$ times more (or less) frequent than in the whole database. The characterization algorithm usually starts with a small set of target objects, selected for instance by a condition on some non-spatial attribute(s) such as "rate of retired people = HIGH" (see figure 4, left). Then, the algorithm expands regions around the target objects, simultaneously selecting those attributes of the regions for which the distribution of values differs significantly from the distribution in the whole database (figure 4, right).

target objects　　　　　　　　**maximally expanded regions**

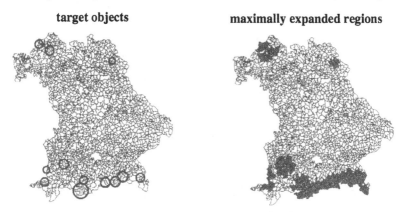

Fig. 4. Characterizing wrt. high rate of retired people [8]

In the last step of the algorithm, the following characterization rule is generated describing the target regions. Note that this rule lists not only some non-spatial attributes but also the neighborhood of mountains (after three extensions) as significant for the characterization of the target regions:

```
community has high rate of retired people ⇒
          apartments per building = very low (0, 9.1) ∧
          rate of foreigners = very low (0, 8.9)∧
          . . . ∧
          object type = mountain (3, 4.1)
```

3.3 Spatial Trend Detection

A *spatial trend* [8] is as a regular change of one or more non-spatial attributes when moving away from a given start object o . Neighborhood paths starting from o are used to model the movement and a regression analysis is performed on the respective attribute values for the objects of a neighborhood path to describe the regularity of change. For the regression, the distance from o is the independent variable and the difference of the attribute values are the dependent variable(s) for the regression. The correlation of the observed attribute values with the values predicted by the regression function yields a measure of confidence for the discovered trend.

Algorithm *global-trends* detects global trends around a start object *o*. The existence of a global trend for a start object *o* indicates that if considering all objects on all paths starting from *o* the values for the specified attribute(s) *in general* tend to increase (decrease) with increasing distance. Figure 5 (left) depicts the result of algorithm *global-trends* for the trend attribute "average rent" and a start object representing the city of Regensburg.

Algorithm *local-trends* detects single paths starting from an object *o* and having a certain trend. The paths starting from *o* may show different pattern of change, for example, some trends may be positive while the others may be negative. Figure 5 (right) illustrates this case again for the trend attribute "average rent" and the start object representing the city of Regensburg.

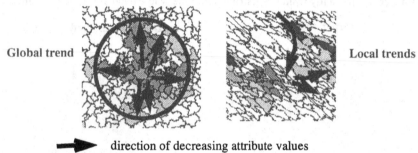

Global trend   Local trends

➤ direction of decreasing attribute values

Fig. 5. Trends of the"average rent" starting from the city of Regensburg

4 Efficient DBMS Support Based on Neighborhood Indices

Typically, spatial index structures, e.g. R-trees [11], are used in an SDBMS to speed up the processing of queries such as region queries or nearest neighbor queries [10]. Therefore, our default implementation of the `neighbors` operations uses an R-tree. If the spatial objects are fairly complex, however, retrieving the neighbors of some object this way is still very time consuming due to the complexity of the evaluation of neighborhood relations on such objects. Furthermore, when creating all neighborhood paths with a given source object, a very large number of `neighbors` operations has to be performed. Finally, many SDBS are rather static since there are not many updates on objects such as geographic maps or proteins. Therefore, materializing the relevant neighborhood graphs and avoiding to access the spatial objects themselves may be worthwhile. This is the idea of the neighborhood indices.

4.1 Neighborhood Indices

Our concept of *neighborhood indices* is related to the work of [15] and [13]. [15] introduced the concept of *spatial join indices* as a materialization of a spatial join with the goal of speeding up spatial query processing. This paper, however, does not deal with the questions of efficient implementation of such indices. [13] extends spatial join indices by associating each pair of objects with their distance. In its basic form, this index requires $O(n^2)$ space because it needs one entry not only for pairs of neighboring objects but for each pair of objects. Therefore, in [13] a hierarchical version of distance associated join indices is proposed. In general, however, we cannot rely on such hierarchies for the purpose of supporting spatial data mining. Our approach, called *neighborhood indices*, extends distance associated join indices with the following new contributions:

- A specified maximum distance restricts the pairs of objects represented in a neighborhood index.
- For each of the different types of neighborhood relations (that is distance, direction, and topological relations), the concrete relation of the pair of objects is stored.

Definition 7: (*neighborhood index*) Let DB be a set of spatial objects and let max and $dist$ be real numbers. Let D be a direction relation and T be a topological relation. Then the *neighborhood index* for DB with maximum distance max, denoted by I_{max}^{DB}, is defined as follows: $I_{max}^{DB} = \{(O_1,O_2,dist,D,T) \mid$

$$O_1, O_2 \in DB \wedge O_1 \, distance_{=dist} \, O_2 \wedge dist \leq max \wedge O_2 D O_1 \wedge O_1 T O_2\}.$$

A simple implementation of a neighborhood index using a B^+-tree on the key attribute *Object-ID* is illustrated in figure 6.

	Object-ID	Neighbor	Distance	Direction	Topology
B^+-tree	o_1	o_2	2.7	southwest	disjoint
	o_1	o_3	0	northwest	overlap
	...	...	...	...	...

Fig. 6. Sample Neighborhood Index

A neighborhood index supports not only one but a set of neighborhood graphs. We call a neighborhood index *applicable* for a given neighborhood graph if the index contains an entry for each of the edges of the graph. To find the neighborhood indices applicable for some neighborhood graph, we introduce the notion of the critical distance of a neighborhood relation. Intuitively, the *critical distance* of a neighborhood relation r is the maximum possible distance for a pair of objects O_1 and O_2 satisfying $O_1 \, r \, O_2$.

Definition 8: (*applicable neighborhood index*) Let G_r^{DB} be a neighborhood graph and let I_{max}^{DB} be a neighborhood index. I_{max}^{DB} is *applicable* for G_r^{DB} iff

$$\forall(O_1 \in DB, O_2 \in DB)O_1 r O_2 \Rightarrow (O_1, O_2, dist, D, T) \in I_{max}^{DB}$$

Definition 9: (*critical distance of a neighborhood relation*) Let r be a neighborhood relation. The *critical distance* of r, denoted by *c-distance(r)*, is defined as follows:

$$c\text{-}distance(r) = \begin{cases} 0 & \text{if } r \text{ is a topological relation except } disjoint \\ c & \text{if } r \text{ is the relation } distance_{<c} \text{ or } distance_{=c} \\ \infty & \text{if } r \text{ is a direction relation, the relation } distance_{>c}, \text{ or } disjoint \\ min(cdistance(r_1), cdistance(r_2)) & \text{if } r = r_1 \wedge r_2 \\ max(cdistance(r_1), cdistance(r_2)) & \text{if } r = r_1 \vee r_2 \end{cases}$$

A neighborhood index with a maximum distance of max is applicable for a neighborhood graph with relation r if the critical distance of r is not larger than max.

Lemma 3: Let G_r^{DB} be a neighborhood graph and let I_{max}^{DB} be a neighborhood index. If $max \geq c\text{-}distance(r)$, then I_{max}^{DB} is applicable for G_r^{DB}.

Obviously, if two neighborhood indices I_{c1}^{DB} and I_{c2}^{DB} with $c_1 < c_2$ are available and applicable, using I_{c1}^{DB} is more efficient because in general it has less entries than I_{c2}^{DB}. The *smallest applicable neighborhood index* for some neighborhood graph is the applicable neighborhood index with the smallest critical distance.

In figure 7, we sketch the algorithm for processing the neighbors operation which makes use of the smallest applicable neighborhood index. If there is no applicable neighborhood index, then the standard approach that uses an R-tree is followed.

---

neighbors (graph G^{DB}, object o, predicate *pred*)
  select as I the smallest applicable neighborhood index for G_r^{DB}; // *Index Selection*
  if such I exists **then**   // *Filter Step*
     use the neighborhood index I to retrieve as *candidates* the set of objects c
     having an entry (o,c,dist, D, T) in I
  else use the R-tree to retrieve as *candidates* the set of objects c satisfying o r c;
  initialize an empty set of *neighbors*;   // *Refinement Step*
  for each c in *candidates* **do**
     if o r c **and** *pred(c)* **then** add c to *neighbors*
  return *neighbors*;

Fig. 7. Algorithm neighbors

The first step of algorithm neighbors, the *index selection*, selects a neighborhood index. The *filter step* returns a set of candidate objects (which may satisfy the specified neighborhood relation) with a cardinality significantly smaller than the database size. In the last step, the *refinement step*, for all these candidates the neighborhood relation as well as the additional predicate *pred* are evaluated and all objects passing this test are returned as the resulting neighbors. The extensions operation can obviously be implemented by iteratively performing neighbors operations. Therefore, it is obvious that the performance of the neighbors operation is very important for the efficiency of our approach.

To create a neighborhood index I_{max}^{DB}, a spatial join on *DB* with respect to the neighborhood relation $(O_1 distance_{\text{-}dist} O_2 \wedge dist \leq max)$ is performed. A spatial join can be efficiently processed by using a spatial index structure, see e.g. [4]. For each pair of objects returned by the spatial join, we then have to determine the exact distance, the direction relation and the topological relation. The resulting tuples of the form $(O_1, O_2, Distance, Direction, Topology)$ are stored in a relation which is indexed by a B+-tree on the attribute O_1.

Updates of a database, i.e. insertions or deletions, require updates of the derived neighborhood indices. Fortunately, the update of a neighborhood index I_{max}^{DB} is restricted to the neighborhood of the respective object defined by the neighborhood relation A $distance_{<\ max} B$. This neighborhood can be efficiently retrieved by using either a neighborhood index (in case of a deletion) or by using a spatial index structure (in case of an insertion).

4.2 Cost Model

We developed a cost model to predict the cost of performing a `neighbors` operation with and without a neighborhood index. We use t_{page}, i.e. the execution time of a page access, and t_{float}, i.e. the execution time of a floating point comparison, as the units for I/O time and CPU time, respectively.

In table 1, we define the parameters of the cost model and list typical values for each of them. The system overhead s includes client-server communication and the overhead induced by several SQL queries for retrieving the relevant neighborhood index and the minimum bounding box of a polygon (necessary for the access of the R-tree). p_{index} and p_{data} denote the probability that a requested index page and data page, respectively, have to be read from disk according to the buffering strategy.

Table 1: Parameters of the cost model

name	meaning	typical values
n	number of nodes in the neighborhood graph	$[10^3 .. 10^5]$
f	average number of edges per node in the graph (fan out)	$[1 .. 10^2]$
v	average number of vertices of a spatial object	$[1 .. 10^3]$
ff	ratio of fanout of the index and fanout (f) of the graph	$[1 .. 10]$
c_{index}	capacity of a page in terms of index entries	128
c_v	capacity of a page in terms of vertices	64
p_{index}	probability that a given index page must be read from disk	$[0..1]$
p_{data}	probability that a given data page must be read from disk	$[0..1]$
t_{page}	average execution time for a page access	$1 * 10^{-2}$ sec
t_{float}	execution time for a floating point comparison	$3 * 10^{-6}$ sec
s	system overhead	depends on DBMS

Table 2 shows the cost for the three steps of processing a `neighbors` operation with and without a neighborhood index. In the R-tree, there is one entry for each of the n nodes of the neighborhood graph whereas the B+-tree stores one entry for each of the $f * n$ edges. We assume that the number of R-tree paths to be followed is proportional to the number of neighboring objects, i.e. proportional to f. A spatial object with v vertices requires v/c_v data pages. We assume a distance relation as neighborhood relation requiring v^2 floating point comparisons. When using a neighborhood index, the filter step returns $ff * f$ candidates. The refinement step has to access their index entries but does not have to access the vertices of the candidates since the refinement test can be directly performed by using the attributes *Distance*, *Direction* and *Topology* of the index entries. This test involves a constant (i.e. independent of v) number of floating point comparisons and requires no page accesses implying that its cost can be neglected.

Table 2: Cost model for the `neighbors` operation

Step	Cost without neighborhood index	Cost with neighborhood index
Selection	s	s
Filter	$f \cdot \left\lceil \log_{c_{index}} n \right\rceil \cdot p_{index} \cdot t_{page}$	$\left\lceil \log_{c_{index}} (f \cdot n) \right\rceil \cdot p_{index} \cdot t_{page}$
Refinement	$(1+f) \cdot \left\lceil v/c_v \right\rceil \cdot p_{data} \cdot t_{page} + f \cdot v^2 \cdot t_{float}$	$ff \cdot f \cdot p_{data} \cdot t_{page}$

4.3 Experimental Results

We implemented the database primitives on top of the commercial DBMS Illustra using its 2D spatial data blade which provides R-trees. A geographic database of Bavaria was used for an experimental performance evaluation and validation of the cost model. This database represents the Bavarian communities with one spatial attribute (polygon) and 52 non-spatial attributes (such as average rent or rate of unemployment). All experiments were run on HP9000/715 (50MHz) workstations under HP-UX 10.10.

The first set of experiments compared the performance predicted by our cost models with the experimental performance when varying the parameters n, f and v. The results show that our cost model is able to predict the performance reasonably well. For instance, figure 8 depicts the results for $n = 2{,}000$, $v = 35$ and varying values for f.

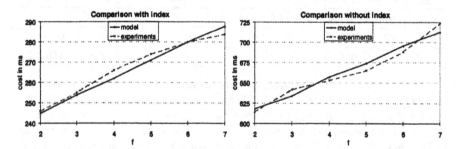

Fig. 8. Comparison of cost model versus experimental results

We used our cost model to compare the performance of the `neighbors` operation with and without neighborhood index for combinations of parameter values which we could not evaluate experimentally with our database. Figure 9 depicts the results (1) for $f = 10$, $v = 100$ and varying n and (2) for $n = 100{,}000$, $f = 10$ and varying v. These results demonstrate a significant speed-up for the `neighbors` operation with compared to without neighborhood index. Furthermore, this speed-up grows strongly with increasing number of vertices of the spatial objects.

The next set of experiments analyzed the system overhead which is rather large. This overhead, however, can be reduced when calling multiple correlated neighbors operations issued by one `extensions` operation, since the client-server communication,

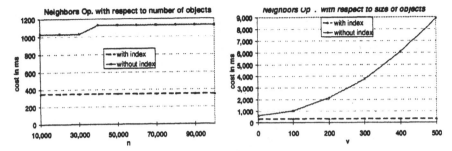

Fig. 9. Comparison with and without neighborhood index

the retrieval of the relevant neighborhood index etc. is necessary only once for the whole extensions operation and not for each of the neighbors operations. In our experiments, we found that the system overhead was typically reduced by 50%, e.g. from 211 ms to 100 ms, when calling multiple correlated neighbors operations.

5 Conclusions

In this paper, we defined neighborhood graphs and paths and a small set of database primitives for spatial data mining. We showed that spatial data mining algorithms such as spatial clustering, characterization, and trend detection are well supported by the proposed operations. Furthermore, we discussed filters restricting the search to such neighborhood paths "leading *away*" from a starting object. An analytical as well as an experimental analysis demonstrated the effectiveness of the proposed filter. Finally, we introduced neighborhood indices to speed-up the processing of our database primitives. Neighborhood indices can be easily created in a commercial DBMS by using standard functionality, i.e. relational tables and index structures. We implemented the database primitives on top of the object-relational DBMS Illustra. The efficiency of the neighborhood indices was evaluated by using an analytical cost model and an extensive experimental study on a geographic database.

So far, the neighborhood relations between two objects depend only on the properties of the two involved objects. In the future, we will extend our approach to neighborhood relations such as "being among the k-nearest neighbors" which depend on more than the two related objects. The investigation of other filters for neighborhood paths with respect to their effectiveness and efficiency in different applications is a further interesting issue. Finally, a tighter integration of the database primitives with the DBMS should be investigated.

References

[1] Agrawal R., Imielinski T., Swami A.: *"Database Mining: A Performance Perspective"*, IEEE Transactions on Knowledge and Data Engineering, Vol. 5, No. 6, 1993, pp. 914-925.

[2] Bavarian State Bureau of Topography and Geodasy, CD-Rom, 1996.

[3] Bill, Fritsch: *"Fundamentals of Geographical Information Systems: Hardware, Software and Data"* (in German), Wichmann Publishing, Heidelberg, Germany, 1991.

[4] Brinkhoff T., Kriegel H.-P., Schneider R., and Seeger B.: *"Efficient Multi-Step Processing of Spatial Joins"*. *Proc.* ACM SIGMOD '94, Minneapolis, MN, 1994, pp. 197-208.

[5] Egenhofer M. J.: *"Reasoning about Binary Topological Relations"*, Proc. 2nd Int. Symp. on Large Spatial Databases, Zurich, Switzerland, 1991, pp. 143-160.

[6] Ester M., Gundlach S., Kriegel H.-P., Sander J.: *"Database Primitives for Spatial Data Mining"*, Proc. Int. Conf. on Databases in Office, Engineering and Science (BTW'99), Freiburg, Germany, 1999.

[7] Ester M., Kriegel H.-P., Sander J.: *"Spatial Data Mining: A Database Approach"*, Proc. 5th Int. Symp. on Large Spatial Databases, Berlin, Germany, 1997, pp. 47-66.

[8] Ester M., Frommelt A., Kriegel H.-P., Sander J.: *"Algorithms for Characterization and Trend Detection in Spatial Databases"*, Proc. 4th Int. Conf. on Knowledge Discovery and Data Mining, New York City, NY, 1998, pp. 44-50.

[9] Fayyad U. M., J., Piatetsky-Shapiro G., Smyth P.: *"From Data Mining to Knowledge Discovery: An Overview"*, in: Advances in Knowledge Discovery and Data Mining, AAAI Press, Menlo Park, 1996, pp. 1 - 34.

[10] Gueting R. H.: *"An Introduction to Spatial Database Systems"*, Special Issue on Spatial Database Systems of the VLDB Journal, Vol. 3, No. 4, October 1994.

[11] Guttman A.: *"R-trees: A Dynamic Index Structure for Spatial Searching"*, Proc. ACM SIGMOD '84, 1984, pp. 47-54.

[12] Koperski K., Adhikary J., Han J.: *"Knowledge Discovery in Spatial Databases: Progress and Challenges"*, Proc. SIGMOD Workshop on Research Issues in Data Mining and Knowledge Discovery, Technical Report 96-08, UBC, Vancouver, Canada, 1996.

[13] Lu W., Han J.: *"Distance-Associated Join Indices for Spatial Range Search"*, Proc. 8th Int. Conf. on Data Engineering, Phoenix, AZ, 1992, pp. 284-292.

[14] Ng R. T., Han J.: *"Efficient and Effective Clustering Methods for Spatial Data Mining"*, Proc. 20th Int. Conf. on Very Large Data Bases, Santiago, Chile, 1994, pp. 144-155.

[15] Rotem D.: *"Spatial Join Indices"*, Proc. 7th Int. Conf. on Data Engineering, Kobe, Japan, 1991, pp. 500-509.

[16] Sander J., Ester M., Kriegel H.-P., Xu X.: *"Density-Based Clustering in Spatial Databases: A New Algorithm and its Applications"*, Data Mining and Knowledge Discovery, an International Journal, Kluwer Academic Publishers, Vol.2, No. 2, 1998.

Cooperative Distributed Vision:
*Dynamic Integration of Visual Perception, Action, and Communication**

Takashi Matsuyama

Department of Intelligence Science and Technology
Kyoto University, Kyoto 606-8501, Japan
e-mail: tm@i.kyoto-u.ac.jp

Abstract. We believe intelligence does not dwell solely in brain but emerges from active interactions with environments through perception, action, and communication. This paper give an overview of our five years project on Cooperative Distributed Vision (CDV, in short). From a practical point of view, the goal of CDV is summarized as follows: Embed in the real world a group of network-connected Observation Stations (real-time image processor with active camera(s)) and mobile robots with vision. And realize 1) wide area dynamic scene understanding and 2) versatile scene visualization. Applications of CDV include real-time wide area surveillance, remote conference and lecturing systems, navigation of (non-intelligent) mobile robots and disabled people. In this paper, we first define the framework of CDV and then present technical research results so far obtained: 1) fixed viewpoint pan-tilt-zoom camera for wide area active imaging, 2) active vision system for real-time moving object tracking and, 3) cooperative moving object tracking by communicating active vision agents.

1 Introduction

This paper gives an overview of our five years project on *Cooperative Distributed Vision* (CDV, in short). From a practical point of view, the goal of CDV is summarized as follows (Fig. 1):
Embed in the real world a group of network-connected *Observation Stations* (real-time image processor with active camera(s)) and mobile robots with vision, and realize
1. Wide area dynamic scene understanding and
2. Versatile scene visualization.

   Applications of CDV include real-time wide area surveillance and traffic monitoring systems, remote conference and lecturing systems, interactive 3D TV and intelligent TV studio, navigation and guidance of (non-intelligent) mobile robots and disabled people, and cooperative mobile robots.

---

* Invited by the DAGM-99 for the common part of the DAGM-99 and KI-99.

Fig. 1. Cooperative distributed vision.

From a scientific point of view, we put our focus upon *dynamic integration of visual perception, action, and communication.* That is, the scientific goal of the project is to investigate how the *dynamics* of these three functions can be characterized and how they should be integrated *dynamically* to realize intelligent systems.

In this paper, we first discuss functionalities of and mutual dependencies among perception, action, and communication to formally clarify the meaning of their integration. Then we present technical research results so far obtained on moving target detection and tracking by cooperative observation stations:
Visual Perception:
Fixed Viewpoint Pan-Tilt-Zoom (FV-PTZ) camera for wide area active imaging
Visual Perception ⊕ Action[1]:
Real-time object detection and tracking by an FV-PTZ camera
Visual Perception ⊕ Action ⊕ Communication:
Cooperative object tracking by communicating active vision agents.

2 Integrating Perception, Action, and Communication

2.1 Modeling Intelligence by Dynamic Interactions

To model intelligence, (classic) Artificial Intelligence employs the scheme

$$Intelligence = Knowledge + Reasoning$$

and puts its major focus upon symbolic knowledge representation and symbolic computation. In this sense, it may be called *Computational Intelligence*[1].

---

[1] The meaning of ⊕ will be explained later.

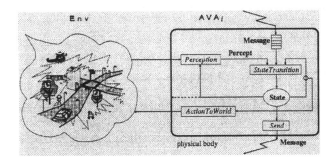

Fig. 2. Model of an Active Vision Agent.

In the CDV project, on the other hand, we propose an idea of *modeling intelligence by dynamic interactions*, which can be represented by the following scheme:

$$Intelligence = Perception \oplus Action \oplus Communication,$$

where $\oplus$ implies dynamic interactions among the functional modules.

That is, we define an *agent* as an intelligent system with perception, action, and communication capabilities and regard these three functions as fundamental modules to realize dynamic interactions between the agent and its outer world (i.e. scene and other agents):

Function		From		To
Perception	:	**World**	$\rightarrow$	**Self**
Action	:	**Self**	$\rightarrow$	**World**
Communication	:	**Self**	$\leftrightarrow$	**Others**

By integrating perception, action, and communication, various dynamic information flows are formed. In our model, reasoning implies the function which dynamically controls such flows of information. We believe intelligence does not dwell solely in brain but emerges from active interactions with environments through perception, action, and communication.

2.2 Model of Active Vision Agent

First we define *Active Vision Agent* (AVA, in short) as a *rational agent* with visual perception, action, and communication capabilities. Let **Sate**$_i$ denote the internal state of ith AVA, AVA$_i$, and **Env** the state of the world. Then, fundamental functions of AVA$_i$ can be defined as follows (Fig. 2):

$$Perception_i : \mathbf{Env} \times \mathbf{Sate}_i \mapsto \mathbf{Percept}_i \qquad (1)$$

$$Action_i : \mathbf{Sate}_i \mapsto \mathbf{Sate}_i \times \mathbf{Env}, \qquad (2)$$

$$Reasoning_i : \mathbf{Percept}_i \times \mathbf{Sate}_i \mapsto \mathbf{Sate}_i, \qquad (3)$$

where **Percept**$_i$ stands for entities perceived by AVA$_i$.

The communication by AVA$_i$ can be defined by the following pair of message exchange functions:

$$Send_i : \mathbf{Sate}_i \mapsto \mathbf{Message}_j \tag{4}$$

$$Receive_i : \mathbf{Message}_i \times \mathbf{Sate}_i \mapsto \mathbf{Sate}_i, \tag{5}$$

where $\mathbf{Message}_i$ and $\mathbf{Message}_j$ denote messages sent out to AVA$_i$ and AVA$_j$ via the communication network, respectively.

Based on the functional definitions given above, we can derive the following observations:

1. $Action_i$ in (2) can be decomposed into

$$Internal Action_i : \mathbf{Sate}_i \mapsto \mathbf{Sate}_i, \tag{6}$$

$$ActionToWorld_i : (\mathbf{Sate}_i \mapsto \mathbf{Sate}_i) \mapsto \mathbf{Env}. \tag{7}$$

The former implies pure internal state changes, such as pan-tilt-zoom controls of an active camera, while the latter means those state transitions whose side-effects incur state changes of $\mathbf{Env}$, e.g. mechanical body-arm controls of a robot.

2. The internal state transition is caused by the (internal) action, the perception followed by the reasoning, and/or the message acception. Thus we can summarize these processes into

$$StateTransition_i : \mathbf{Percept}_i \times \mathbf{Message}_i \times \mathbf{Sate}_i \mapsto \mathbf{Sate}_i. \tag{8}$$

3. The above described model merely represents static functional dependencies and no dynamic properties are taken into account. A straightforward way to introduce dynamics into the model would be to incorporate time variable t into the formulae. For example, equation (8) can be augmented to

$$StateTransition_i(\mathbf{Percept}_i(t), \mathbf{Message}_i(t), \mathbf{Sate}_i(t)) = \mathbf{Sate}_i(t + \Delta t). \tag{9}$$

This type of formulation is widely used in control systems. In fact, Asada[2] used linearized state equations to model vision-based behaviors of a mobile robot. We believe, however, that more flexible models are required to implement the dynamics of an AVA;

1) Asynchronous Dynamics: Communications between AVAs are asynchronous in their nature.

2) Conditional Dynamics: Cooperations among AVAs require conditional reactions.

In what follows,

1. we first introduce *Fixed Viewpoint Pan-Tilt-Zoom (FV-PTZ) camera*, with which camera actions can be easily correlated with perceived images. (Section 3).

2. Then, a real-time object tracking system with an FV-PTZ camera is presented, where a novel dynamic interaction mechanism between perception and action is proposed (Section 4).

3. Finally, we present a cooperative object tracking system, where a state transition network is employed to realize asynchronous and conditional dynamics of

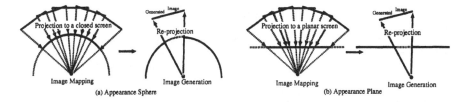

Fig. 3. Appearance sphere and plane.

an AVA, i.e. dynamic interactions among perception, action, and communication modules. (Section 5).

3 Fixed Viewpoint Pan-Tilt-Zoom Camera for Wide Area Active Imaging

The pan-tilt-zoom camera control can be modeled by $Internal Action_i$ in (6), which transforms AVA_i's internal state from $\mathbf{State}_i(before)$ to $\mathbf{State}_i(after)$. This state change is reflected on $\mathbf{Percept}_i$ by $Perception_i$ in (1). As is well know in Computer Vision, however, it is very hard to find the correlation between $\mathbf{State}_i(before) \to \mathbf{State}_i(after)$ and $\mathbf{Percept}_i(before) \to \mathbf{Percept}_i(after)$; complicated 3D $\to$ 2D geometric and photometric projection processes are involved in $Perception_i$ even if $\mathbf{Env}$ is fixed.

We devised a sophisticated pan-tilt-zoom camera, with which camera actions can be easily correlated with perceived images.

3.1 Fixed Viewpoint Pan-Tilt-Zoom Camera

In general, a pan-tilt camera includes a pair of geometric singularities: 1) the projection center of the imaging system and 2) the rotation axes. In ordinary active camera systems, no deliberate design about these singularities is incorporated and the discordance of the singularities causes photometric and geometric appearance variations during the camera rotation: varying highlights and motion parallax. In other words, 2D appearances of a scene change dynamically depending on the (unknown) 3D scene geometry.

Our idea to overcome the above problem is simple but effective:
1. Make pan and tilt axes intersect with each other.
2. Place the projection center at the intersecting point.
3. Design such a zoom lens system whose projection center is fixed irrespectively of zooming.

We call the above designed active camera the *Fixed Viewpoint Pan-Tilt-Zoom camera* (in short, FV-PTZ camera)[3].

3.2 Image Representation for the FV-PTZ Camera

While images observed by an FV-PTZ camera do not include any geometric and photometric variations depending on the 3D scene geometry, object shapes

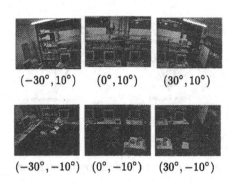

$(-30°, 10°)$ $(0°, 10°)$ $(30°, 10°)$

$(-30°, -10°)$ $(0°, -10°)$ $(30°, -10°)$

Fig. 4. Observed images with (pan, tilt).   **Fig. 5.** Generated APP image.

in the images vary with the camera motion. These variations are caused by the movement of the image plane, which can be rectified by projecting observed images onto a common virtual screen. On the virtual screen, the projected images form a seamless wide panoramic image.

For the rectification, we can use arbitrarily shaped virtual screens:

APS: When we can observe the 360° panoramic view, a spherical screen can be used (Fig. 3 (a)). We call the omnidirectional image on the spherical screen *APpearance Sphere* (APS in short).

APP: When the rotation angle of the camera is limited, we can use a planar screen (Fig. 3 (b)). The panoramic image on the planar screen is called *APpearance Plane* (APP in short).

As illustrated in the right sides of Figs. 3(a)(b), once an APS or an APP is obtained, images taken with arbitrary combinations of pan-tilt-zoom parameters can be generated by re-projecting the APS or APP onto the corresponding image planes. This enables the virtual look around of the scene.

We developed a sophisticated camera calibration method for an off-the-shelf active video camera, SONY EVI G20, which we found is a good approximation of an FV-PTZ camera ($-30° \leq$ pan $\leq 30°$, $-15° \leq$ tilt $\leq 15°$, and zoom: $15° \leq$ horizontal view angle $\leq 44°$) [4]. Figs. 4 and 5 show a group of observed images and the APP panoramic image synthesized.

4   Dynamic Integration of Visual Perception and Action for Real-Time Moving Object Tracking

As shown in (1), (2), and (3), $Perception_i$ and $Action_i$ are mutually dependent on each other and their integration has been studied in Active Vision and Visual Servo[5]. To implement an AVA system, moreover, we have to integrate three modules with different intrinsic dynamics:

Visual Perception : video rate periodic cycle

Action   : mechanical motions involving variable (large) delays

Communication : asynchronous message exchanges, in which variable delays are incurred depending on network activities.

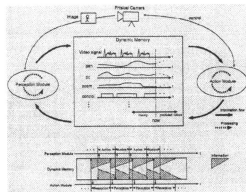

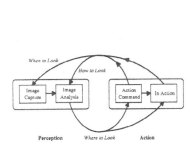

Fig. 6. Information flows in dynamic vision.

Fig. 7. Dynamic memory architecture.

4.1 Dynamic Memory

We are proposing a novel scheme named *Dynamic Vision*, whose distinguishing characteristics are as follows.

1. In a dynamic vision system, complicated information flows are formed between perception and action modules to solve *When to Look* and *How to Look* problems as well as ordinary *Where to Look* problem (Fig. 6):

Where to Look : Geometric camera motion planning based on image analysis

When to Look : Image acquisition timing should be determined depending on the camera motion as well as the target object motion, because quick camera motion can degrade observed images (i.e. motion blur).

How to Look : Depending on camera parameters (e.g. motion speed, iris, and shutter speed), different algorithms and/or parameter values should be used to realize robust image processing, because the quality of observed images is heavily dependent on the camera parameters.

2. The system dynamics is represented by a pair of parallel time axes, on which the dynamics of perception and action modules are represented respectively (See the lower diagram in Fig. 7). That is, the modules run in parallel dynamically exchanging data.

To implement a dynamic vision system, the *dynamic memory architecture* illustrated in Fig. 7 has been proposed, where perception and action modules share what we call the *dynamic memory*. It records temporal histories of state variables such as pan-tilt angles of the camera and the target object location. In addition, it stores their predicted values in the future (dotted lines in the figure). Perception and action modules are implemented as parallel processes which dynamically read from and write into the memory according to their own intrinsic dynamics.

While the above architecture looks similar to the "whiteboard architecture" proposed in [6], the critical difference rests in that the dynamic memory main-

82

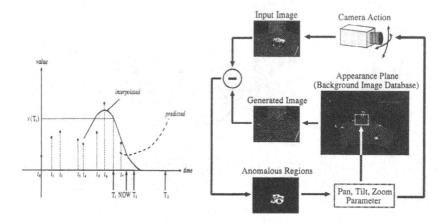

Fig. 8. Representation of a time
varying variable.

Fig. 9. Basic scheme for object tracking.

tains dynamically varying variables whose temporal periods are *continuously* spanning from the past to the future. That is, each entity in the dynamic memory is associated with the following temporal interpolation and prediction functions (Fig. 8):

1. *Temporal Interpolation*: Since the perception and action modules can write data only intermittently, the dynamic memory interpolates such discrete data. With this function, a module can read a value at any temporal moment, for example at T_1 in Fig. 8.
2. *Future Prediction*: Some module runs fast and may require data which are not written yet by another module (for example, the value at T_3 in Fig. 8). The dynamic memory generates predicted values based on those data so far recorded. Note that multiple values may be defined by the interpolation and prediction functions, for example, at NOW and T_2 in Fig. 8. We have to define the functions to avoid such multiple value generation.

With these two functions, each module can get any data along the time axis freely without waiting (i.e. wasting time) for synchronization with others. That is, the dynamic memory integrates parallel processes into a unified system while decoupling their dynamics; each module can run according to its own dynamics without being disturbed by the others. Moreover, prediction-based processing can be easily realized to cope with various delays involved in image processing and physical camera motion.

4.2 Prototype System Development

To embody the idea of Dynamic Vision, we developed a prototype system for real-time moving object detection and tracking with the FV-PTZ camera [7]. Fig. 9 illustrates its basic scheme:

frame:0 frame:20 frame:40 frame:60 frame:80 frame:100 frame:120 frame:140

Fig. 10. Partial image sequence of tracking (Upper: captured images, Lower: detected target regions).

1. Generate the APP image of the scene.

2. Extract a window image from the APP according to the current pan-tilt-zoom parameters and regard it as the background image.

3. Compute difference between the background image and an observed image.

4. If anomalous regions are detected in the difference image, select one and control the camera parameters to track the selected target.

5. Otherwise, move the camera along the predefined trajectory to search for an object.

Fig. 10 illustrates a partial sequence of human tracking by the system. Fig. 11 illustrates object and camera motion dynamics which were written into and read from the dynamic memory:

1. The action module reads pan-tilt angles (P',T') from the camera and writes them as CAM data into the dynamic memory.

2. When necessary, the perception module reads the CAM data from the dynamic memory: i.e. $(Cp(t),Ct(t))$ in the figure. Note that since the latter module runs faster, the frequency of reading operations of $(Cp(t),Ct(t))$ is much higher than that of writing operations of (P',T') by the former. (Compare two upper graphs.)

3. Then, the perception module conducts the object detection as illustrated in Fig. 9, whose output, i.e. the detected object centroid $(Op(t),Ot(t))$, is written back to the dynamic memory as OBJ data.

4. The action module, in turn, reads the OBJ data to control the camera.

Fig. 12 shows the read/write access timing to the dynamic memory by the perception and action modules. Note that both modules work asynchronously keeping their own intrinsic dynamics. Note also that the perception module runs almost twice faster than the action module (about 66msec/cycle).

These figures verify that the smooth real-time dynamic interactions between the perception and action modules are realized without introducing any interruption or idle time for synchronization.

5 Cooperative Object Tracking by Communicating Active Vision Agents

5.1 Task Specification

This section addresses a multi-AVA system which cooperatively detects and tracks a focused target object. The task of the system is as follows: 1) Each AVA is equipped with the FV-PTZ camera and mutually connected via the communication network. 2) Initially, it searches for a moving object independently of the

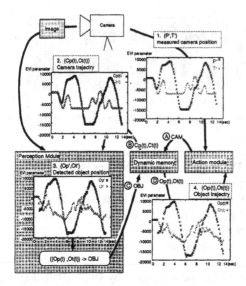

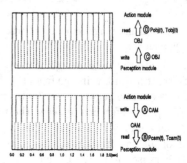

Fig. 11. Dynamic data exchanged between the perception and action modules (Large amplitude: pan, Small amplitude: tilt).

Fig. 12. Access timing to the dynamic memory by the perception and action modules (Upper: Object information, Lower: Camera information).

others. 3) When an AVA detects an object, it navigates the gazes of the other AVAs toward that object (Fig. 13). 4) All AVAs keep tracking the focused target cooperatively without being disturbed by obstacles or other moving objects (Fig. 14). 5) When the target goes out of the scene, the system returns back to the initial search mode.

All FV-PTZ cameras are calibrated and the object detection and tracking by each AVA is realized by the same method as described in Section 4.

5.2 Cooperative Object Tracking Protocol

Target Object Representation The most important ontological issue in the cooperative object tracking is how to represent the target object being tracked. In our multi-AVA system, "agent" means an AVA with visual perception, action, and communication capabilities. The target object is tracked by a group of such AVAs. Thus, we represent the target object by an *agency*, a group of those AVAs that are observing the target at the current moment.

Agency Formation Protocol When AVA_i detects an object, it broadcasts the object detection message. If no other AVAs detect objects, then AVA_i generates an agency consisting of itself alone (Fig. 13). When multiple object detection messages are broadcast simultaneously, AVA_i can generate an agency only if it has the highest priority among those AVAs that have detected objects.

Suppose AVA_i has generated an agency.

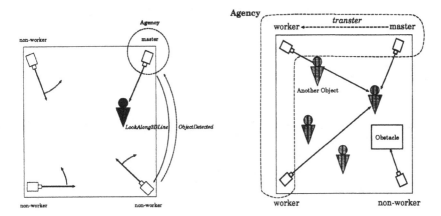

Fig. 13. Agency formation.   Fig. 14. Role assignment.

Gaze Navigation :First, AVA$_i$ broadcasts the 3D line, L_i, defined by the projection center of its camera and the object centroid in the observed image. Then, the other AVAs search for the object along this 3D line by controlling their cameras respectively (Fig. 13).

Object Identification : Those AVAs which can successfully detect the same object as AVA$_i$ are allowed to join into the agency. This object identification is done by the following method. Suppose AVA$_j$ detects an object and let L_j denote the 3D view line directed toward that object from AVA$_j$. AVA$_j$ reports L_j to AVA$_i$, which then examines the nearest 3D distance between L_i and L_j. If the distance is less than the threshold, a pair of detected objects by AVA$_i$ and AVA$_j$ are considered as the same object and AVA$_j$ is allowed to join the agency.

Object Tracking in 3D : Once multiple AVAs join the agency and their perception modules are synchronized, the 3D object location can be estimated by computing the intersection point among 3D view lines emanating from the member AVAs. Then, the 3D object location is broadcast to the other AVAs which have not detected the object.

Role Assignment Protocol Since the agency represents the target object being tracked, it has to maintain the object motion history, which is used to guide the search of non-member AVAs. Such object history maintenance should be done exclusively by a single AVA in the agency to guarantee the consistency. We call the right of maintaining the object history the *master authority* and the AVA with this right the *master* AVA. The other member AVAs are called *worker* AVAs and AVAs outside the agency *non-worker* AVAs (Fig. 14).

When an AVA first generates the agency, it immediately becomes the master. The master AVA conducts the object identification described before to allow other AVAs to join the agency, and maintains the object history. All these processings are done based on the object information observed by the master AVA. Thus, the reliability of the information observed by the master AVA is crucial to

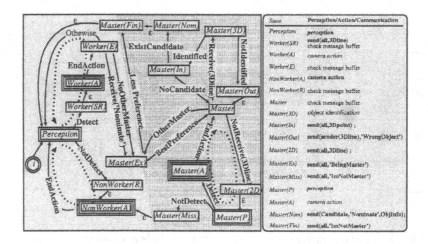

Fig. 15. State transition network for the cooperative object tracking.

realize robust and stable object tracking. In the real world, however, no single AVA can keep tracking the object persistently due to occluding obstacles and interfering moving objects.

The above discussion leads us to introducing the dynamic master authority transfer protocol. That is, the master AVA always checks the reliability of the object information observed by each member, and transfers the master authority to such AVA that gives the most reliable object information (Fig. 14).

The prototype system employs a simple method: the master AVA transfers the authority to such member AVA whose object observation time is the latest, since the latest object information may be the most reliable.

5.3 Prototype System Development

Fig. 15 illustrates the state transition network designed to implement the above mentioned cooperative object tracking protocols. The network specifies event driven asynchronous interactions among perception, action, and communication modules as well as communication protocols with other AVAs, through which behaviors of an AVA emerge.

State i in the double circles denotes the initial state. Basically the states in rectangular boxes represent the roles of an AVA: master, worker, and non-worker. Since the master AVA conducts several different types of processing depending on situations, its state is subdivided into many substates. Those states in the shaded area show the states with the master authority. Each arrow connecting a pair of states is associated with the condition under which that state transition is incurred. ε means the unconditional state transition.

The right side of the figure shows what kind of processing, i.e. perception, action, receive, or send, is executed at each state. Double rectangular boxes denote states where perception is executed, while in triple rectangular boxes,

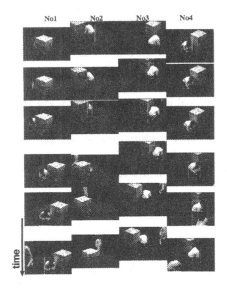

No1   No2   No3   No4

time

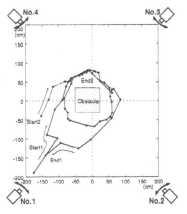

Fig. 16. Partial image sequences observed by four cameras. The vertical length of an image represents 0.5 sec.

Fig. 17. 3D target motion trajectories.

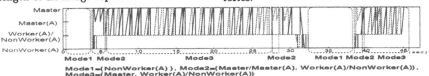

Mode1={NonWorker(A)}, Mode2={Master/Master(A), Worker(A)/NonWorker(A)},
Mode3={Master, Worker(A)/NonWorker(A)}

Fig. 18. State transition histories of the four AVAs.

camera action is executed. Thus, each state has its own dynamics and dynamic behaviors of an AVA are fabricated by state transitions.

We conducted experiments to verify the system performance. Two persons walked around a large box located at the center of the room (5m × 6m). Four FV-PTZ cameras (i.e. four AVAs) are placed at the four corners of the room respectively, looking downward obliquely from about 2.5m above the floor. The person who first entered in the scene was regarded as the target. He crawled around the box not to be detected by the cameras. The other person walked around the box to interfere the camera views to the target person. Then, both went out from the scene and after a while, a new person came into the scene.

Fig. 16 illustrates partial image sequences observed by the cameras, where the vertical axis represents the time when each image is captured. Each detected object is enclosed by a rectangle. Note that while some images include two objects and others nothing, the gaze of each camera is directed toward the crawling target person. Note also that the image acquisition timings of the cameras are almost synchronized. This is because the master AVA broadcasts the 3D view line or the 3D position of the target to the other AVAs, by which their perception processes are activated. This synchronized image acquisition by multiple cameras enables the computation of the 3D target motion trajectory (Fig. 17).

Fig. 18 illustrates the state transition histories of the four AVAs. We can see that the system exhibits well coordinated behaviors as designed:

Mode 1: All AVAs are searching for an object.

Mode 2: The master AVA itself tracks the object since the others are still searching for the object.

Mode 3: All AVAs form the agency to track the object under the master's guidance.

The zigzag shape in the figure shows the continuous master authority transfer is conducted inside the agency.

6 Concluding Remarks

This paper describes the idea and goal of our five years project on cooperative distributed vision and shows technical research results so far obtained on moving object detection and tracking by cooperative active vision agents. Currently, we are studying

1. Robust object detection in complex natural scenes
2. Communication protocols for cooperative multi targets tracking
3. Application system developments such as remote lecturing and intelligent TV studio systems.

For detailed activities of the CDV project, see our homepage at
URL: http://vision.kuee.kyoto-u.ac.jp/CDVPRJ/.

This work was supported by the Research for the Future Program of the Japan Society for the Promotion of Science (JSPS-RFTF96P00501).

References

1. Poole, D., Mackworth, A., and Goebel, R.: Computational Intelligence, Oxford University Press, 1998.
2. Uchibe, E., Asada, M., and Hosoda, K.: Behavior Coordination for a Mobile Robot Using Modular Reinforcement Learning, Proc. of IROS'96, pp.1329-1336, 1996.
3. Wada, T. and Matsuyama, T.: Appearance Sphere: Background Model for Pan-Tilt-Zoom Camera, Proc. of ICPR, Vol. A, pp. 718-722, 1996.
4. Wada, T., Ukita, N., and Matsuyama, T.: Fixed Viewpoint Pan-Tilt-Zoom Camera and its Applications, Trans. IEICE, Vol.J81D-II, No.6, pp.1182-1193, 1998 (in Japanese).
5. Aloimonos, Y. (ed.): Active Perception, Lawrence Erlbaum Associates Publisher, 1993
6. Shafer, S.A., Stentz, A., and Thorpe, C.E.: "An Architecture for Sensor Fusion in a Mobile Robot," Proc. of IEEE Conf. on Robotics and Automation, pp.2002-2011, 1986.
7. Murase, K., Hiura, S., and Matsuyama, T.: Dynamic Integration of Perception and Action for Real-Time Object Detection and Tracking, IPSJ SIG-CVIM115-20, 1999 (in Japanese).

Computing Probabilistic Least Common Subsumers in Description Logics

Thomas Mantay, Ralf Möller, and Alissa Kaplunova

Artificial Intelligence Lab, Department of Computer Science, University of Hamburg

Abstract. Computing least common subsumers in description logics is an important reasoning service useful for a number of applications. As shown in the literature, it can, for instance, be used for similarity-based information retrieval where information retrieval is performed on the basis of the similarities of user-specified examples. In this article, we first show that, for crisp DLs, in certain cases the set of retrieved information items can be too large. Then we propose a probabilistic least common subsumer operation based on a probabilistic extension of the description logic language $\mathcal{ALN}$. We show that by this operator the amount of retrieved data can be reduced avoiding information flood.

1 Introduction

Knowledge representation languages based on description logics (DLs) have proven to be a useful means for representing the terminological knowledge of an application domain in a structured and formally well understood way. In DLs, knowledge bases are formed out of *concepts* representing sets of individuals. Using the concept constructors provided by the DL language, complex concepts are built out of atomic concepts and atomic roles. Roles represent binary relations between individuals. For example, in the context of a TV information system, the set of all football broadcasts can be described with a concept term using the atomic concepts teamsports-broadcast and football and the atomic role has-sports-tool: teamsports-broadcast ⊓ ∀ has-sports-tool.football.

A central feature of knowledge representation systems based on DLs is a set of reasoning services with the possibility to deduce implicit knowledge from explicitly represented knowledge. For instance, the subsumption relation between two concepts can be determined. Intuitively speaking, a concept C *subsumes* a concept D if the set of individuals represented by C is a superset of the set of individuals represented by D, i.e., if C is more general than D. Furthermore, *retrieval* describes the problem of determining all individuals which are instances of a given concept.

As another reasoning service, the *least common subsumer* (LCS) operation, applied to concepts $C_1, \ldots, C_m$, computes the most specific concept which subsumes $C_1, \ldots, C_m$. The LCS operation is an important and non-trivial reasoning service useful for a number of applications. In [2], an LCS operator is considered for the DL $\mathcal{ALN}$ including feature chains in order to approximate a disjunction operation which is not explicitly included in $\mathcal{ALN}$. In addition, the operator is used as a subtask for the "bottom-up" construction of knowledge bases based on the DLs $\mathcal{ALN}$ with cyclic concept definitions [1]. See also [3] for a similar application concerning the constructive induction of a P-CLASSIC KB from data.

In our applications, the LCS operation is used as a subtask for similarity-based information retrieval [5]. The goal is to provide a user of an information system with an example-based query mechanism. The data of an information system are modeled as DL individuals. For instance, a specific TV broadcast about football could be modeled as an instance of the concept for football broadcasts introduced above. The "commonalities" of the selected examples of interest to the user are formalized by a DL concept of which (*i*) the user-selected examples are instances and (*ii*) which is the most specific concept (w.r.t. subsumption) with property (*i*). A concept fulfilling properties (*i*) and (*ii*) will then be used as a retrieval filter. The task of similarity-based information retrieval can be split into three subtasks: First, the most specific concepts of a finite set of individuals are computed yielding a finite set of concepts. Then, the LCS of these concepts is computed. Finally, by determining its instances the LCS concept is used as a retrieval concept. For the purpose of similarity-based information retrieval, the first task is fulfilled by the well-known realization inference service. The third subtask, determining the instances of the LCS concept, is accomplished by the instance retrieval inference service of the knowledge representation system.

In certain cases, computing the LCS of concepts yields a very general concept. As a consequence, a large set of information items are retrieved resulting in an information flood if all items are displayed at once. Thus, at least a ranking is needed or we have to define a new operator for computing the commonalities between concepts. In this paper, we pursue the second approach and define an LCS operator that takes additional domain knowledge into account.

The main contribution of this paper is the proposal of a probabilistic LCS operation for a probabilistic extension of the DL $\mathcal{ALN}$ which has been introduced in [4] for the knowledge representation system P-CLASSIC. The probabilistic LCS operator makes use of P-CLASSIC's ability to model the degree of overlap between concepts. With the probabilistic LCS operator we investigate an example-based retrieval approach in which well known information retrieval techniques are integrated with formally well investigated inference services of DLs.

2 Preliminaries

In this section, we introduce syntax and semantics of the underlying knowledge representation language $\mathcal{ALN}$ and give formal definitions of relevant terms.

Definition 1 (Syntax). *Let C be a set of atomic concepts and $\mathcal{R}$ a set of atomic roles disjoint from C. $\mathcal{ALN}$ concepts are recursively defined as follows:*

- *The symbols $\top$ and $\bot$ are $\mathcal{ALN}$ concepts (top concept, bottom concept).*
- *A and $\neg A$ are $\mathcal{ALN}$ concepts for each $A \in C$ (atomic concept, negated atomic concept).*
- *Let C and D be $\mathcal{ALN}$ concepts, $R \in \mathcal{R}$ an atomic role, and $n \in \mathbb{N} \cup \{0\}$. Then $C \sqcap D$ (concept conjunction), $\forall R.C$ (universal role quantification), $(\geq n R)$ ($\geq$-restriction), and $(\leq n R)$ ($\leq$-restriction) are also concepts.*

We set $(= n R)$ as an abbreviation for $(\geq n R) \sqcap (\leq n R)$. The semantics of concepts is given in terms of an interpretation.

Definition 2 (Interpretation, Model, Consistency). *An interpretation* $\mathcal{I} = (\Delta^{\mathcal{I}}, \cdot^{\mathcal{I}})$ *of an $\mathcal{ALN}$ concept consists of a non-empty set $\Delta^{\mathcal{I}}$ (the* domain *of $\mathcal{I}$) and an* interpretation function $\cdot^{\mathcal{I}}$. *The interpretation function maps every atomic concept A to a subset $A^{\mathcal{I}} \subseteq \Delta^{\mathcal{I}}$ and every role R to a subset $R^{\mathcal{I}} \subseteq \Delta^{\mathcal{I}} \times \Delta^{\mathcal{I}}$. The interpretation function is recursively extended to complex $\mathcal{ALN}$ concepts as follows. Assume that $A^{\mathcal{I}}, C^{\mathcal{I}}, D^{\mathcal{I}}$, and $R^{\mathcal{I}}$ are already given and $n \in \mathbb{N} \cup \{0\}$. Then*

- $\top^{\mathcal{I}} := \Delta^{\mathcal{I}}$, $\bot^{\mathcal{I}} := \emptyset$, $(\neg A)^{\mathcal{I}} := \Delta^{\mathcal{I}} \setminus A^{\mathcal{I}}$, $(C \sqcap D)^{\mathcal{I}} := C^{\mathcal{I}} \cap D^{\mathcal{I}}$,
- $\forall R.C^{\mathcal{I}} := \{d \in \Delta^{\mathcal{I}} | \forall d' : (d, d') \in R^{\mathcal{I}} \Rightarrow d' \in C^{\mathcal{I}}\}$,
- $(\geq n R)^{\mathcal{I}} := \{d \in \Delta^{\mathcal{I}} | \sharp\{d' | (d, d') \in R^{\mathcal{I}}\} \geq n\}$, *and*
- $(\leq n R)^{\mathcal{I}} := \{d \in \Delta^{\mathcal{I}} | \sharp\{d' | (d, d') \in R^{\mathcal{I}}\} \leq n\}$.

An interpretation $\mathcal{I}$ is a model *of an $\mathcal{ALN}$ concept C iff $C^{\mathcal{I}} \neq \emptyset$. C is called* consistent *iff C has a model.*

Note that both constructors $\top$ and $\bot$ are expressible by $(\geq 0 R)$ and $A \sqcap \neg A$, respectively.

Definition 3 (Subsumption, Equivalence, Instance). *A concept C is* subsumed *by a concept D ($C \sqsubseteq D$) iff $C^{\mathcal{I}} \subseteq D^{\mathcal{I}}$ for all interpretations $\mathcal{I}$. Two concepts C and D are* equivalent *iff $C^{\mathcal{I}} = D^{\mathcal{I}}$ holds for all interpretations $\mathcal{I}$. The interpretation function $\cdot^{\mathcal{I}}$ is extended to individuals by mapping them to elements of $\Delta^{\mathcal{I}}$ such that $a^{\mathcal{I}} \neq b^{\mathcal{I}}$ if $a \neq b$. An individual $d \in \Delta^{\mathcal{I}}$ is an* instance *of a concept C iff $d^{\mathcal{I}} \in C^{\mathcal{I}}$ holds for all interpretations $\mathcal{I}$.*

Definition 4 (Depth). *The* depth *of a concept is recursively defined as follows:*

- *If $C = A$, $C = \neg A$, $C = (\geq n R)$, or $(\leq n R)$, then $depth(C) := 0$.*
- *If $C = \forall R.C'$, then $depth(C) := 1 + depth(C')$.*

Note that, in contrast to usual definitions of the concept depth, we define the depth of number restrictions as 0.

Definition 5 (Canonical form). *Let $C_1, \ldots, C_m$ be concepts and $\{R_1, \ldots, R_M\}$ the set of all roles occuring in $C_1, \ldots, C_m$. Then C_i is in* canonical form *iff*

$$C_i = \alpha_{i1} \sqcap \cdots \sqcap \alpha_{in_i} \sqcap \beta_{iR_1} \sqcap \cdots \sqcap \beta_{iR_{j_i}}$$

where $j_i \in \{0, \ldots, M\}$, α_{ik} is an atomic concept or negated atomic concept with no atomic concept appearing more than once and $\beta_{iR_j} = (\geq l_{iR_j} R_j) \sqcap (\leq m_{iR_j} R_j) \sqcap \forall R_j.C'_{iR_j}$ with C'_{iR_j} also being in canonical form.

It is easy to see that any concept can be transformed into an equivalent concept in canonical form in linear time.

Definition 6 (LCS). *Let $C_1, \ldots, C_m$ be concepts. Then we define the set of* least common subsumers *(LCSs) of $C_1, \ldots, C_m$ as*

$$lcs(C_1, \ldots, C_m) := \{E | C_1 \sqsubseteq E \wedge \cdots \wedge C_m \sqsubseteq E \wedge$$
$$\forall E' : C_1 \sqsubseteq E' \wedge \cdots \wedge C_m \sqsubseteq E' \Rightarrow E \sqsubseteq E'\}.$$

In [2], it is shown that, for the DL $\mathcal{ALN}$, all elements of $lcs(C_1, \ldots, C_m)$ are equivalent. Therefore, we will consider $lcs(C_1, \ldots, C_m)$ as a single concept instead of a set of concepts.

The following example shows that the concept computed by the LCS is sometimes too general and, thus, might not always be a useful retrieval concept. Let sports-broadcast (SB), team-sports-broadcast (TSB), individual-sports-broadcast (ISB), basketball (B), football (FB), and tennis-racket (TR) be atomic concepts, has-sports-tool an atomic role, and

$$\text{basketball-broadcast (BB)} := \text{team-sports-broadcast} \sqcap (= 1 \text{ has-sports-tool}) \sqcap$$
$$\forall \text{ has-sports-tool.basketball},$$

$$\text{football-broadcast (FB)} := \text{team-sports-broadcast} \sqcap (= 1 \text{ has-sports-tool}) \sqcap$$
$$\forall \text{ has-sports-tool.football, and}$$

$$\text{tennis-broadcast (TB)} := \text{individual-sports-broadcast} \sqcap (= 1 \text{ has-sports-tool}) \sqcap$$
$$\forall \text{ has-sports-tool.tennis-racket}$$

be concepts. Subsequently, we will use the concept abbreviations given in brackets. Let us consider a user interested in TV broadcasts similar to FB and BB. Then, computing the LCS of FB and BB would result in a useful retrieval concept: $\text{TSB} \sqcap (= 1 \text{ has-sports-tool}) \sqcap \forall \text{has-sports-tool.ST}$. However, let us consider a user whose interests are expressed by FB and TB. The LCS computation then yields the retrieval concept $A := \text{SB} \sqcap (= 1 \text{has-sports-tool}) \sqcap \forall \text{has-sports-tool.ST}$ denoting the set of *all* sports broadcasts with a sports tool. Since A is a very general concept, using A as a retrieval concept would result in a large amount of TV broadcasts, which might not be acceptable on the part of the user. A more suitable result would be to allow for $B := \text{TSB} \sqcap (= 1 \text{has-sports-tool}) \sqcap \forall \text{has-sports-tool.ST}$ and $C := \text{ISB} \sqcap (= 1 \text{has-sports-tool}) \sqcap \forall \text{has-sports-tool.ST}$ as alternative retrieval concepts. This is plausible because in Davis Cup matches, for instance, teams of tennis players compete against each other. Hence, in our intuition, there is a non-empty overlap between the concepts TSB and ISB which cannot be adequately quantified in $\mathcal{ALN}$. In order to model the degree of overlap between concepts by probabilities, the knowledge representation system P-CLASSIC was introduced in [4]. The DL underlying P-CLASSIC is a probabilistic extension of $\mathcal{ALN}$ augmented by functional roles (attributes).

One of the goals of P-CLASSIC is to compute probabilistic subsumption relationships of the form $P(D|C)$ denoting the probability of an individual to be an instance of D given that it is an instance of C. In case $C \equiv \top$, we write $P(D)$. In order to fully describe a concept, its atomic concept components and the properties of number restrictions and universal role quantifications need to be described. Therefore, a set $\mathcal{P}$ of probabilistic classes (p-classes) is introduced describing a probability distribution over the properties of individuals conditioned on the knowledge that the individuals occur on the right-hand side of a role. Each p-class is represented by a Bayesian network and one of the p-classes $P^* \in \mathcal{P}$ is the root p-class. The root p-class describes the distribution over all individuals and all other p-classes describe the distribution over role successors assuming independence between distinct individuals. The Bayesian networks are

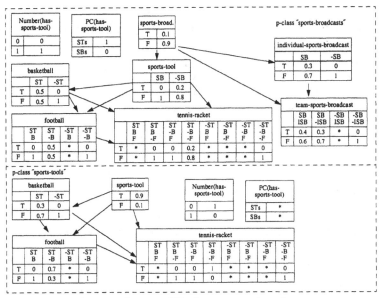

Fig. 1. P-CLASSIC KB about sports broadcasts.

modeled as DAGs whose nodes represent atomic concepts, number restrictions $[Number(R)]$, and the p-class from which role successors are drawn $[PC(R)]$. In addition to P-CLASSIC, we introduce extra nodes for negations of atomic concepts. Dependencies are used to model conditional probabilities and are modeled by edges in the network. For instance, for an individual, we can state the probability of this individual to be an instance of ISB under the condition that it is an instance of SB. The range of the variables of a node representing an atomic or negated atomic concept can be either *true* or *false* and for $Number(R)$ it is a subset of $\mathbb{N}$. In order to guarantee termination of the inference algorithm for computing $P(D|C)$, this subset must be finite. Thus, the number of role successors for a role is bounded. The function $bound(R)$ indicates the maximum number of role successors for R. The range of a $PC(R)$ node is the set of p-classes $\mathcal{P}$ indicating the p-classes the R-successors are drawn from. The reason for introducing special nodes for negations of atomic concepts is that this extension enables us to evaluate expressions of the form $P(A \sqcap \neg A)$ as 0 which will be a necessary property subsequently. In order to demonstrate the advantages of the probabilistic LCS operator, we will now create a P-CLASSIC KB with overlapping concepts. Figure 1 shows a knowledge base about sports broadcasts enriched by probability information. For instance, it is stated that a broadcast is considered to be about team-sports (TSB) with probability 0.3 given that it is a broadcast about sports (SB) but no individual-sports (-ISB). Two p-classes are represented. The concept **sports-broadcasts** is the root p-class and the role successors for the role **has-sports-tool** are drawn from the p-class **sports-tools**. For each concept C, the probability $P_{P*}(C)$ with which an individual is an instance of C can then be computed by a standard inference algorithm for Bayesian networks. For example, the probability of $P_{P*}(\text{TSB} \sqcap (= 1\,\text{has-sports-tool}) \sqcap \forall\,\text{has-sports-tool.B})$ is computed by setting the nodes for TSB and B to *true*, $Number(\text{has-sports-tool}) = 1$, and $PC(\text{has-sports-tool}) = \text{STs}$. By Bayesian network propagation we yield a value of 0.015. With the formalism for computing expressions of the form $P(D|C)$ it is possible to express the degree of overlap between C and D by a probability.

Based on the probabilistic description logic summarized in this section, it is possible to define a probabilistic LCS operator which takes into account the degree of overlap between concepts.

3   A Probabilistic Extension of the LCS Operator

Intuitively, given concepts $C_1, \ldots, C_m$, the key idea is to allow those concepts for candidates of a probabilistic least common subsumer (PLCS) of $C_1, \ldots, C_m$ which have a non-empty overlap with $C_1, \ldots, C_m$. In order to keep the set of candidates finite, we consider only concepts whose depth is not larger than $\max\{depth(C_i)|i \in \{1, \ldots, m\}\}$. From the viewpoint of information retrieval this is no severe restriction, since in practical applications deeply nested concepts usually do not have any relevant individuals as instances (e.g., the concept FB $\sqcap$ $\forall$has-sports-tool.$\forall$has-sports-tool.F in our example).

Definition 7. *Let* $C_1, \ldots, C_m$ *be* $\mathcal{ALN}$ *concepts and* P^* *the root p-class of a* P-CLASSIC *KB. Then we define the set of* PLCS *concept candidates of* $C_1, \ldots, C_m$ *as*

$$Can(C_1, \ldots, C_m) := \{E | P_{P^*}(E \sqcap C_1) > 0 \land \cdots \land P_{P^*}(E \sqcap C_m) > 0 \land$$
$$depth(E) \leq \max\{depth(C_i)|i = 1, \ldots, m\}\}.$$

Definition 7 induces the following observation.

Proposition 1. *Let* $C_1, \ldots, C_m$ *be* $\mathcal{ALN}$ *concepts. Then, in the worst case, the cardinality of* $Can(C_1, \ldots, C_m)$ *is exponential in* m.

Proof. Given a P-CLASSIC KB in which $C_1, \ldots, C_m$ are all atomic concepts with $\forall i, j \in \{1, \ldots, m\} : P(C_i \sqcap C_j) > 0$, we can bound $\sharp Can(C_1, \ldots, C_m)$ by the exponential function 2^m. □

In the next step, we want to measure the effectiveness of using a certain PLCS candidate for retrieval. It will be helpful to be able to express the probability of an individual to be an instance of a concept disjunction. Since this language operator is not contained in $\mathcal{ALN}$, we use the following definition which is essentially taken from [6].

Definition 8. *Let* $C_1, \ldots, C_m$ *be* $\mathcal{ALN}$ *concepts. Then we define*

$$P(C_1 \sqcup \cdots \sqcup C_m) := (-1)^2 \sum_{k=1, \ldots, m} P(C_k) + (-1)^3 \sum_{k_1 < k_2} P(C_{k_1} \sqcap C_{k_2}) +$$
$$(-1)^4 \sum_{k_1 < k_2 < k_3} P(C_{k_1} \sqcap C_{k_2} \sqcap C_{k_3}) + \ldots$$
$$+ (-1)^{m+1} P(C_1 \sqcap \cdots \sqcap C_m).$$

It should be noted that by Definition 8 we do not extend the syntax of the underlying DL.

Proposition 2. *Let $C_1, \ldots, C_m$ be concepts. Then computing $P(C_1 \sqcup \cdots \sqcup C_m)$ is exponential in m.*

The proof is obvious and is omitted here.

In many retrieval environments, it is customary to introduce two real numbers: recall and precision. Both values indicate the quality of a concept E to function as an appropriate PLCS. By these measures the qualities of potential PLCSs can be compared to one another. The comparison will be formalized by the notion of dominance between triples $(E, r_{E,C_1,\ldots,C_m}, p_{E,C_1,\ldots,C_m})$ and $(E', r_{E',C_1,\ldots,C_m}, p_{E',C_1,\ldots,C_m})$.

Definition 9 (Recall). *Let E and $C_1, \ldots, C_m$ be $\mathcal{ALN}$ concepts. Then we define E's recall of $C_1, \ldots, C_m$ as*

$$r_{E,C_1,\ldots,C_m} := P(C_1 \sqcup \cdots \sqcup C_m | E) = \frac{P(E \sqcap (C_1 \sqcup \cdots \sqcup C_m))}{P(E)}.$$

According to this definition, the larger the recall measure of a concept E, the more specific it is w.r.t. probabilistic subsumption of $C_1, \ldots, C_m$. For a concept E, a perfect recall is yielded iff $r_{E,C_1,\ldots,C_m} = 1$. For example, if E is a PLCS candidate and A an atomic concept such that $A \sqsubseteq E$, then $r_{E,A,\neg A} = 1$. Unlike in the definition of the (crisp) LCS, a concept expression does not necessarily need to subsume $C_1, \ldots, C_m$ (completely) in order to be a PLCS candidate. This motivates the introduction of the precision measure.

Definition 10 (Precision). *Let E and $C_1, \ldots, C_m$ be $\mathcal{ALN}$ concepts. Then we define E's precision of $C_1, \ldots, C_m$ as*

$$p_{E,C_1,\ldots,C_m} := P(E | C_1 \sqcup \cdots \sqcup C_m) = \frac{P(E \sqcap (C_1 \sqcup \cdots \sqcup C_m))}{P(C_1 \sqcup \cdots \sqcup C_m)}.$$

The precision measures the probability with which a randomly chosen individual, which is an instance of any of the C_i, $i \in \{1, \ldots, m\}$, is also an instance of the PLCS candidate E. As a consequence of Definition 10, if $E = lcs(C_1, \ldots, C_m)$, we have $q_{E,C_1,\ldots,C_m} = 1$.

Figure 2 illustrates the meaning of both measures given four concepts represented as areas in the 2D space. The recall of E_1, $r_{E_1,C,D}$, corresponds to the ratio of the size of the hatched area and the size of E_1. E_1's precision, $p_{E_1,C,D}$, is the ratio of the size of E_1 and the size of the union of E_1, C, and D. Given the appropriate values for E_2 we see that $r_{E_2,C,D}$ is smaller than $r_{E_1,C,D}$ but $p_{E_2,C,D}$ is larger than $p_{E_1,C,D}$.

Proposition 3. *Let E and $C_1, \ldots, C_m$ be $\mathcal{ALN}$ concepts. Then, computing $r_{E,C_1,\ldots,C_m}$ and $p_{E,C_1,\ldots,C_m}$ takes time exponential in the length of $E, C_1, \ldots, C_m$.*

Proof. Since $P(E \sqcap (C_1 \sqcup \cdots \sqcup C_m)) = P(E) - (P(E \sqcup C_1 \sqcup \cdots \sqcup C_m) - P(C_1 \sqcup \cdots \sqcup C_m))$, the claim follows from Proposition 2. $\qquad\square$

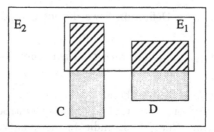

Fig. 2. Scenario of four concepts illustrating the meaning of "recall" and "precision".

With the above considerations, we will define the set of PLCSs of concepts $C_1, \ldots, C_m$ as a set of triples where the first component is a concept $E \in Can(C_1, \ldots, C_m)$ and the other components are E's recall and precision. In a concrete application, a user should be able to specify minimum values for at least one of the measures that he is willing to accept. For example, he could specify a recall of 0.8 preventing him from obtaining too general PLCS concepts and, thus, restricting the amount of retrieved data.

With the notion of dominance between candidates we can define the set of probabilistic least common subsumers.

Definition 11 (Dominance). *Let E, E' and $C_1, \ldots, C_m$ be $\mathcal{ALN}$ concepts. Then $(E, r_{E,C_1,\ldots,C_m}, p_{E,C_1,\ldots,C_m})$ dominates $(E', r_{E',C_1,\ldots,C_m}, p_{E',C_1,\ldots,C_m})$ iff $r_{E,C_1,\ldots,C_m} > r_{E',C_1,\ldots,C_m} \wedge p_{E,C_1,\ldots,C_m} > p_{E',C_1,\ldots,C_m}$.*

Definition 12 (Set of Probabilistic Least Common Subsumers). *Let $C_1, \ldots, C_m$ be $\mathcal{ALN}$ concepts. Then we define the set of probabilistic least common subsumers of $C_1, \ldots, C_m$ as*

$$p\text{-}lcs(C_1, \ldots, C_m) := \{(E, r_{E,C_1,\ldots,C_m}, p_{E,C_1,\ldots,C_m}) \in Can(C_1, \ldots, C_m) \times \mathbb{R} \times \mathbb{R}|$$
$$\neg \exists (E', r_{E',C_1,\ldots,C_m}, p_{E',C_1,\ldots,C_m}):$$
$$(E', r_{E',C_1,\ldots,C_m}, p_{E',C_1,\ldots,C_m}) \text{ dominates}$$
$$(E, r_{E,C_1,\ldots,C_m}, p_{E,C_1,\ldots,C_m})\}.$$

$p\text{-}lcs(C_1, \ldots, C_m)$ *is called* minimal *iff* $\forall (E, r_{E,C_1,\ldots,C_m}, p_{E,C_1,\ldots,C_m}), (E', r_{E',C_1,\ldots,C_m}, p_{E',C_1,\ldots,C_m}) \in p\text{-}lcs(C_1, \ldots, C_m) : E \not\equiv E'$.

In Definition 12 we formalize the ideas of Fig. 2 conditioned on the general case of m concepts. When defining $p\text{-}lcs(C_1, \ldots, C_m)$ we consider only concepts with a non-empty overlap with each of the $C_1, \ldots, C_m$. We only accept triples with the best quality measures and, therefore, accept only dominating triples in $p\text{-}lcs(C_1, \ldots, C_m)$. From this definition we can derive the following statement.

Proposition 4. *The set $p\text{-}lcs(C_1, \ldots, C_m)$ has the following properties:*

(i) $p\text{-}lcs(C_1, \ldots, C_m)$ is finite.

(ii) Minimality: $(E, r_{E,C_1,\ldots,C_m}, p_{E,C_1,\ldots,C_m}) \in p\text{-}lcs(C_1, \ldots, C_m) \Longrightarrow$
$\forall i \in \{1, \ldots, m\} : P(E \sqcap C_i) > 0 \wedge \neg \exists (E', r_{E',C_1,\ldots,C_m}, p_{E',C_1,\ldots,C_m}) :$
$r_{E',C_1,\ldots,C_m} > r_{E,C_1,\ldots,C_m} \wedge p_{E',C_1,\ldots,C_m} > p_{E,C_1,\ldots,C_m} \wedge depth(E') \leq depth(E)$.

Proof. (i) is obvious since the maximum depth of the concepts in $p\text{-}lcs(C_1, \ldots, C_m)$ is limited by the maximum depth of the $C_1, \ldots, C_m$ and the number of concept components of $C_1, \ldots, C_m$ is finite ensuring the number of PLCS candidates to be finite. Hence, $p\text{-}lcs(C_1, \ldots, C_m)$ is finite as well. For $(E, r_{E,C_1,\ldots,C_m}, p_{E,C_1,\ldots,C_m}) \in p\text{-}lcs(C_1, \ldots, C_m)$, the fact that $P(E \sqcap C_i) > 0$, for all $i \in \{1, \ldots, m\}$, follows immediately by the definition of $Can(C_1, \ldots, C_m)$. $\neg\exists(E', r_{E',C_1,\ldots,C_m}, p_{E',C_1,\ldots,C_m})$: $r_{E',C_1,\ldots,C_m} > r_{E,C_1,\ldots,C_m} \wedge p_{E',C_1,\ldots,C_m} > p_{E,C_1,\ldots,C_m} \wedge depth(E') \leq depth(E)$ also follows since $p\text{-}lcs(C_1, \ldots, C_m)$ contains only dominating triples $(E, r_{E,C_1,\ldots,C_m}, p_{E,C_1,\ldots,C_m})$ with $depth(E) \leq \max\{depth(C_i)|i \in \{1, \ldots, m\}\}$. $\square$

The minimal set $p\text{-}lcs(C_1, \ldots, C_m)$ can be computed in three steps: First, we must find the set of concepts which have a non-empty overlap with each of the $C_1, \ldots, C_m$. Proposition 4 (i) states a necessary criterion for a corresponding algorithm to terminate since the set of concepts E which have a non-empty overlap with each of the C_i is finite. Then, for each concept E in this set, we have to compute the parameters $r_{E,C_1,\ldots,C_m}$ and $p_{E,C_1,\ldots,C_m}$ and then build the set of dominant triples $p\text{-}lcs(C_1, \ldots, C_m)$. Proposition 4 (ii) guarantees that there is no relevant retrieval concept with better recall *and* precision than the corresponding measures of the triples in $p\text{-}lcs(C_1, \ldots, C_m)$. Finally, we must determine the minimal set $p\text{-}lcs(C_1, \ldots, C_m)$. This can be done by successively eliminating a triple $(E, r_{E,C_1,\ldots,C_m}, p_{E,C_1,\ldots,C_m})$ from $p\text{-}lcs(C_1, \ldots, C_m)$ as long as the following condition holds:

$$\forall(E, r_{E,C_1,\ldots,C_m}, p_{E,C_1,\ldots,C_m}) \in p\text{-}lcs(C_1, \ldots, C_m) :$$
$$\neg\exists(E', r_{E',C_1,\ldots,C_m}, p_{E',C_1,\ldots,C_m}) \in p\text{-}lcs(C_1, \ldots, C_m) \text{ with } E \equiv E'.$$

The necessary equivalence test can be performed by structural comparisons since the involved concepts are in normal form. In general, a minimal $p\text{-}lcs(C_1, \ldots, C_m)$ is not unique since there is no rule stating which triple to eliminate in case two triples with equivalent concepts are present. However, in our similarity-based information retrieval application this is no problem because the sets of instances of equivalent concepts are equal.

Algorithm 1 computes the set of PLCS candidates given concepts $C_1, \ldots, C_m$ and the KB as a Bayesian network BN. In the first step, all atomic concepts and negated atomic concepts in the Bayesian network are collected in the set X_1 if there is a non-empty overlap with each of the $C_1, \ldots, C_m$. Computing the concept candidates for our example, *compute-concept-candidate*(FB, TB), we get $X_1 = \{\text{SB}, \text{TSB}, \text{ISB}\}$. Secondly, we build the set of all conjunctions of concepts of X_1 which have a non-empty overlapping with each of the $C_1, \ldots, C_m$ including the ones consisting of only one conjunct. In our case, we yield $X_2 = \{\text{SB}, \text{TSB}, \text{ISB}, \text{SB} \sqcap \text{TSB}, \text{SB} \sqcap \text{ISB}, \text{ISB} \sqcap \text{TSB}, \text{SB} \sqcap \text{TSB} \sqcap \text{ISB}\}$. In the next part of the algorithm, we collect all number restrictions having a non-empty overlap with each of the $C_1, \ldots, C_m$ in the set X_3. Since the maximum number of role successors is bounded, we can guarantee finiteness of X_3. Let X be an abbreviation for $(= 1\,\text{has-sports-tool})$ and Y an abbreviation for $(\geq 0\,\text{has-sports-tool}) \sqcap (\leq 1\,\text{has-sports-tool})$. Then, in our example, we have $X_3 = \{X, Y\}$. Subsequently,

---

Algorithm 1 compute-concept-candidates($C_1, \ldots, C_m, BN$)

$X_1 := \emptyset, X_2 := \emptyset, X_3 := \emptyset, X_4 := \emptyset, X_5 := \emptyset, X_6 := \emptyset, X_7 := \emptyset$

for all nodes in BN representing an atomic or negated atomic concept A **do**
    add A to X_1 if $\forall i \in \{1, \ldots, m\} : P_{PCL_i}(A \sqcap C_i) > 0$
end for
for all $K \in 2^{X_1}$ **do**
    add $E := \sqcap_{D \in K} D$ to X_2 if $\forall i \in \{1, \ldots, m\} : P_{PCL_i}(E \sqcap C_i) > 0$
end for
for $i \in \{1, \ldots, M\}$ and subexpressions of the form $(\geq i\, R) \sqcap (\leq j\, R)$ in $C_1, \ldots, C_m$
do
    add $(\geq k\, R) \sqcap (\leq l\, R)$ to X_3 if $0 \leq k \leq j \wedge bound(R) \geq l \geq i \wedge k \leq l$
end for
for $i \in \{1, \ldots, M\}$ and subexpressions of the form $\forall R_i.C_1', \ldots, \forall R_i.C_m'$ in $C_1, \ldots, C_m$
do
    add the concepts resulting from the invocation
    compute-concept-candidates($C_1', \ldots, C_m', BN$) to X_4
end for
$X_5 := \{C \sqcap \forall R_i.D | i \in \{1, \ldots, M\}$ and $C \in X_3$ and $D \in X_4$ and C refers to role $R_i\}$
for all $K \in 2^{X_5}$ **do**
    add $E := \sqcap_{D \in K} D$ to X_6 if $\forall i \in \{1, \ldots, m\} : P_{PCL_i}(E \sqcap C_i) > 0$
end for
$X_7 := X_2 \cup X_6 \cup \{C \sqcap D | C \in X_2 \wedge D \in X_6 \wedge \forall i \in \{1, \ldots, m\} : P_{PCL_i}(C \sqcap D \sqcap C_i) > 0\}$
return X_7

---

for all roles R_i and all $\forall$-quantifications occurring in $C_1, \ldots, C_m$ and involving R_i, we add those concepts to X_4 which have a non-empty overlap with each of the R_i quantifiers ($C_1', \ldots, C_m'$ in the algorithm). In our example, we compute $X_4 := \{\mathsf{ST}\}$. Now, in X_5 we collect all conjunctions of number restrictions from X_3 involving role R and $\forall R.D$ where D is a concept overlapping with R's quantifiers $C_1', \ldots, C_m'$. Let X' be an abbreviation for $X \sqcap \forall\, \mathsf{has\text{-}sports\text{-}tool}.\mathsf{ST}$ and Y' an abbreviation for $Y \sqcap \forall\, \mathsf{has\text{-}sports\text{-}tool}.\mathsf{ST}$. Then, in our example, we have $X_5 = \{X', Y'\}$. In X_6, we collect the conjunctions of elements of X_5 over all occurring roles if a conjunction has a non-empty overlap with each of the $C_1, \ldots, C_m$. Since, in our example, we have only one role, we get $X_6 = X_5$. Finally, we combine the results in X_2 (conjunctions of atomic and negated atomic concepts) and the ones in X_6 (conjunctions of number restrictions and $\forall$-quantifications) into X_7 which is returned by the algorithm. In our example, X_7 consists of 21 concepts from which we will only list the ones which are unique w.r.t. to equivalence: $\{E_1, \ldots, E_{12}\} = \{\mathsf{SB}, \mathsf{TSB}, \mathsf{ISB}, \mathsf{TSB} \sqcap \mathsf{ISB}, \mathsf{SB} \sqcap X', \mathsf{TSB} \sqcap X', \mathsf{ISB} \sqcap X', \mathsf{TSB} \sqcap \mathsf{ISB} \sqcap X', \mathsf{SB} \sqcap Y', \mathsf{TSB} \sqcap Y', \mathsf{ISB} \sqcap Y', \mathsf{TSB} \sqcap \mathsf{ISB} \sqcap Y'\}$ as desired.

Theorem 1. *For concepts $C_1, \ldots, C_m$ and a Bayesian network BN representing a* P-Classic *KB, algorithm* compute-concept-candidates *returns the set* $Can(C_1, \ldots, C_m)$.

Proof. We give only a sketch of the proof. Algorithm *compute-concept-candidates* terminates because the maximum number of iterations is bounded by the maximum depth of $C_1, \ldots, C_m$. It is sound since every output concept has a non-empty overlap with $C_1, \ldots, C_m$. It is also complete because the algorithm recursively checks all possible concepts resulting from the concept-forming operators of Definition 1 for a non-empty overlap with $C_1, \ldots, C_m$. □

The set of concept candidates computed by Algorithm 1 can easily be transformed into a set in which all pairs of concepts are not equivalent. Therefore, later no additional algorithm for transforming $p\text{-}lcs(C_1, \ldots, C_m)$ into a minimal $p\text{-}lcs(C_1, \ldots, C_m)$ will be necessary. Now recall and precision must be determined for each candidate by means of the formulae given in Definitions 9 and 10. This can be done straightforwardly by algorithms taking concepts E and $C_1, \ldots, C_m$ as input parameters and returning $r_{E,C_1,\ldots,C_m}$ and $p_{E,C_1,\ldots,C_m}$, respectively. The set $p\text{-}lcs(C_1, \ldots, C_m)$ contains only those triples whose quality measures dominate those of other triples.

Algorithm 2 compute-minimal-plcs($(E_1, r_1, p_1), \ldots, (E_n, r_n, p_n)$)

$p\text{-}lcs(C_1, \ldots, C_m) := sort(((E_1, r_1, p_1), \ldots, (E_n, r_n, p_n)), p_i)$
for $i = 1$ to n **do**
    eliminate all (E', r', p') from $p\text{-}lcs(C_1, \ldots, C_m)$ with $r' < r_i$ and $p' < p_i$
end for.

Algorithm 2 computes the largest subset of dominant triples of $\{(E_1, r_{E_1,C_1,\ldots,C_m}, p_{E_1,C_1,\ldots,C_m}), \ldots, (E_n, r_{E_n,C_1,\ldots,C_m}, p_{E_n,C_1,\ldots,C_m})\}$. In the example, we get $p\text{-}lcs(\mathsf{FB}, \mathsf{TB}) = \{(\mathsf{SB} \sqcap X', 0.22, 1), (\mathsf{SB} \sqcap Y', 0.22, 1), (\mathsf{TSB} \sqcap X', 0.24, 0.354), (\mathsf{TSB} \sqcap Y', 0.24, 0.354), (\mathsf{ISB} \sqcap X', 0.26, 0.345), (\mathsf{ISB} \sqcap Y', 0.26, 0.345)\}$. As a result we get six possible retrieval concepts. $\mathsf{SB} \sqcap X'$ is the (crisp) LCS of FB and TB. Naturally, this concept has a precision of 1.0 since, according to Definition 6, $lcs(\mathsf{FB}, \mathsf{TB})$ is a concept which (completely) subsumes FB and TB. Alternatively, the result suggests the use of $\mathsf{TSB} \sqcap X'$ or $\mathsf{ISB} \sqcap X'$ as retrieval concepts. Both concepts have a better recall measure, and using them for retrieval results in a smaller set of information items. On the other hand, $\mathsf{TSB} \sqcap X'$ and $\mathsf{ISB} \sqcap X'$ have a worse precision measure than $\mathsf{SB} \sqcap X'$. Hence, the probability of meeting an individual which does not incorporate the commonalities represented by the concepts FB and TB is higher. The three concepts involving Y' have the same quality measures than the ones involving X'. The reason is that from our P-CLASSIC KB it follows that $P(Number(\mathsf{has\text{-}sports\text{-}tool}) = 0) = 1.0$, i.e, we do not need to consider them.

Theorem 2. *Let $C_1, \ldots, C_m$ be $\mathcal{ALN}$ concepts. Then, in the worst case, computing $p\text{-}lcs(C_1, \ldots, C_m)$ takes time exponential in m.*

Proof. This result follows from Proposition 1 since computing the set of PLCS candidates of $C_1, \ldots, C_m$ is a subtask of computing $p\text{-}lcs(C_1, \ldots, C_m)$. □

Propositions 2, 1, and 3 show the sources of complexity for the presented inference task. Due to the subterms $P(C_1 \sqcup \cdots \sqcup C_m)$ and $P(E \sqcap (C_1 \sqcup \cdots \sqcup C_m))$

occuring in Definitions 9 and 10, the computation of the precision and the recall measure take time exponential in the number of m. Also the computation of the set of PLCS candidates takes time exponential in the number of concepts. In practice, however, the exponential behavior of the computation comes into effect only for knowledge bases with many overlapping concepts. Thus, when building a KB, the number of concept overlaps should be kept small.

4 Conclusion

In this article, we contributed to the problem of similarity-based information retrieval on the basis of the DL $\mathcal{ALN}$. It is shown that in certain cases the computation of commonalities with the (crisp) LCS operation yields too general retrieval concepts which can result in an information flood in a retrieval context. In order to circumvent this problem, we introduced a probabilistic LCS for a probabilistic extension of the DL $\mathcal{ALN}$. It is proved that the retrieval concepts provided by this operation are in some sense optimal and can be used as an alternative to retrieval concepts computed by a crisp LCS operation. By demonstrating the performances of the PLCS operator with an example we showed that meaningful retrieval results can be achieved with this operator. In the retrieval approach we integrated known information retrieval techniques with formally investigated inference services of DLs. Further research can be done on extending the expressivity of the underlying DL—especially integrating a disjunction operator. It is not clear if the disjunction $C \sqcup D$ of two concepts C and D should also belong to the set of PLCS candidates since it is questionable if it sufficiently represents the commonalities of C and D. Another problem is that the number of PLCS candidates will dramatically increase in the presence of an or-operator.

References

1. F. Baader and R. Küsters. Computing the Least Common Subsumer and the Most Specific Concept in the Presence of Cyclic $\mathcal{ALN}$-concept Descriptions. In O. Herzog and A. Günter, editors, *Proc. of the 22nd KI-98*, volume 1504, pages 129–140, 1998.
2. W. W. Cohen, A. Borgida, and H. Hirsh. Computing Least Common Subsumers in Description Logics. In *Proceedings of the National Conference on Artificial Intelligence AAAI'92*, pages 754–760. AAAI Press/The MIT Press, 1992.
3. J.U. Kietz and K. Morik. A polynomial approach to the constructive induction of structural knowledge. *Machine Learning*, 14:193–217, 1994.
4. D. Koller, A. Levy, and A. Pfeffer. P-Classic: A tractable probabilistic description logic. In *Proc. of AAAI 97*, pages 390–397, Providence, Rhode Island, 1997.
5. R. Möller, V. Haarslev, and B. Neumann. Semantics-based Information Retrieval. In *Int. Conf. on Inf. Techn. and Knowl. Systems*, Vienna, Budapest, 1998.
6. V. K. Rohatgi. *An Introduction to Probability Theory and Mathematical Statistics*. Wiley Series in Probability and Mathematical Statistics, 1976.

Revising Nonmonotonic Theories: The Case of Defeasible Logic

D. Billington, G. Antoniou, G. Governatori, and M. Maher

School of Computing and Information Technology
Griffith University, QLD 4111, Australia
{db,ga,guido,mjm}@cit.gu.edu.au

Abstract. The revision and transformation of knowledge is widely recognized as a key issue in knowledge representation and reasoning. Reasons for the importance of this topic are the fact that intelligent systems are gradually developed and refined, and that often the environment of an intelligent system is not static but changes over time. Traditionally belief revision has been concerned with revising first order theories. Nonmonotonic reasoning provides rigorous techniques for reasoning with incomplete information. Until recently the dynamics of nonmonotonic reasoning approaches has attracted little attention. This paper studies the dynamics of defeasible logic, a simple and efficient form of nonmonotonic reasoning based on defeasible rules and priorities. We define revision and contraction operators and propose postulates. Our postulates try to follow the ideas of AGM belief revision as far as possible, but some AGM postulates clearly contradict the nonmonotonic nature of defeasible logic, as we explain. Finally we verify that the operators satisfy the postulates.

1 Introduction

The revision and transformation of knowledge is widely recognized as a key problem in knowledge representation and reasoning. Reasons for the importance of this topic are the fact that intelligent systems are gradually developed and refined, and that often the environment of an intelligent system is not static but changes over time.

Belief revision [1, 7] studies reasoning with changing information. Traditionally belief revision techniques have been concerned with the revision of knowledge expressed in classical logic. The approach taken is to study *postulates for operators*, the most well-known operators being revision and contraction.

Until recently little attention was devoted to the revision of more complex kinds of knowledge. But in the past few years there has been an increasing amount of work on revising nonmonotonic knowledge, in particular default logic theories [17, 9, 18]. These works were motivated by the ability of default reasoning to maintain inconsistent knowledge, and the use of default reasoning in various application domains. For example, the use of default rules has been proposed for the maintenance of software [5, 14]. In requirements engineering the use of default rules has been proposed [2, 19] to identify and trace inconsistencies among single requirements. One key issue in requirements engineering is the evolution of requirements, which technically translates to the evolution of default theories.

Default logics are known to be computationally complex [10, 12]. In this paper we will study a simple, efficient default reasoning approach, defeasible logic [16]. It is a sceptical reasoning approach based on the use of defeasible rules and priorities between them. Its usefulness has been demonstrated in several domains [3, 6].

In studying the revision of knowledge in defeasible logic, first we formulate postulates for revision and contraction operators in defeasible logic. We chose to be guided by the AGM postulates for classical belief revision [1]. Some of the AGM postulates can be readily adopted. Others need to be slightly modified, but we can demonstrate a close link to the *motivation* of the postulate as expressed, say, by Gardenfors [7]. But some AGM postulates contradict the nonmonotonic nature of defeasible logic. This contradiction is not surprising since AGM belief revision was designed for the revision of (monotonic) classical logical theories.

Once we establish the postulates we define concrete revision and contraction operators for defeasible logic, and show that they satisfy the proposed postulates.

2   Defeasible Logic

In this paper we use a simplified version of defeasible logic, in that strict rules are not considered; for the description of the full logic see [15]. We also consider only an essentially propositional version of the logic: the language does not contain function symbols and every expression with variables represents the finite set of its variable-free instances. A knowledge base consists of facts, rules (defeasible rules and defeaters), and a superiority relation among rules.

Facts denote simple pieces of information that are deemed to be true regardless of other knowledge items. Thus, facts are not revisable. A typical fact is that Tweety is a bird: *bird(tweety)*.

There are two kinds of rules, defeasible rules and defeaters. A rule r consists of its *antecedent* $A(r)$ (written on the left) which is a finite set of literals, an arrow, and its *consequent* $C(r)$ which is a literal. In examples we will often omit set notation for $A(r)$.

Defeasible rules are rules that can be defeated by contrary evidence. An example of such a rule is "Birds typically fly"; written formally:

$$bird(X) \Rightarrow flies(X).$$

The idea is that if we know that something is a bird, then we may conclude that it flies, unless there is other evidence suggesting that it may not fly. Defeasible rules with an empty antecedent are a little like facts, but they are defeatable and revisable.

Defeaters are rules that cannot be used to draw any conclusions. Their only use is to prevent some conclusions. In other words, they are used to defeat some defeasible rules by producing evidence to the contrary. An example is "If an animal is heavy then it may not be able to fly". Formally:

$$heavy(X) \rightsquigarrow \neg flies(X)$$

The main point is that the information that an animal is heavy is not sufficient evidence to conclude that it doesn't fly. It is only evidence that the animal *may* not be able to fly. In other words, we don't wish to conclude $\neg flies(X)$ if *heavy*(X), we simply want to prevent a conclusion $flies(X)$.

The *superiority relation* among rules is used to define priorities among rules, that is, where one rule may override the conclusion of another rule. For example, given the defeasible rules

$$r : \quad republican \Rightarrow \neg pacifist$$
$$r' : \qquad quaker \Rightarrow pacifist$$

which contradict one another, no conclusive decision can be made about the pacifism of a person who is both a republican and a quaker. But if we introduce a superiority relation $>$ with $r > r'$, then we can indeed conclude $\neg pacifist$.

A *defeasible theory* T is a triple $(F, R, >)$ where F is a finite consistent set of literals (called *facts*), R a finite set of rules, and $>$ an acyclic superiority relation on R. $R[q]$ denotes the set of rules in R with head q, and $R_d[q]$ denotes the set of defeasible rules in R with head q

A *conclusion* of T is a tagged literal and can have one of the following three forms: (i) $+\partial q$, which means that q is defeasibly provable in T; (ii) $-\partial q$, which means that we have proved that q is not defeasible provable in T; and (iii) Σq, which means that there is a reasoning chain supporting q.

Provability is defined below. It is based on the concept of a proof in $T = (F, R, >)$. A *proof* or *derivation* is a finite sequence $P = (P(1), \ldots P(n))$ of tagged literals satisfying the following conditions ($P(1..i)$ denotes the initial part of the sequence P of length i, and $\sim p$ the complement of a literal p):

$+\partial$: If $P(i + 1) = +\partial q$ then either
    (1) $q \in F$ or
    (2) (2.1) $\exists r \in R_d[q] \; \forall a \in A(r) : +\partial a \in P(1..i)$ and
        (2.2) $\sim q \notin F$ and
        (2.3) $\forall s \in R[\sim q]$ either
            (2.3.1) $\exists a \in A(s) : -\partial a \in P(1..i)$ or
            (2.3.2) $\exists t \in R_d[q]$
                 $\forall a \in A(t) : +\partial a \in P(1..i)$ and
                 $t > s$

$-\partial$: If $P(i + 1) = -\partial q$ then
    (1) $q \notin F$ and
    (2) (2.1) $\forall r \in R_d[q] \; \exists a \in A(r) : -\partial a \in P(1..i)$ or
        (2.2) $\sim q \in F$ or
        (2.3) $\exists s \in R[\sim q]$ such that

(2.3.1) $\forall a \in A(s) : +\partial a \in P(1..i)$ and
(2.3.2) $\forall t \in R_d[q]$ either
$\quad \exists a \in A(t) : -\partial a \in P(1..i)$ or
$\quad$ not $t > s$

Σq: If $P(i+1) = \Sigma q$ then
$\quad$ (1) $q \in F$ or
$\quad$ (2) $\exists r \in R_d[q] \ \forall a \in A(r) : \Sigma a \in P(1..i)$

The elements of a proof are called *lines* of the proof. We say that a tagged literal L is *provable* in $T = (F, R, >)$, denoted $T \vdash L$, iff there is a proof P in T such that L is a line of P. We say that literal q is *provable in T* iff $T \vdash +\partial q$, and that q is *supported in T* iff $T \vdash \Sigma q$.

Even though the definition seems complicated, it follows ideas which are intuitively appealing. For example, the condition $+\partial$ states the following: One way of establishing that q is defeasibly provable is to show that it is a fact. The other way is to find a rule with conclusion q, all antecedents of which are defeasibly provable. In addition, it must be established that $\sim q$ is not a fact (to do otherwise would be counterintuitive – to derive q defeasibly, although there might be a definite reason against it), and for every rule s which might provide evidence for $\sim q$, either one of its antecedents is provably not derivable, or there is a rule with conclusion q which is stronger than s and can be applied (that is, all its antecedents are defeasibly provable). Essentially the defeasible rules with head q form a *team* which tries to counterattack any rule with head $\sim q$. If the rules for q win then q is derived defeasibly; otherwise q cannot be derived in this manner.

Support means simple forward chaining reasoning without considering counterarguments.

Finally let $T = (F, R, >)$ be a defeasible theory, M a set of literals and $X = M \cup \{\sim p \mid p \in M\}$. We define $T_M^\ominus = (F, R - R[X], >')$, where $R[X]$ denotes the set of rules in R with head a literal in X, and $>'$ is the reduct of $>$ on $R - R[X]$.

3 Postulates for Belief Change Operators

In this section we will formulate reasonable postulates for belief revision operators in a defeasible reasoning framework. We will be considering the classical AGM postulates and will be proposing necessary modifications.

3.1 Belief Bases and Belief Sets

First we need to specify the kinds of belief sets we will consider. In classical logic, a belief set is supposed to be a deductively closed set of formulas. In our framework it is natural to study sets of conclusions from a given defeasible theory T. We note that in the defeasible logic variant we are studying here the conclusions are single literals. But we can easily reason with a somewhat more general set of conclusions: conjunctions of literals. Other types of conclusions can also be treated but we would need more complex technical means, so we defer their discussions to later papers. Technically we will consider a conjunction of literals $p_1 \wedge \ldots \wedge p_n$ to be a conclusion of T iff $T \vdash +\partial p_i$ for all $i = 1, \ldots, n$.

A set of literals BB is a *belief base* iff there is a defeasible theory T such that $BB = \{p \mid T \vdash +\partial p\}$. Equivalently, a belief base is any finite consistent set of literals. We say that BB *is generated by* T.

Now we define the belief set $B(T)$ generated by the defeasible theory T to be the conjunctive closure of the corresponding belief base. That is, a set of conjunctions of literals B is a *belief set* iff there is a defeasible theory T such that $B = \{p_1 \wedge \ldots \wedge p_n \mid T \vdash +\partial p_1, \ldots, T \vdash +\partial p_n\}$.

In the following we assume that $T = (F, R, >)$ is a defeasible theory, BB the belief base generated by T, and $B(T)$ the belief set generated by T. ΣT denotes the set of beliefs supported by T. Finally p_i and q_j denote literals, and c, c_k denote conjunctions of literals.

3.2 The Change Operators

As in classical belief revision, we will distinguish between three kinds of belief changes:

(i) *expansion* $^+$, which seeks to add a new formula φ to the belief set B if the negation of φ is not included in B. This is the original motivation of expansion as explained by Gardenfors in [7].

(ii) *revision* *, which adds a formula φ to the belief set B even in cases where the negation of φ is in B. To achieve this outcome the operation may need to delete formulas in B.

(iii) *contraction* $^-$, which seeks to retract a sentence φ without adding new conclusions.

In practice, revision and contraction have attracted the greatest attention. Therefore we will not be addressing postulates for expansion.

In accordance with the defeasible logic we are studying in this paper, the sentences φ will be either literals or conjunctions of literals. In future work we will study change with full propositional formulas in variants of defeasible logic allowing for disjunction.

3.3 Postulates for Revision

The first postulate in classical belief revision states that B_p^* is closed under logical consequences. According to the discussion in section 3.1, its counterpart is the following which, under our definitions, holds trivially:

($*1$) $B(T_c^*)$ is a belief set.

The second AGM postulate in classical belief revision guarantees that the sentence φ is added to the belief set. Basically we adopt the same idea, with one difference: we forbid the addition of a contradiction to the belief base. There are two kinds of contradictions. The obvious one is to have a complementary pair of literals in the conjunction. For the other kind of contradiction we note that we consider facts in a defeasible theory to be undisputed information. Thus if we attempt to add a literal p_i but $\sim p_i$ is a fact, then the addition of p_i is rejected, as is the addition of any conjunction of literals that contains p_i.

(∗2) If $\sim p_i \notin F$ and $p_i \neq \sim p_j$ (for all i, j) then $p_1 \wedge \ldots \wedge p_n \in B(T^*_{p_1 \wedge \ldots \wedge p_n})$.

According to Gardenfors [7], the aim of the third and fourth AGM postulates is to identify revision with expansion in case the negation of the sentence φ to be added is not in B. The next two postulates achieve the same for defeasible logic. (The use of two postulates by [1] is for technical reasons [7], which do not seem to hold for defeasible logic. We use two postulates only to be consistent with the AGM numbering.)

(∗3) If $\sim p_1, \ldots, \sim p_n \notin B(T)$ then $B(T^*_{p_1 \wedge \ldots \wedge p_n}) \subseteq B(T^+_{p_1 \wedge \ldots \wedge p_n})$.

(∗4) If $\sim p_1, \ldots, \sim p_n \notin B(T)$ then $B(T^+_{p_1 \wedge \ldots \wedge p_n}) \subseteq B(T^*_{p_1 \wedge \ldots \wedge p_n})$.

The fifth AGM postulate states that the result of a revision by φ is the absurd belief set iff $\neg\varphi$ is logically valid. Since, by definition, there is no absurd belief set in defeasible logic (due to its sceptical nature, p and $\sim p$ cannot be proven together), establishing an exact counterpart to the fifth AGM postulate is not straightforward. We note, though, that in AGM postulates 2 and 5 are related: it is postulate 2 which admits a contradiction and creates the possibility of an absurd belief set, as specified in postulate 5. In our case we mentioned already that the addition of a contradictory sentence should be rejected. The types of contradiction were discussed before the definition of (∗2). As in AGM, our fifth postulate defines the behaviour of revision in case we try to add a contradictory formula.

(∗5) If $\sim p_i \in F$ or $\sim p_i = p_j$ (for some i, j) then $p_1 \wedge \ldots \wedge p_n \notin B(T^*_{p_1 \wedge \ldots \wedge p_n})$.

In fact this postulate can be strengthened by providing more information on what happens if addition of a contradictory formula is rejected. We formulate two variations. (∗5a) states that in case $p_1 \wedge \ldots \wedge p_n$ is self-contradictory, or at least one $\sim p_i \in F$, then revision by $p_1 \wedge \ldots \wedge p_n$ should not cause any change. Note that in case $\sim p_i \in F$ for some i, the entire conjunction contradicts the facts.

Condition (∗5b) is weaker in that *all* $p_1, \ldots, p_n$ are required to contradict F for the revision of $p_1 \wedge \ldots \wedge p_n$ to have no effect.

(∗5a) If $\sim p_i \in F$ or $\sim p_i = p_j$ (for some i, j) then $B(T^*_{p_1 \wedge \ldots \wedge p_n}) = B(T)$.

(∗5b) If $\sim p_i \in F$ (for all i) or $\sim p_i = p_j$ (for some i, j) then $B(T^*_{p_1 \wedge \dots \wedge p_n}) = B(T)$.

The sixth AGM postulate expresses syntax-independence, and has a natural counterpart in defeasible logic.

(∗6) If the set of literals in the conjunctions c_1 and c_2 is the same, then $B(T^*_{c_1}) = B(T^*_{c_2})$.

This concludes the adaptation of the basic AGM postulates. Now we consider the composite postulates, the seventh and eighth, which regard the conjunction of sentences. According to [7] by the principle of minimal change revision with both φ and ψ ought to be the same as the expansion of $B(T^*_\varphi)$ by ψ, provided that ψ does not contradict the beliefs in $B(T^*_\varphi)$. Again for technical reasons, two AGM postulates were formulated. Below we state two postulates for defeasible logic revision to maintain the correspondence to the AGM postulates for classical belief revision.

(∗7) If $\sim q_1, \dots, \sim q_m \notin B(T^*_{p_1 \wedge \dots \wedge p_n})$ then
$B(T^*_{p_1 \wedge \dots \wedge p_n \wedge q_1 \wedge \dots \wedge q_m}) \subseteq B((T^*_{p_1 \wedge \dots \wedge p_n})^+_{q_1 \wedge \dots \wedge q_m})$.

(∗8) If $\sim q_1, \dots, \sim q_m \notin B(T^*_{p_1 \wedge \dots \wedge p_n})$ then
$B((T^*_{p_1 \wedge \dots \wedge p_n})^+_{q_1 \wedge \dots \wedge q_m}) \subseteq B(T^*_{p_1 \wedge \dots \wedge p_n \wedge q_1 \wedge \dots \wedge q_m})$.

3.4 Postulates for Contraction

The first postulate we formulate is a natural adaptation of the first AGM postulate.

(−1) $B(T^-_c)$ is a belief set.

The second AGM postulate states that we contract a formula *only* by deleting some formulas, but not by adding new ones. This postulate cannot be adopted in our framework because it contradicts the sceptical nonmonotonic nature of defeasible logic. To see this, suppose that we know a, and we have rules $\Rightarrow p$ and $a \Rightarrow \neg p$. Then a is sceptically provable and p is not. But if we decide to contract a then p becomes sceptically provable. We note that this behaviour is not confined to the specifics of defeasible logic but holds in any sceptical nonmonotonic formalism (e.g. in the sceptical interpretation of default logic).

But this example also suggests what proportion of the original idea of the second AGM postulate we can maintain: even though p was not in the original belief set, its appearance in the result of the contraction is not due to the addition of new rules. Stated differently, even though p was not *provable* in the original nonmonotonic knowledge base, it was nevertheless *supported*.

(−2) $B(T^-_c) \subseteq \Sigma T$.

The principle of minimal change requires that if the sentence to be retracted is not included in the belief set, no change is necessary.

(−3) If $c \notin B(T)$ then $B(T_c^-) = B(T)$.

The fourth AGM postulate states that the sentence to be retracted is not included in the outcome of the contraction operation, unless it is a tautology. In our framework this situation cannot arise since a conjunction of literals is never a tautology. But if we follow our earlier idea that the negations of facts are contradictions (thus facts may be viewed as being always true), it is natural to state the following:

(−4) If $p_1 \wedge \ldots \wedge p_n \in B(T_{p_1 \wedge \ldots \wedge p_n}^-)$ then $p_1, \ldots, p_n \in F$.

The fifth AGM postulate expresses the possibility of recovery: if we retract a sentence and then add it again, then we do not lose any of the original beliefs. This idea is naturally expressed in defeasible logic below.

The condition $c \in B(T)$ is necessary: in case $c \notin B(T)$, the addition of c may remove a literal $\sim p_i$ which contradicts a conjunct p_i in c. Due to the nonmonotonic nature of defeasible logic, this may cause the removal of further literals from $B(T)$. Thus without the condition $c \in B(T)$ the postulate cannot be reasonably expected to hold.

(−5) If $c \in B(T)$ then $B(T) \subseteq B((T_c^-)_c^+)$.

Syntax independence is again straightforward.

(−6) If the set of literals in the conjunctions c_1 and c_2 is the same, then $B(T_{c_1}^-) = B(T_{c_2}^-)$.

The seventh and eighth AGM postulates are simply adopted.

(−7) $B(T_{c_1}^-) \cap B(T_{c_2}^-) \subseteq B(T_{c_1 \wedge c_2}^-)$
(−8) If $c_1 \notin B(T_{c_1 \wedge c_2}^-)$ then $B(T_{c_1 \wedge c_2}^-) \subseteq B(T_{c_1}^-)$

4 Revising Defeasible Theories

Having formulated postulates for change operators motivated by the original AGM postulates, the question remains what *concrete operators* might satisfy these sets of postulates, indeed whether such operators exist. In the following we define such concrete operators and prove that indeed they satisfy our postulates.

We will use a common notation for all operators. Let $c = p_1 \wedge \ldots \wedge p_n$ be the formula to be added/deleted.

4.1 Expansion

Gardenfors [7] states that expansion is meant to add a formula φ to $B(T)$ *only if* $\neg\varphi \notin B(T)$. In this sense, the case where $\neg\varphi \in B(T)$ is irrelevant. However AGM decided to also add φ in this case. We will keep T unchanged, following [7] rather than [1].

$$T_c^+ = \begin{cases} T & \text{if } \sim p_i \in B(T) \text{ for some } i \in \{1, \ldots, n\} \\ T & \text{if } \sim p_i = p_j \text{ for some } i, j \in \{1, \ldots, n\} \\ (F, R', >') & \text{otherwise} \end{cases}$$

where

$R' = R \cup \{\Rightarrow p_1, \ldots, \Rightarrow p_n\}$
$>' = (> \cup \{\Rightarrow p_i > r \mid i \in \{1, \ldots, n\}, r \in R[\sim p_i]\}) - \{r > \Rightarrow p_i \mid i \in \{1, \ldots, n\}, r \in R[\sim p_i]\}.$

Thus we add rules that prove each of the literals p_i, and ensure that these are strictly stronger than any possibly contradicting rules.

4.2 Revision

AGM revision works in the same way as AGM expansion when the formula to be added does not cause an inconsistency, but revision adds a formula even if its negation is in the belief set. In our framework the definition of revision looks as follows:

$$T_c^* = \begin{cases} T & \text{if } p_1 \wedge \ldots \wedge p_n \in B(T) \\ (F, R', >') & \text{otherwise} \end{cases}$$

where

$R' = R \cup \{\Rightarrow p_1, \ldots, \Rightarrow p_n\}$
$>' = (> \cup \{\Rightarrow p_i > r \mid i \in \{1, \ldots, n\}, r \in R[\sim p_i]\}) - \{r > \Rightarrow p_i \mid i \in \{1, \ldots, n\}, r \in R[\sim p_i]\}.$

The difference to our previous definition of $^+$ is that now we make the modifications to the rules and the superiority relation even if the negation of one of the p_i was in B. Note that due to the sceptical nature of defeasible logic, we still do not get a contradiction (in the sense of being able to prove both p_i and $\sim p_i$. Theorem 1 below (equivalently, the postulates $(*1) - (*8)$) specifies when we will be able to prove p_i and when $\sim p_i$.

4.3 Contraction

We define a concrete contraction operator as follows.

$$T_c^- = \begin{cases} T & \text{if } p_1 \wedge \ldots \wedge p_n \notin B(T) \\ (F, R', >') & \text{otherwise} \end{cases}$$

where

$R' = R \cup \{p_1, \ldots, p_{i-1}, p_{i+1}, \ldots, p_n \rightsquigarrow \sim p_i \mid i \in \{1, \ldots, n\}\}$
$>' = > - \{s > r \mid r \in R' - R\}.$

Intuitively we wish to prevent the proof of $p_1 \wedge \ldots \wedge p_n$, that is, the proof of all the p_i. We achieve this by ensuring that at least one of the p_i will not be proven. The new rules in R' ensure that if all but one p_i have been proven, a defeater with head $\sim p_j$ will fire. Having made the defeaters not weaker than any other rules, the defeater cannot be "counterattacked" by another rule, and p_j will not be proven, as an inspection of the condition $+\partial$ in section 2 shows.

4.4 Results

Here we formulate some results on the revision concepts introduced in this paper. The main theme is to investigate which of the stated postulates are satisfied by the concrete change operations.

Neither of the two composite contraction AGM postulates (7 and 8) translate naturally into our framework, because they contradict the sceptical nonmonotonic nature of our underlying logical machinery. For example, the seventh AGM postulate would suggest $B(T_p^-) \cap B(T_q^-) \subseteq B(T_{p \wedge q}^-)$. Now consider a nonmonotonic knowledge base with defeasible rules $\Rightarrow p, \Rightarrow q, p \Rightarrow a$ and $q \Rightarrow a$. Then a is provable in both $B(T_p^-)$ and $B(T_q^-)$, but not in $B(T_{p \wedge q}^-)$. As for the eighth AGM postulate, it is motivated by the idea that if we need to remove more (say p and $p \wedge q$), then we will get a smaller belief set than just removing p. But already in our discussion of (-2) we saw that a similar property cannot be reasonably expected in our framework.

In our postulates we maintain the following idea of the last two AGM postulates: we try to express an upper and a lower bound for a composite contraction operation. For the lower bound we expect that $B(T_{p \wedge q}^-)$ contains at least the conclusions that can be proven if we remove from the underlying defeasible theory all rules with head $p, \sim p, q,$ or $\sim q$. For the upper bound we use the same idea as for (-2), that is, by contracting more information we may not add beliefs that are not *supported* in the result of contracting less.

$(-7a)$ $B(T_{\{p1, \ldots, pn, q_1, \ldots, q_m\}}^{\ominus}) \subseteq B(T_{p_1 \wedge \ldots \wedge p_n \wedge q_1 \wedge \ldots \wedge q_m}^-)$.

$(-8a)$ If $p_1 \wedge \ldots \wedge p_n \notin B(T_{p_1 \wedge \ldots \wedge p_n \wedge q_1 \wedge \ldots \wedge q_m}^-)$ then $B(T_{p_1 \wedge \ldots \wedge p_n \wedge q_1 \wedge \ldots \wedge q_m}^-) \subseteq \Sigma T_{p_1 \wedge \ldots \wedge p_n}^-$.

The following result connects the concrete change operators defined above to the postulates established in the previous section.

Theorem 1. *Let* $^+, ^*$ *and* $^-$ *be the concrete change operators defined in the previous subsections. Then*

- *Postulates* $(*1) - (*8)$ *are satisfied. Moreover,* $(*5b)$ *is satisfied but not* $(*5a)$.
- *Postulates* $(-1) - (-6)$ *are satisfied. Postulates* (-7) *and* (-8) *are not satisfied, but* $(-7a)$ *and* $(-8a)$ *are satisfied.*

The proof can be found in the full version of this paper.

The Levi and Harper Identities [13, 11] have been proposed as ways to define revision and contraction in terms of the other two operators. These identities are consistent with the AGM postulates for classical belief revision. The Levi Identity holds for our concrete operators.

Theorem 2. *Let* $^+$,* *and* $^-$ *be the change operators defined above. Then the following is true:*

$$B(T_p^*) = B((T_{\sim p}^-)_p^+). \qquad\qquad Levi\ Identity$$

On the other hand, the counterpart of the Harper Identity does not hold, nor can it be reasonably expected:

$$B(T_p^-) = B(T) \cap B(T_{\sim p}^*). \qquad\qquad Harper\ Identity$$

As we have already seen in the discussion of (-2), $B(T_p^-)$ cannot be expected to be a subset of $B(T)$, but clearly the right hand side of the Harper Identity is a subset of $B(T)$.

Note that we are only able to formulate these identities for literals. For more complex formulas we need to incorporate disjunction in the belief sets and the deductive machinery of defeasible logic (to be able to express the negation of a conjunction).

5 Conclusion

In this paper we studied the revision of knowledge in defeasible logic. In particular, we formulated desirable postulates for revision and contraction operators. In some cases we followed the intuition and formalization of the original AGM postulates. In other cases we explained why certain AGM postulates are inappropriate because they contradict the nonmonotonic nature of defeasible logic, and proposed reasonable alternatives.

To our knowledge this is the first complete set of postulates for the revision of explicit nonmonotonic knowledge (of course we are aware of the close relationship between belief revision and nonmonotonic reasoning [8], as well as work showing that belief revision can be achieved using default reasoning [4]).

Then we defined concrete revision and contraction operators, and showed that they fulfill the postulates, as well as adaptations of the Levi Identity.

Our future work in the area will include the study of revision in defeasible logics with disjunction, and the examination of the relevance of revision to the evolution of knowledge in particular application domains.

References

1. C.E. Alchourron, P. Gardenfors and D. Makinson. On the Logic of Theory Change: Partial Meet Contraction and Revision Functions. *Journal of Symbolic Logic* 50 (1985), 510–530.

2. G. Antoniou. The role of nonmonotonic representations in requirements engineering. *International Journal of Software Engineering and Knowledge Engineering* 8,3 (1998): 385–399.

3. G. Antoniou, D. Billington and M. Maher. On the analysis of regulations using defeasible rules. In *Proc. 32nd Hawaii International Conference on Systems Science*, IEEE Press 1999.

4. G. Brewka. Preferred Subtheories: An Extended Logical Framework for Default Reasoning. In *Proc. 11th IJCAI*, 1989, 1043–1048.

5. D.E. Cooke and Luqi. Logic Programming and Software Maintenance. *Annals of Mathematics and Artificial Intelligence* 21 (1997): 221–229.

6. M.A. Covington. Defeasible Logic on an Embedded Microcontroller. In *Proc. 10th International Conference on Industrial and Engineering Applications of Artificial Intelligence and Expert Systems*, June 1997.

7. P. Gardenfors. *Knowledge in Flux: Modeling the Dynamics of Epistemic States.* The MIT Press 1988.

8. P. Gardenfors and D. Makinson. Nonmonotonic Inference Based on Expectations. *Artificial Intelligence* 65 (1994): 197–245.

9. A. Ghose and R. Goebel. Belief states as default theories: Studies in non-prioritized belief change. In *Proc. ECAI'98*.

10. G. Gottlob. Complexity Results for Nonmonotonic Logics. *Journal of Logic and Computation* 2,3 (1992): 397–425.

11. W.L. Harper. Rational Conceptual Change. In *PSA 1976*, Philosophy of Science Association, Vol. 2, 462–494.

12. H.A. Kautz and B. Selman. Hard problems for simple default logics. *Artificial Intelligence* 49 (1991):243–279.

13. I. Levi. Subjunctives, dispositions and chances. *Synthese* 34, 423–455, 1977.

14. Luqi and D.E. Cooke. How to Combine Nonmonotonic Logic and Rapid Prototyping to Help Maintain Software. *International Journal on Software Engineering and Knowledge Engineering* 5,1 (1995): 89–118.

15. M.J. Maher, G. Antoniou and D. Billington. A Study of Provability in Defeasible Logic. In *Proc. 11th Australian Joint Conference on Artificial Intelligence*, LNAI 1502, Springer 1998, 215–226.

16. D. Nute. Defeasible Reasoning. In *Proc. 20th Hawaii International Conference on Systems Science*, IEEE Press 1987, 470–477.

17. M.A. Williams and N. Foo. Nonmonotonic Dynamics of Default Logics. In *Proc. 9th European Conference on Artificial Intelligence*, 1990, 702–707.

18. M.A. Williams and G. Antoniou. A Strategy for Revising Default Theory Extensions. in *Proc. 6th International Conference on Principles of Knowledge Representation and Reasoning*, Morgan Kaufmann 1998.

19. D. Zowghi and R. Offen. A Logical Framework for Modeling and Reasoning about the Evolution of Requirements. In *Proc. Third International Symposium on Requirements Engineering*, IEEE Press 1997.

On the Translation of Qualitative Spatial Reasoning Problems into Modal Logics

Werner Nutt

German Research Center for Artificial Intelligence GmbH (DFKI)
Stuhlsatzenhausweg 3, 66123 Saarbrücken, Germany
Werner.Nutt@dfki.de

Abstract. We introduce topological set constraints that express qualitative spatial relations between regions. The constraints are interpreted over topological spaces. We show how to translate our constraints into formulas of a multimodal propositional logic and give a rigorous proof that this translation preserves satisfiability. As a consequence, the known algorithms for reasoning in modal logics can be applied to qualitative spatial reasoning. Our results lay a formal foundation to previous work by Bennett, Nebel, Renz, and others on spatial reasoning in the RCC8 formalism.

1 Introduction

An approach to qualitative spatial reasoning that has received considerable attention is the so-called *Region Connection Calculus* (RCC), which has been introduced by Randell, Cui, and Cohn [9]. A specialization of RCC is the calculus RCC8. Similar to Allen's calculus for temporal reasoning [1], which is based on 13 elementary relations that can hold between time intervals, in RCC8, there are eight elementary topological relations that can hold between regions. The base relations are used to specify qualitative constraints that hold between regions. Then, the reasoning tasks are to determine whether a set of constraints is consistent or whether it entails another one.

Different semantics of RCC8 have been considered, which essentially differ in the way that "regions" are interpreted. Applications in geographical information systems, for instance, motivate to interpret a region as a connected subset of the plane whose boundary is a continuous curve or, more specifically, as a polygon [3]. For this semantics, however, it is not known whether reasoning is decidable.

As an alternative, Bennett has proposed to interpret regions as subsets of a topological space [2]. There is a close connection between propositional languages describing sets in topological spaces and modal propositional logics, which has already been pointed out by McKinsey and Tarski [7, 8]. Bennett gave a translation of RCC8 constraints to formulas in a multimodal logic. However, in his work a thorough semantical foundation of RCC8 and of the translation into modal logic is missing. In particular, it is unclear whether the translation preserves the satisfiability of constraints. Nonetheless, other researchers have taken

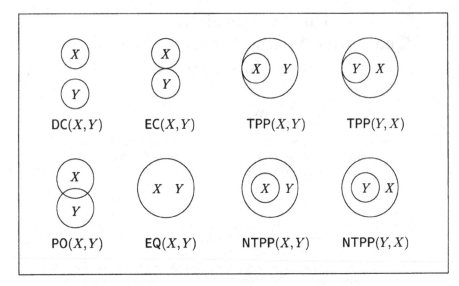

Fig. 1. A graphical representation of the RCC8 relations

the correctness of Bennett's translation for granted and based their own work upon it. For instance, Renz and Nebel applied it to identify maximal tractable fragments of RCC8 [10], and Haarslev et al. [4] used it to develop algorithms for spatioterminological reasoning.

In the present paper, we give for the first time a rigorous foundation for reasoning in RCC8. To represent topological relationships between regions, we introduce a language of *topological set constraints,* which generalizes RCC8. For example, we can express in the new language relations that involve more than two regions. We formulate a semantics for our language that interprets regions as subsets of topological spaces. Using McKinsey and Tarski's topological interpretation of the modal propositional logic **S4**, we reduce reasoning about topological constraints to reasoning in that logic. Algorithms for the latter are well-known (see e.g. [5]). Through our translation, they become applicable to spatial reasoning problems. We show that reasoning in the general language is PSPACE-complete. As a special case, we obtain also a reduction of constraint solving in RCC8. Interestingly, the work by Renz and Nebel on the complexity of RCC8 still applies to our translation, although it differs from the one given by Bennett.

Figure 1 illustrates each base relation of RCC8 by a pair X, Y of planar regions with continuous boundary curve such that the relation holds between them. The "interior" of such a region is the region area minus the boundary curve. The regions are "disconnected," written $DC(X,Y)$, if they are disjoint. They are "externally connected," written $EC(X,Y)$, if their interiors are disjoint, but the regions are not. They "partially overlap," written $PO(X,Y)$, if their interiors intersect, but none is a subset of the other. They are "equal," written $EQ(X,Y)$ if they are identical. The region X is a "tangential proper part" of Y, written $\mathsf{TPP}(X,Y)$, if X is a subset of Y, but not of the interior of Y. Finally, X is a "non-tangential proper part" of Y, written $\mathsf{NTPP}(X,Y)$, if X is contained in the interior of Y. Since tangential and non-tangential proper part are asymmetric relations, their converses can also hold between X and Y.

In the rest of the paper, we remind the reader of the basic notions of point set topology so far as they are relevant for our subject (Section 2), define topological set constraints and show how to formally express with them the RCC8-relations (Section 3), recall the topological interpretation of modal logic (Section 4), and then reduce reasoning about set constraints to reasoning in **S4** (Sections 5 and 6). Finally, we translate set constraints into formulas of a multimodal logic with an additional **K**-operator such that satisfiability is preserved (Section 7).

2   Reminder on Point Set Topology

Point set topology—or simply topology—is a mathematical theory that deals with properties of space that are independent of size and shape. In topology, one can define concepts such as interior, exterior, and isolated points of regions, boundaries and connected components of regions, and connected regions. Point set topology is therefore a possible framework for qualitative spatial reasoning.

In this section, we review the basic concepts of topology, with particular emphasis on interior operators. For further information on the topic we refer to standard textbooks like [6].

2.1   Definitions

A topological space is a pair $\mathcal{T} = (U, \mathcal{O})$ where U is a nonempty set, called the *universe*, and $\mathcal{O}$ is a set of subsets of U, called the *topology* of $\mathcal{T}$, such that the following holds:

1. $\emptyset \in \mathcal{O}$ and $U \in \mathcal{O}$;
2. if $O_1 \in \mathcal{O}$ and $O_2 \in \mathcal{O}$, then $O_1 \cap O_2 \in \mathcal{O}$;
3. if O_i, $i \in I$, is a (possibly infinite) family of elements of $\mathcal{O}$, then

$$\bigcup_{i \in I} O_i \in \mathcal{O}.$$

The elements of $\mathcal{O}$ are called the *open sets* of $\mathcal{T}$. For example, in Euclidean space, a set O is open if for each point $p \in O$ there is a circle surrounding p that is contained in O.

Let $\mathcal{T} = (U, \mathcal{O})$ be a topological space, $S \subseteq U$ a subset of U, and $p \in U$ a point. Then p is an *interior point* of S if there is an open set $O \in \mathcal{O}$ such that $p \in O$ and $O \subseteq S$. We denote the *set of interior points* of S as $i(S)$. The set $i(S)$ is the largest open set contained in S.

Proposition 1. *Let $\mathcal{T} = (U, \mathcal{O})$ be a topological space and $i: 2^U \to 2^U$ be the mapping that associates to every subset of U its set of interior points. Then for all $S, T \subseteq U$ we have*

1. $i(U) = U;$
2. $i(S) \subseteq S;$
3. $i(S) \cap i(T) = i(S \cap T);$
4. $i(i(S)) = i(S).$

The point p is a *touching point* of S if every open set containing p has a nonempty intersection with S. Finally, p is a *boundary point* of S if p is a touching point of S and of its complement $U \setminus S$. We denote the set of touching points of S as $cl(S)$, and the set of boundary points of S as $bd(S)$. We call $i(S)$ the *interior* of S, $cl(S)$ the *closure* of S, and $bd(S)$ the *boundary* of S. We say that a set S is *closed* if it contains all its touching points. Closure and boundary, openness and closedness, can all be expressed in terms of interiors of sets.

Proposition 2. *Let $\mathcal{T} = (U, \mathcal{O})$ be a topological space and $S \subseteq U$. Then*

- $cl(S) = U \setminus i(U \setminus S)$, *and* $bd(S) = cl(S) \cap cl(U \setminus S)$
- S *is open iff* $S = i(S)$, *and* S *is closed iff* $S = cl(S)$.

2.2 Interior Operators

In the preceding subsection, we have defined a topological space in terms of its open sets. Now we show that it is also possible to define it through the mapping i.

Let U be a set and $i: 2^U \to 2^U$ an operator that maps subsets of U to subsets of U. We say that i is an *interior operator* if it satisfies the properties (1) to (4) of Proposition 1. Every topology determines an interior operator.

Proposition 3. *Let $\mathcal{T} = (U, \mathcal{O})$ be a topological space and i be the operator that maps every subset of U to its interior. Then*

1. i *is an interior operator;*
2. *a set $S \subseteq U$ is open if and only if it is a fixpoint of i, that is, $i(S) = S$.*

The second statement of the above proposition says that not only determines the topology of $\mathcal{T}$ an interior operator, but the topology can also be reconstructed from the interior operator. We will show that also the converse is true. An interior operator determines a topology, from which the operator can be reconstructed. For an interior operator i on a set U we define $\mathcal{O}_i$ as the set of fixpoints of i, that is,

$$\mathcal{O}_i := \{O \subseteq U \mid O = i(O)\}.$$

Proposition 4. *If i is an interior operator, then $\mathcal{O}_i$ is a topology.*

Proposition 5. *Let i be an interior operator and $i_{\mathcal{O}_i}$ be the interior operator corresponding to the topology $\mathcal{O}_i$. Then $i = i_{\mathcal{O}_i}$.*

This shows that point set topology can equivalently be based on open sets and on interior operators. In the sequel, we will therefore talk about topological spaces as pairs (U, i), where i is an interior operator on U.

3 Set Expressions and Set Constraints

We now introduce a formal language in which we can make statements about relationships between subsets of a topological space. We assume that there is a countably infinite set of *variables* (written X, Y, Z). *Set expressions* (written s, t) are built up according to the syntax rule

$$s, t \quad \longrightarrow \quad X \mid \top \mid \bot \mid s \sqcap t \mid s \sqcup t \mid \bar{s} \mid \mathrm{I} s.$$

A *topological interpretation* $\mathcal{I}$ is a triple

$$\mathcal{I} = (U, i, d),$$

where (U, i) is a topological space, and d is a function that maps every variable to a subset of U. The function d can be extended in a unique way to set expressions such that the following identities hold:

$$d(\bot) = \emptyset \qquad\qquad d(\top) = U$$
$$d(s \sqcap t) = d(s) \cup d(t) \qquad d(s \sqcup t) = d(s) \cap d(t)$$
$$d(\bar{s}) = U \setminus d(s) \qquad\qquad d(\mathrm{I} s) = i(d(s)).$$

If no misunderstanding can arise, we will refer to topological interpretations simply as interpretations.

Elementary set constraints have the form

$$s \doteq t \qquad \text{or} \qquad s \not\doteq t,$$

where s, t are set expressions. Constraints of the form $s \doteq t$ are *positive*, while those of the form $s \not\doteq t$ are *negative*. *Complex set constraints* (written C, D) are obtained from elementary ones using the propositional connectives conjunction, disjunction, and negation. Thus, if C, D are set constraints, so are $C \wedge D$, $C \vee D$, and $\neg C$.

Next we define when an interpretation $\mathcal{I} = (U, i, d)$ *satisfies* a set constraint C (written $\mathcal{I} \models C$). For elementary constraints we have

$$\mathcal{I} \models s \doteq t \quad \text{iff} \quad d(s) = d(t)$$
$$\mathcal{I} \models s \not\doteq t \quad \text{iff} \quad d(s) \neq d(t).$$

For conjunctions, disjunctions, and negations of constraints, satisfaction is defined as one would expect.

An interpretation $\mathcal{I}$ is a *model* of C if $\mathcal{I}$ satisfies C. A constraint is *satisfiable* if it has a model. Two constraints are *equivalent* if they have the same models. If $\mathcal{C}$ is a set of constraints, then $\mathcal{I}$ is a model of $\mathcal{C}$ if $\mathcal{I}$ is a model of every constraint in $\mathcal{C}$. The set $\mathcal{C}$ *entails* a constraint D (written $\mathcal{C} \models D$) if every model of $\mathcal{C}$ is also a model of D. As usual, entailment can be reduced to (un)satisfiability, since $\mathcal{C} \models D$ if and only if $\mathcal{C} \cup \{\neg D\}$ is unsatisfiable.

With our topological set constraints, we can now give a semantics to the RCC8-relations. The constraints range over variables that represent regions. Which sets are regions? First, regions should be nonempty. Moreover, among subsets of the plane, most authors consider lines or sets with cracks as pathological and rule them out as regions. Formally, such sets can be excluded by requiring that regions are *regular*, that is, they coincide with the closure of their interior. That a set X is regular and nonempty can be expressed by the constraint

$$X \doteq \overline{\overline{IIX}} \wedge X \neq \bot. \tag{1}$$

We define now each base relation as a shorthand for a certain set constraint:

$$
\begin{aligned}
\mathsf{DC}(X,Y) &:= X \sqcap Y \doteq \bot \\
\mathsf{EC}(X,Y) &:= X \sqcap Y \neq \bot \wedge IX \sqcap IY \doteq \bot \\
\mathsf{PO}(X,Y) &:= IX \sqcap IY \neq \bot \wedge X \sqcap \overline{Y} \neq \bot \wedge \overline{X} \sqcap Y \neq \bot \\
\mathsf{EQ}(X,Y) &:= X \doteq Y \\
\mathsf{TPP}(X,Y) &:= X \sqcap \overline{Y} \doteq \bot \wedge X \sqcap \overline{IY} \neq \bot \\
\mathsf{NTPP}(X,Y) &:= X \sqcap \overline{IY} \doteq \bot.
\end{aligned}
$$

It is easy to verify that the eight base relations are mutually exclusive and cover all possible cases. That is, any two subsets of a topological space, regardless of whether they are regions or not, satisfy one and only one base relation.

A *basic RCC8 constraint system* over a set of region variables is a conjunction that contains for each variable the regularity constraint (1) and for some pairs of variables a base relation. In a *general constraint system*, there can be disjunctions of base relations between variables. Satisfiability and entailment of RCC8 constraints are defined as for arbitrary set constraints. Methods to decide the satisfiability of arbitrary topological set constraints are therefore also applicable to RCC8.

4 The Topological Interpretation of Modal Logics

In propositional logic, formulas are constructed from variables (again denoted as X, Y) and the constants $\top$ and $\bot$ (standing for *true* and *false*) with the connectives $\wedge$, $\vee$, and $\neg$.[1] In *modal* propositional logic, there is in addition a unary operator, which is often written as $\Box$. We denote such formulas with the letters ϕ, ψ.

The semantics of such formulas is usually given in terms of Kripke-interpretations. Such an interpretation consists of a set, whose elements are called *worlds*. The worlds are connected by a binary relation, the *reachability* relation. To each world, a propositional interpretation is associated that assigns truth values to the variables. A set of worlds with a reachability relation is called a *frame*.

One singles out variants of modal logics by admitting only interpretations whose reachability relation has certain properties. The most general logic is the logic **K**, where all possible interpretations are allowed. For the logic **S4**, only interpretations are admitted whose reachability relation is *reflexive* and *transitive*. For reasons that will become obvious later on we write the modal operator in **S4** as **I**. The following schemas are valid in **S4**, that is, they hold in every **S4**-interpretation and for every modal formula ϕ:

---

[1] We also use, as shorthands, the connectives $\rightarrow$ and $\leftrightarrow$.

1. $\mathbf{I}\top \leftrightarrow \top$;
2. $\mathbf{I}\phi \rightarrow \phi$;
3. $\mathbf{I}(\phi \wedge \psi) \leftrightarrow \mathbf{I}\phi \wedge \mathbf{I}\psi$;
4. $\mathbf{II}\phi \leftrightarrow \mathbf{I}\phi$.

These formulas are also characteristic for **S4** in the sense that a frame with the property that all interpretations over it satisfy all formulas of the above kind is reflexive and transitive.

There is an obvious analogy between the above formulas and the equations in Proposition 1 that characterize interior operators on topological spaces. It has been shown that the analogy is not accidental. To see this, let π be the bijection that maps set expressions s to modal propositional formulas $\pi(s)$ by mapping variables, $\top$ and $\bot$ to themselves, and by recursively replacing $s \sqcap t$, $s \sqcup t$, $\overline{s}$, and $\mathbf{I}s$ with $\pi(s) \wedge \pi(t)$, $\pi(s) \vee \pi(t)$, $\neg\pi(s)$, and $\mathbf{I}\pi(s)$, respectively. The bijection π abstracts from the fact that set expressions and propositional modal formulas are notational variants of each other. We write the inverse of π as π^{-1}.

A set term s is a *topological tautology* if $d(s) = U$ for all topological interpretations (U, i, d), that is, if s always denotes the entire space. A modal formula ϕ is *topologically valid* if $\pi^{-1}(\phi)$ is a topological tautology. The fundamental result that links the topological interpretation of set terms to modal logics is due to McKinsey and Tarski [7].

Theorem 1 (McKinsey and Tarski). *A modal propositional formula is topologically valid if and only if it is S4-valid.*

5 A Deduction Theorem for Set Constraints

By Theorem 1, the satisfiability of an elementary constraint $s \neq \bot$ and the **S4**-satisfiability of $\pi(s)$ are equivalent properties. To make use of the theorem also for complex constraints, we prove a deduction theorem which will allow us to reduce entailment of elementary constraints to validity, and thus disentailment to satisfiability of elementary set constraints, which is equivalent to satisfiability in **S4**.

Let $\mathcal{I} = (U, i, d)$ be a topological interpretation and $U' \subseteq U$ be a nonempty subset of U. Then the *restriction* of $\mathcal{I}$ to U' is the interpretation $\mathcal{I}' = (U', i', d')$ that is defined by

- $i'(S) := i(S)$ for all $S \subseteq U'$;[2]
- $d'(X) := d(X) \cap U'$ for all variables X.

We denote the unique extension of d' to arbitrary set expressions again with the letter d'.

Lemma 1. *Let $\mathcal{I} = (U, i, d)$ be a topological interpretation, $U' \subseteq U$ be a nonempty subset of U, and $\mathcal{I}' = (U', i', d')$ be the restriction of $\mathcal{I}$ to U'. If U' is open, then for all set expressions s*

$$d'(s) := d(s) \cap U'.$$

---

[2] It is easy to check that i' is again an interior operator.

Proof. The proof is by induction over the structure of set expressions. By definition of d', the claim holds for variables. The induction step for expressions of the form $\top$, $\bot$, $s \sqcap t$, $s \sqcup t$ and $\overline{s}$ involves only the properties of the Boolean connectives and is therefore omitted. We show the claim for expressions of the form Is:

$$d'(Is) = i'(d'(s)) \tag{2}$$
$$= i'(d(s) \cap U') \tag{3}$$
$$= i(d(s) \cap U') \tag{4}$$
$$= i(d(s)) \cap i(U') \tag{5}$$
$$= i(d(s)) \cap U' \tag{6}$$
$$= d(Is) \cap U', \tag{7}$$

where (2) holds because of the definition of d' for expressions, (3) because of the induction hypothesis, (4) because of the definition of i', (5) because of Property 3 of interior operators (see Proposition 1), (6) because U' is open, that is, a fixpoint of i, and (7) because of the definition of d for expressions. Note that in (6) we indeed make use of the fact that U' is open.

For set expressions, we write $s \sqsubseteq t$ as a shorthand for $\overline{s} \sqcup t \doteq \top$.

Lemma 2 (Deduction Lemma). *Let s, t be set expressions. Then*

$$s \doteq \top \models t \doteq \top \quad \text{iff} \quad \models Is \sqsubseteq t.$$

Proof. "$\Leftarrow$" Suppose that $\mathcal{I} \models Is \sqsubseteq t$ for all topological interpretations $\mathcal{I}$. Let $\mathcal{I} = (U, i, d)$ be an interpretation such that $\mathcal{I} \models s \doteq \top$, that is, $d(s) = U$. This implies $d(Is) = i(d(s)) = i(U) = U$, and therefore $d(\overline{Is}) = \emptyset$. Since $\mathcal{I} \models \overline{Is} \sqcup t \doteq \top$, we have $d(t) = U$. Hence, $\mathcal{I} \models t \doteq \top$.

"$\Rightarrow$" Suppose that $s \doteq \top \models t \doteq \top$. We want to show that $\mathcal{I} \models \overline{Is} \sqcup t \doteq \top$ for every interpretation $\mathcal{I} = (U, i, d)$. We distinguish whether $d(Is)$ is empty or not.

If $d(Is) = \emptyset$, then $d(\overline{Is}) = U$, and the claim holds. Suppose therefore that $d(Is) \neq \emptyset$. Then $U' := d(Is) = i(d(s))$ is a nonempty open subset of U. Consider the restriction $\mathcal{I}' = (U', i', d')$ of $\mathcal{I}$ to U'. Then we have

$$d'(s) = d(s) \cap U' \tag{8}$$
$$= d(s) \cap i(d(s)) \tag{9}$$
$$= i(d(s)) = d(Is) = U', \tag{10}$$

where (8) holds because of Lemma 1, (9) because of the definition of U', and (10) holds, since $i(d(s)) \subseteq d(s)$ because of Property 2 of interior operators (see Proposition 1).

Since $d'(s) = U'$, we have that $\mathcal{I}' \models s \doteq \top$. Hence, $\mathcal{I}' \models t \doteq \top$, and therefore, $U' = d'(t) = d(t) \cap U'$, where the second equation holds because of Lemma 1. This implies that $d(Is) = U' \subseteq d(t)$, which means that $\mathcal{I} \models Is \sqsubseteq t$.

One may conjecture that a stronger version of the Deduction Lemma holds, stating that $s \doteq \top \models t \doteq \top$ if and only if $\models s \sqsubseteq t$. However, this statement is not correct. For instance, we have that $X \doteq \top \models IX \doteq \top$, because of Property 1 of interior operators in Proposition 1. But $X \sqsubseteq IX$ does not hold in all interpretations.

Theorem 2 (Deduction Theorem for Topological Set Constraints). *Let s, t be set expressions. Then*

$$s \doteq \top \models t \doteq \top \qquad \text{iff} \qquad \models Is \sqsubseteq It.$$

Proof. The theorem is an immediate consequence of the Deduction Lemma because $\models Is \sqsubseteq It$ holds if and only if $\models Is \sqsubseteq t$ holds.

The "if" statement is true because $i(T) \subseteq T$ for every set T in a topological space. The "only if" statement is true, since for sets S, T in a topological space, $i(S) \subseteq T$ implies $i(S) \subseteq i(T)$ because $i(S)$ is open and $i(T)$ is the largest open subset of T.

6 Checking the Satisfiability of Set Constraints

We define conjunctive set constraints and translate them equivalently into sets of **S4**-formulas. Then we show that this is sufficient to decide satisfiability of arbitrary set constraints.

A set constraint is in *normal form* if every elementary constraint occurring in it is either of the form $s \doteq \top$ or $s \not\doteq \top$. Obviously, every set constraint can be rewritten in polynomial time into an equivalent one in normal form, making use of the fact that $s \doteq t$ is equivalent to $(\bar{s} \sqcup t) \sqcap (s \sqcup \bar{t}) \doteq \top$. A set constraint in normal form is *conjunctive* if it is a conjunction of elementary constraints, that is, if it has the form

$$s_1 \doteq \top \wedge \ldots \wedge s_m \doteq \top \wedge t_1 \not\doteq \top \wedge \ldots \wedge t_n \not\doteq \top. \tag{11}$$

As has been noted in [2], the negative conjuncts in a conjunctive constraint are independent of each other. That is, to determine satisfiability of the constraint it is sufficient to consider conjunctions of the positive conjuncts and one negative conjunct at a time.

Lemma 3 (Convexity). *The conjunctive set constraint (11) is satisfiable if and only for each $j \in 1..n$ the constraint*

$$s_1 \doteq \top \wedge \ldots \wedge s_m \doteq \top \wedge t_j \not\doteq \top \tag{12}$$

is satisfiable.

Proof. We denote the constraint (11) as C and the constraints (12) as C_j. Clearly, if C is satisfiable, then each C_j is satisfiable. Suppose, conversely, that each C_j is satisfiable. Then there are interpretations $\mathcal{I}_j = (U_j, i_j, d_j)$ such that $d_j(t_i) = U_j$ for all $i \in 1..m$ and $d_j(t_j) \neq U_j$. Without loss of generality, we can assume that the universes U_j are pairwise disjoint. We define a new interpretation $\mathcal{I} = (U, i, d)$ through

$$U := U_1 \cup \cdots \cup U_n$$
$$i(S) := i_1(S \cap U_1) \cup \cdots \cup i_n(S \cap U_n) \quad \text{for every } S \subseteq U, \text{ and}$$
$$d(X) := d_1(X) \cup \cdots \cup d_n(X) \quad \text{for every variable } X.$$

It is easy to check that i is in fact an interior operator on U. Then $\mathcal{I}$ is an interpretation such that $\mathcal{I} \models s_i \doteq \top$ for all $i \in 1..m$ and $\mathcal{I} \models t_j \not\equiv \top$ for all $j \in 1..n$. Thus, $\mathcal{I}$ satisfies C.

Theorem 3 (Translation). *The conjunctive set constraint* (11) *is satisfiable if and only for each $j \in 1..n$ the modal formula*

$$\mathbf{I}\,\pi(s_1) \wedge \ldots \wedge \mathbf{I}\,\pi(s_m) \wedge \neg \mathbf{I}\,\pi(t_j) \tag{13}$$

is **S4**-*satisfiable.*

Proof. Let $s = s_1 \sqcap \cdots \sqcap s_m$. We observe that the constraint $s_1 \doteq \top \wedge \ldots \wedge s_m \doteq \top$ is equivalent to $s \doteq \top$, and that the formula $\mathbf{I}\,\pi(s_1) \wedge \ldots \wedge \mathbf{I}\,\pi(s_m)$ is **S4**-equivalent to $\mathbf{I}\,\pi(s)$. Thus, by Lemma 3 it suffices to show that $s \doteq \top \wedge t_j \not\equiv \top$ is satisfiable if and only $\mathbf{I}\,\pi(s) \wedge \neg \mathbf{I}\,\pi(t_j)$ is **S4**-satisfiable.

Now, $s \doteq \top \wedge t_j \not\equiv \top$ is satisfiable if and only if $s \doteq \top \not\models t_j \doteq \top$. By the Deduction Theorem (Theorem 2), this is the case if and only if $\not\models \mathbf{I}s \sqsubseteq \mathbf{I}t_j$, that is, $\overline{\mathbf{I}s} \sqcup \mathbf{I}t_j$ is not a topological tautology. Then, by McKinsey and Tarski (Theorem 1), $\neg \mathbf{I}\,\pi(s) \vee \mathbf{I}\,\pi(t_j)$ is not **S4**-valid, that is, $\mathbf{I}\,\pi(s) \wedge \neg \mathbf{I}\,\pi(t_j)$ is **S4**-satisfiable.

Satisfiability checking of arbitrary constraints in normal form can be reduced to satisfiability checking of conjunctive constraints. To see this, note that we can equivalently rewrite a constraint into its disjunctive normal form, that is, into a disjunction of conjunctive constraints. The latter is satisfiable if and only if one of its disjuncts is satisfiable.

Alternatively, we can associate to each constraint a propositional formula by viewing the elementary constraints as propositions. Then we guess a truth assignment that satisfies the propositional formula. Next, we check whether our truth assignment can arise from a topological interpretation. To do so, we form a conjunction whose conjuncts are those elementary constraints for which we have guessed "true" and inverses of those for which we have guessed "false." (The inverse of $s \doteq \top$ is $s \not\equiv \top$ and vice versa.) The truth assignment can arise from a topological interpretation if and only if this conjunctive constraint is satisfiable.

From this consideration, we infer the complexity of satisfiability for arbitrary set constraints. In addition, we obtain an an upper bound for RCC8 by an argument that is much simpler than the one in [10].

Proposition 6 (Complexity). *Satisfiability of arbitrary topological set constraints is* PSPACE-*complete, while satisfiability of* RCC8 *constraints systems is* NP-*complete.*

Proof. Satisfiability of arbitrary set constraints is at least as hard as satisfiability in **S4**, which is PSPACE-complete [5]. On the other hand, as shown above, the satisfiability of a set constraint can be reduced in nondeterministic polynomial time to a linear number of **S4**-satisfiability tests. This gives us the upper bound.

As regards RCC8, we observe that translating RCC8 constraint systems results in formulas where the maximal nesting depth of the modal operator **I** is two. Satisfiability of such formulas can be decided in nondeterministic polynomial time (cf. the algorithm in [5]). The lower bound has been proved in [10].

7 A Multimodal Encoding

Bennett has proposed a translation of RCC8 constraint systems into a multimodal logic with an **S4**-operator and a "strong **S5**"-operator, however, without proving the correctness of the translation [2].

We show now, more generally, that arbitrary topological set constraints can be translated into multimodal formulas that have an **S4**-operator **I** and a **K**-operator $\Box$ such that equivalence is preserved. We define the translation π as an extension of the mapping π between set expressions and **S4**-formulas by:

$$\pi(s \doteq \top) := \Box \mathbf{I}\, \pi(s) \qquad \pi(s \doteq \bot) := \Box \mathbf{I}\, \neg \pi(s)$$
$$\pi(s \not\doteq \top) := \Diamond \neg \mathbf{I}\, \pi(s) \qquad \pi(s \not\doteq \bot) := \Diamond \neg \mathbf{I}\, \neg \pi(s),$$

and for Boolean combinations of constraints as one would expect ($\Diamond$ is the standard shorthand for $\neg \Box \neg$). Under this mapping, the RCC8-relations are translated into the following formulas:

$$
\begin{aligned}
\pi(\mathsf{DC}(X,Y)) &= \Box \mathbf{I}\, \neg (X \wedge Y) \\
\pi(\mathsf{EC}(X,Y)) &= \Box \mathbf{I}\, \neg (\mathbf{I}\,X \wedge \mathbf{I}\,Y) \wedge \Diamond \neg \mathbf{I}\, \neg (X \wedge Y) \\
\pi(\mathsf{PO}(X,Y)) &= \Diamond \neg \mathbf{I}\, \neg (\mathbf{I}\,X \wedge \mathbf{I}\,Y) \wedge \\
&\qquad \Diamond \neg \mathbf{I}\, \neg (X \wedge \neg Y) \wedge \Diamond \neg \mathbf{I}\, \neg (\neg X \wedge Y) \\
\pi(\mathsf{EQ}(X,Y)) &= \Box \mathbf{I}\, (\neg X \vee Y) \wedge \Box \mathbf{I}\, (X \vee \neg Y) \\
\pi(\mathsf{TPP}(X,Y)) &= \Box \mathbf{I}\, (\neg X \vee Y) \wedge \Diamond \neg \mathbf{I}\, \neg (X \wedge \neg \mathbf{I}\,Y) \\
\pi(\mathsf{NTPP}(X,Y)) &= \Box \mathbf{I}\, (\neg X \vee \mathbf{I}\,Y).
\end{aligned}
$$

The intuition behind this encoding is that we need the modality $\Box$ to appropriately combine the positive and negative constraints. The following elementary proposition (see e.g. [5]) shows that the condition for satisfiability of a conjunction of formulas with boxes and diamonds is analogous to the one for satisfiability of a conjunction of positive and negative elementary set constraints in Lemma 3.

Proposition 7. *Let $\phi_1, \dots, \phi_m, \psi_1, \dots, \psi_n$ be multimodal formulas. Then*

$$\Box \phi_1 \wedge \cdots \wedge \Box \phi_m \wedge \Diamond \psi_1 \wedge \cdots \wedge \Diamond \psi_n$$

is satisfiable if and only if for each $j \in 1..m$ the formula

$$\phi_1 \wedge \cdots \wedge \phi_m \wedge \psi_j$$

is satisfiable.

Interestingly, Proposition 7 would not be true if we used I instead of $\square$. Thus, indeed a second modal operator is needed.

Together with the Translation Theorem 3, Proposition 7 yields our claim for conjunctive constraints, from which it can be generalized to arbitrary constraints.

Theorem 4. *Let C be a topological set constraint. Then C is satisfiable if and only if $\pi(C)$ is satisfiable.*

8 Conclusion

We have defined syntax and semantics of topological set constraints, which generalize the RCC8 constraints, a prominent formalism for qualitative spatial reasoning. For instance, we can say now that a certain region consists of at least two nonempty disconnected components, which is not possible in RCC8. We have proved that reasoning about our constraints can be reduced to reasoning in a multimodal logic with an **S4** and additional **K**-operator. Thus, we have provided for the first time a rigorous theoretical foundation for the research into topological constraint languages like RCC8.

Acknowledgement This research was supported by the Esprit Long Term Research Project 22469 "Foundations of Data Warehouse Quality" (DWQ).

I am grateful to H.-J. Bürckert, D. Hutter, B. Nebel, and J. Renz for discussions about the subject of the paper. In particular, Nebel and Renz helped me to understand the area and to eliminate errors in earlier drafts.

References

1. J.F. Allen. Maintaining knowledge about temporal intervals. *Communications of the ACM*, 26(11):832–843, 1983.
2. B. Bennett. Modal logics for qualitative spatial reasoning. *Journal of the Interest Group in Pure and Applied Logic*, 4(1):23–45, 1996.
3. M. Grigni, D. Papadias, and Chr.H. Papadimitriou. Topological inference. *Proc. IJCAI'95*, 901–907, August 1995.
4. V. Haarslev, C. Lutz, and R. Möller. Foundations of spatioterminological reasoning with Description Logics. *Proc. KR'98*, 112–123, June 1998.
5. J.Y. Halpern and Y. Moses. A guide to completeness and complexity for modal logics of knowledge and belief. *Artificial Intelligence*, 54(3):319–380, 1992.
6. J.L. Kelley. *General Topology*. Van Nostrand, Princeton (New Jersey, USA), 1960.
7. J.C.C. McKinsey and A. Tarski. The algebra of topology. *Annals of Mathematics*, 45:141–191, 1944.
8. J.C.C. McKinsey and A. Tarski. Some theorems about the sentential calculi of Lewis and Heyting. *J. Symbolic Logic*, 13(1):1–15, 1948.
9. D.A. Randell, Zh. Cui, and A.G. Cohn. A spatial logic based on regions and connection. In *Proc. KR'92*, 165–176, October 1992.
10. J. Renz and B. Nebel. On the complexity of qualitative spatial reasoning: A maximal tractable fragment of the region connection calculus. *Artificial Intelligence*, 108(1-2):69–123, 1999.

Following Conditional Structures of Knowledge

Gabriele Kern-Isberner

FernUniversität Hagen
Dept. of Computer Science, LG Prakt. Informatik VIII
P.O. Box 940, D-58084 Hagen, Germany
`gabriele.kern-isberner@fernuni-hagen.de`

Abstract. Conditionals are not only central items for knowledge representation, but also play an important part in belief revision, in particular when dealing with iterated belief revision. To handle conditional beliefs appropriately, epistemic states instead of propositional belief sets have to be considered. Within this framework, the preservation of conditional beliefs turns out to be a principal concern, corresponding to the paradigm of minimal propositional change to be found in AGM theory.
In this paper, we deal with the revision of epistemic states by conditional beliefs, thus extending the usual framework of only taking propositional or factual beliefs as new information. We present a thorough formalization of the *principle of conditional preservation* under revision for ordinal conditional functions, commonly regarded as appropriate representations of epistemic states. Though apparently quite simple and elementary, this formal principle will be shown to imply the postulates dealing with conditional preservation stated in other papers.

Keywords: Belief revision, conditionals, Ramsey test, ordinal conditional functions, iterated revision, AGM theory

1 Introduction

Knowledge typically undergoes change - additional information has to be incorporated into a knowledge base, errors should be corrected without giving up anything that is known, changes over time or caused by actions must be taken into regard. While human beings handle these changes with an amazing ease, machine intelligence has to make use of ingenious belief revision techniques to arrive at consistent and satisfactory results.

The field of belief revision is mainly influenced by the so-called *AGM theory*, named after Alchourron, Gärdenfors and Makinson who set up a framework of postulates for a reasonable change of propositional beliefs (cf. [1], [7]). However, this theory conceived for propositional beliefs proved to be inadequate to handle conditionals; the famous *triviality result* of Gärdenfors illustrates this incompatibility (cf. [7]). As a consequence it was emphasized that conditional beliefs are fundamentally different from propositional beliefs and need the richer structure of an epistemic state to be handled appropriately (cf. e.g. [4]). Moreover, conditionals and epistemic states provide an excellent framework to study *iterated belief revision* (cf. [4]).

The AGM postulates are generalized in [4] (cf. also section 3) to apply to revisions of epistemic states. These postulates constrain revisions of the form $\Psi \star A$, the revision operator $\star$ connecting an epistemic state Ψ, as initial state of belief, and a propositional formula A, as new information. But again they focus on the propositional beliefs held in Ψ. So a further and essential extension of the AGM-theory is necessary to also cover conditional beliefs. Here the preservation of conditional beliefs as a principal concern corresponds to the paradigm of minimal propositional change guiding the AGM-postulates. Darwiche and Pearl [4] advanced four postulates in addition to the AGM axioms to model conditional preservation under revision by propositional beliefs.

In [10], the framework of revising epistemic states is extended to its full complexity by also considering *revisions by conditional beliefs*. This allows not only to take factual information into account but also to incorporate changed or newly acquired revision policies into one's state of belief. A scheme of eight postulates appropriate to guide the revision of epistemic states by conditional beliefs is presented in [10]. These postulates are motivated by following the specific, non-classical nature of conditionals, and the aim of preserving conditional beliefs is achieved by studying specific interactions between conditionals, represented properly by two relations. Because one of the postulates claims propositional belief revision to be a special case of conditional belief revision, they also cover the topic of Darwiche and Pearl's work [4], and we showed that all four postulates stated there may be derived from our postulates.

There are several different ways to represent epistemic states. One of the most sophisticated methods is to make use of probability distributions, requiring, however, a strictly probabilistic structure of knowledge. A more general approach assumes epistemic states Ψ to be equipped with a *plausibility pre-ordering* on worlds, resp. on propositions (cf. e.g. [2], [4]). The most plausible propositions are believed in Ψ, and a conditional $(B \mid A)$ is accepted in Ψ iff, given A, B is more plausible than $\neg B$, i.e. iff $A \wedge B$ is more plausible than $A \wedge \neg B$ (cf. e.g. [2]). So the conditional beliefs are given implicitly by a plausibility pre-ordering in Ψ.

A well-appreciated means to represent epistemic states are Spohn's *ordinal conditional functions (OCF's)* (cf. [11]) assigning non-negative integers as degrees of plausibility to worlds. So they not only induce a *ranking* of worlds but also allow to quantify the relative distances of worlds with respect to plausibility. Ordinal conditional functions may be considered as qualitative abstractions of probabilistic beliefs (cf. [11], [4], [8]). Using them, an appropriate but appealingly simple representation of epistemic states is obtained. Revising such an ordinal conditional function means to compute a new ranking of worlds.

In this paper, we present a thorough formalization of what may be called a *principle of conditional preservation* for ordinal conditional functions, mainly supported by elementary and intuitive arguments involving conditionals. Though apparently quite simple, this formal principle will be shown to imply the postulates dealing with conditional preservation stated in [10] and thus also those in [4]. So in fact, it reveals a fundamental preservation mechanism for conditional beliefs under revision. And because propositional beliefs may be considered as a special kind of conditional beliefs, this principle seems to be of importance for the whole area of belief revision.

This paper is organized as follows: The following section brings some formal properties of conditionals and introduces the relations $\sqsubseteq$ and $\perp$ crucial to study interactions between conditionals. Section 3 explains the formal connection between conditionals and the process of revising epistemic states by way of the Ramsey test and summarizes some basic results for the revision of ordinal conditional functions. The framework is extended to revision by conditional beliefs in section 4. Here we repeat the axioms and the representation theorems of [10], and we give an example. In section 5, the formal principle of conditional preservation is presented, and a characterization for ordinal conditional functions satisfying it is proved. Section 6 concludes this paper with a short summary and with outlining further research.

2 Conditionals

Conditionals $(B \mid A)$ represent expressions of the form "*If A then B*", conjoining two formulas A and B of a propositional language $\mathcal{L}$. Formally, a conditional $(B \mid A)$ may be considered an object of a three-valued (non-boolean) nature, partitioning the set of worlds Ω into three parts: those worlds satisfying $A \wedge B$ and thus confirming the conditional, those worlds satisfying $A \wedge \neg B$, thus refuting the conditional, and those worlds not fulfilling the premise A and so which the conditional may not be applied to at all. Therefore a conditional is adequately represented as a *generalized indicator function* on Ω (cf. e.g. [5], [3])

$$(B \mid A)(\omega) = \begin{cases} 1 & \text{if} \quad \omega \models A \wedge B \\ 0 & \text{if} \quad \omega \models A \wedge \neg B \\ u & \text{if} \quad \omega \models \neg A \end{cases} \tag{1}$$

where u means *undefined*. To simplify notations, we will replace a conjunction by juxtaposition and indicate the negation of a proposition by barring it, i.e. $AB = A \wedge B$ and $\overline{B} = \neg B$.

Two conditionals are considered to be equivalent iff they are identical as indicator functions, i.e. $(B \mid A) \equiv (D \mid C)$ iff $A \equiv C$ and $AB \equiv CD$ (cf. [3]). Usually, a propositional fact $A \in \mathcal{L}$ is identified with the conditional $(A \mid \top)$, where $\top$ is tautological.

For a conditional $(B \mid A)$, we define the *affirmative set* $(B \mid A)^+$ and the *conflicting set* $(B \mid A)^-$ of worlds as

$$(B \mid A)^+ = \{\omega \in \Omega \mid \omega \models AB\} (= Mod(AB))$$
$$(B \mid A)^- = \{\omega \in \Omega \mid \omega \models A\overline{B}\} (= Mod(A\overline{B}))$$

where for any propositional formula $A \in \mathcal{L}$, $Mod(A)$ denotes the set of all A-worlds, i.e. $Mod(A) = \{\omega \in \Omega \mid \omega \models A\}$.

Lemma 1. *Two conditionals $(B \mid A), (D \mid C)$ are equivalent iff their corresponding affirmative and contradictory sets are equal, i.e. $(B \mid A) \equiv (D \mid C)$ iff $(B \mid A)^+ = (D \mid C)^+$ and $(B \mid A)^- = (D \mid C)^-$.*

It is difficult to capture interactions between conditionals. In [3] and [6], logical connectives and implications between conditionals are defined and investigated. Here we will pursue a different idea of interaction. Having the effects of conditionals on worlds in mind, we define two relations $\sqsubseteq$ and $\perp$ between conditionals by

$$(D \mid C) \sqsubseteq (B \mid A) \text{ iff } (D \mid C)^+ \subseteq (B \mid A)^+ \text{and } (D \mid C)^- \subseteq (B \mid A)^-,$$
$$(D \mid C) \perp (B \mid A) \text{ iff } Mod(C) \subseteq M \in \{(B \mid A)^+, (B \mid A)^-, Mod(\overline{A})\}.$$

Thus $(D \mid C) \sqsubseteq (B \mid A)$ if the effect of the former conditional on worlds is in line with the latter one, but $(D \mid C)$ applies to fewer worlds. Thus $(D \mid C)$ is be called a *subconditional* of $(B \mid A)$ in this case.

In contrast to this, the second relation $\perp$ symbolizes a kind of *independency between conditionals*. We have $(D \mid C) \perp (B \mid A)$ if $Mod(C)$, i.e. the range of application of the conditional $(D \mid C)$, is completely contained in one of the sets $(B \mid A)^+, (B \mid A)^-$ or $Mod(\overline{A})$. So for all worlds which $(D \mid C)$ may be applied to, $(B \mid A)$ has the same effect and yields no further partitioning. Note, however, that $\perp$ is not a symmetric independence relation; $(D \mid C) \perp (B \mid A)$ rather expresses that $(D \mid C)$ is *not affected* by $(B \mid A)$.

Both relations may be expressed using the standard ordering $\leq$ between propositional formulas: $A \leq B$ iff $A \models B$, i.e. iff $Mod(A) \subseteq Mod(B)$.

Lemma 2. *Let $(B \mid A), (D \mid C)$ be two conditionals over $\mathcal{L}$.*

(i) $(D \mid C) \sqsubseteq (B \mid A)$ iff $CD \leq AB$ and $C\overline{D} \leq A\overline{B}$;
       in particular, if $(D \mid C) \sqsubseteq (B \mid A)$ *then* $C \leq A$.

(ii) $(D \mid C) \sqsubseteq (B \mid A)$ and $(B \mid A) \sqsubseteq (D \mid C)$ iff $(D \mid C) \equiv (B \mid A)$.

(iii) $(D \mid C) \perp (B \mid A)$ iff $C \leq E$, where E is one of $AB, A\overline{B}$ or $\overline{A}$.

3   Epistemic states and ordinal conditional functions

An epistemic state Ψ represents the cognitive state of an individual, or more generally, of an intelligent agent at a given time. The propositions the agent takes as most plausible are believed in Ψ and thus are elements of the *belief set* $Bel(\Psi)$. Furthermore, a conditional $(B \mid A)$ is accepted by the agent iff, supposed that A were to hold, B would appear more plausible than $\neg B$. So we adopt a subjunctive interpretation of conditionals, implicitly referring to a revision process: A conditional $(B \mid A)$ is accepted in the epistemic state Ψ iff revising Ψ by A yields belief in B. This relationship between conditionals and belief revision is known as the *Ramsey test* (cf. e.g. [2], [7]):

(RT) $\Psi \models (B \mid A)$ iff $Bel(\Psi \star A) \models B$

where $\star$ is a revision operator, taking an epistemic state Ψ and some new belief A as inputs and yielding a revised epistemic state $\Psi \star A$ as output. In this framework, conditionals $(B \mid A)$ play an important part as representations of revision policies, reflecting the beliefs (B) an intelligent agent is inclined to hold if new information (A) becomes obvious.

The revision of epistemic states cannot be reduced to revising its propositional part $Bel(\Psi)$ because two *different* epistemic states Ψ_1, Ψ_2 may have *equivalent* belief sets $Bel(\Psi_1) \equiv Bel(\Psi_2)$, and revising Ψ_1 and Ψ_2 by the same new information A may result in different revised belief sets $Bel(\Psi_1 \star A) \not\equiv Bel(\Psi_2 \star A)$ proving that $Bel(\Psi \star A)$ is not uniquely determined by $Bel(\Psi)$ and A.

Example 1. Two physicians have to make a diagnosis when confronted with a patient showing certain symptoms. They both agree that disease A is by far the most plausible diagnosis, so they both hold belief in A. Moreover, as the physicians know, diseases B and C might also cause the symptoms, but here the experts disagree: One physician regards B to be a possible diagnosis, too, but excludes C, whereas the other physician is inclined to take C into consideration, but not B. That is, the epistemic states Ψ_1, Ψ_2 of the two physicians may be assumed to contain the following propositional and conditional beliefs:

$$\Psi_1 : A, (B \mid \overline{A}), (\overline{C} \mid \overline{A}), \quad \Psi_2 : A, (C \mid \overline{A}), (\overline{B} \mid \overline{A})$$

Suppose now that a specific blood test definitely proves that the patient is not suffering from disease A. So both experts have to revise their beliefs by $\overline{A}$. Using the conditionals above as revision policies, the first physician now takes B to be the correct diagnosis, the second one takes C for granted: $Bel(\Psi_1 \star \overline{A}) \models B, \overline{C}$, $Bel(\Psi_2 \star \overline{A}) \models C, \overline{B}$. Though initially the physicians' opinions may be described by the same belief set, they end up with different belief sets after revision.

Darwiche and Pearl [4] consider the revision of epistemic states with propositional beliefs, mainly concerned with handling iterated revisions. They generalize the AGM-postulates for belief revision to the framework of revising epistemic states (cf. [4]):

Suppose Ψ, Ψ_1, Ψ_2 to be epistemic states and $A, A_1, A_2, B \in \mathcal{L}$;

(R*1) A is believed in $\Psi \star A$: $Bel(\Psi \star A) \models A$.
(R*2) If $Bel(\Psi) \wedge A$ is satisfiable, then $Bel(\Psi \star A) \equiv Bel(\Psi) \wedge A$.
(R*3) If A is satisfiable, then $Bel(\Psi \star A)$ is also satisfiable.
(R*4) If $\Psi_1 = \Psi_2$ and $A_1 \equiv A_2$, then $Bel(\Psi_1 \star A_1) \equiv Bel(\Psi_2 \star A_2)$.
(R*5) $Bel(\Psi \star A) \wedge B$ implies $Bel(\Psi \star (A \wedge B))$.
(R*6) If $Bel(\Psi \star A) \wedge B$ is satisfiable then $Bel(\Psi \star (A \wedge B))$ implies $Bel(\Psi \star A) \wedge B$.

These postulates are nearly exact reformulations of the AGM postulates, as stated in [9], with belief sets replaced throughout by belief sets of epistemic states, with one important exception: Postulate (R*4) now claims that only *identical* epistemic states are supposed to yield equivalent revised belief sets. This is a clear but adequate weakening of the corresponding AGM-postulate referring to *equivalent* belief sets. As we explained above, such a reduction to propositional beliefs is inappropriate when handling epistemic states. $\Psi_1 = \Psi_2$ now requires the two epistemic states to be thoroughly identical, i.e. to incorporate in particular the same propositional beliefs as well as the same conditional beliefs. Obeying the difference between $Bel(\Psi_1) \equiv Bel(\Psi_2)$ and $\Psi_1 = \Psi_2$ circumvents Gärdenfors' famous triviality result [7] and makes the Ramsey test compatible with the AGM theory for propositional belief revision. So, due to their close similarity to the AGM-postulates, the postulates above ensure that the revision of epistemic states is in line with the AGM theory as long as the revision of the corresponding belief sets is considered.

Any revision satisfying (R*1) to (R*6) induces for each epistemic state Ψ a plausibility pre-ordering $\leq_\Psi$ on the set of worlds Ω so that

$$Mod(\Psi \star A) = \min_{\leq_\Psi} Mod(A) \qquad (2)$$

(cf. [4]), where $Mod(\Psi) := Mod(Bel(\Psi))$ denotes the set of models of an epistemic state Ψ. That is, the worlds satisfying $Bel(\Psi \star A)$ are precisely those worlds satisfying A that are minimal (i.e. most plausible) with respect to $\leq_\Psi$. In particular, the models of Ψ are the most plausible worlds in Ω: $Mod(\Psi) = \min_{\leq_\Psi}(\Omega)$.

A particularly adequate representation of epistemic states is obtained by using *ordinal conditional functions, OCF,* (sometimes simply called *ranking functions*) κ from the set of worlds Ω to ordinals, i.e. to non-negative integers, representing degrees of plausibility. For the sake of consistency, it is generally assumed that some worlds are mapped to the minimal element 0 (cf. [11]). Ordinal conditional functions allow not only to compare worlds according to their plausibility but also to take the relative distances between them into account. Each OCF κ induces a plausibility ordering $\leq_\kappa$ by $\omega_1 \leq_\kappa \omega_2$ iff $\kappa(\omega_1) \leq \kappa(\omega_2)$. So the smaller $\kappa(\omega)$ is, the more plausible the world ω appears, and what is believed (for certain) in the epistemic state represented by κ is described precisely by the set $\{\omega \in \Omega \mid \kappa(\omega) = 0\} =: Mod(\kappa)$, and consequently,

$$Bel(\kappa) = \{A \in \mathcal{L} \mid \omega \models A \text{ for all } \omega \in Mod(\kappa)\}$$

For a propositional formula $A \not\equiv \bot$, we set $\kappa(A) = \min\{\kappa(\omega) \mid \omega \models A\}$ so that $\kappa(A \vee B) = \min\{\kappa(A), \kappa(B)\}$. In particular, $0 = \min\{\kappa(A), \kappa(\overline{A})\}$, so that at least one of A or $\overline{A}$ is considered mostly plausible. A proposition A is believed iff $\kappa(\overline{A}) > 0$ (which implies $\kappa(A) = 0$). We abbreviate this by $\kappa \models A$.

For a conditional $(B \mid A)$, we define

$$\kappa(B \mid A) = \kappa(AB) - \kappa(A)$$

Let $\star$ be a (propositional) revision operator on ordinal conditional functions (or on the corresponding epistemic states, respectively) obeying postulates (R*1) to (R*6) (for examples of such revision operators, cf. [11], [4]). Then, according to (2),

$$Mod(\kappa \star A) = \min_\kappa Mod(A) \qquad (3)$$

for $A \in \mathcal{L}$, and the Ramsey test (RT) yields

$$\kappa \models (B \mid A) \quad \text{iff} \quad \kappa \star A \models B, \tag{4}$$

for any conditional $(B \mid A)$ over $\mathcal{L}$. Using (3) and (4), the following lemma is evident:

Lemma 3. *Let $(B \mid A)$ be a conditional over $\mathcal{L}$, let κ be an OCF. The following three statements are equivalent:*
(i) $\quad \kappa \models (B \mid A)$.
(ii) $\quad \kappa(AB) < \kappa(A\overline{B})$.
(iii) $\quad \kappa(\overline{B} \mid A) > 0$.

4 Revising by conditionals

As Darwiche and Pearl [4] emphasized, the AGM postulates are too weak to capture iterated revisions. The minimality of propositional change, implicitly claimed by the postulates (R*1) to (R*6) above (cf. section 3), has to be supplemented by the paradigm of preserving conditional beliefs within epistemic states. So in [4], Darwiche and Pearl set up four additional postulates, (C1) – (C4), aiming at making this idea more concrete, but only dealing with conditional preservation under propositional revisions. Their postulates are generalized in [10] by the following set of axioms, describing reasonable revisions of epistemic states by conditionals. Revising an epistemic state Ψ by a conditional $(B \mid A)$ becomes necessary if a new conditional belief resp. a new revision policy should be included in Ψ, yielding a changed epistemic state $\Psi' = \Psi \star (B \mid A)$ such that $\Psi' \models (B \mid A)$, i.e. $\Psi' \star A \models B$ (cf. example 2 below).

Postulates for conditional revision:

Suppose Ψ is an epistemic state and $(B \mid A), (D \mid C)$ are conditionals. Let $\Psi \star (B \mid A)$ denote the result of revising Ψ by $(B \mid A)$.

(CR0) $\Psi \star (B \mid A)$ is an epistemic state.
(CR1) $\Psi \star (B \mid A) \models (B \mid A)$.
(CR2) $\Psi \star (B \mid A) = \Psi$ iff $\Psi \models (B \mid A)$.
(CR3) $\Psi \star B = \Psi \star (B \mid \top)$ is a propositional AGM-revision operator.
(CR4) $\Psi \star (B \mid A) = \Psi \star (D \mid C)$ whenever $(B \mid A) \equiv (D \mid C)$.
(CR5) If $(D \mid C) \perp (B \mid A)$ then $\Psi \models (D \mid C)$ iff $\Psi \star (B \mid A) \models (D \mid C)$.
(CR6) If $(D \mid C) \sqsubseteq (B \mid A)$ and $\Psi \models (D \mid C)$ then $\Psi \star (B \mid A) \models (D \mid C)$.
(CR7) If $(D \mid C) \sqsubseteq (\overline{B} \mid A)$ and $\Psi \star (B \mid A) \models (D \mid C)$ then $\Psi \models (D \mid C)$.

Postulates (CR0) and (CR1) are self-evident. (CR2) postulates that Ψ should be left unchanged precisely if it already entails the conditional. (CR3) says that the corresponding propositional revision operator should be in accordance with the AGM postulates (so it is coherent to use the same symbol $\star$ for propositional as well as for conditional revision). (CR4) requires the result of the revision process to be independent of the syntactical representation of conditionals.

The next three postulates aim at preserving the conditional structure of knowledge. The rationale behind these postulates is not to minimize conditional change, as e.g. in the work of Boutilier and Goldszmidt [2], but to preserve the *conditional structure* of the knowledge, as far as possible, which is made obvious by studying interactions between conditionals.

(CR5) claims that revising by a conditional should preserve all conditionals that are independent of that conditional, in the sense given by the relation ⊥. This postulate is motivated as follows: The validity of a conditional $(B \mid A)$ in an epistemic state Ψ depends on the relation between (some) worlds in $Mod(AB)$ and (some) worlds in $Mod(A\overline{B})$. So incorporating $(B \mid A)$ to Ψ may require a shift between $Mod(AB)$ on one side and $Mod(A\overline{B})$ on the other side, but should leave intact any relations between worlds within $Mod(AB)$, $Mod(A\overline{B})$, or $Mod(\overline{A})$. These relations may be captured by conditionals not affected by $(B \mid A)$, i.e. by conditionals $(D \mid C) \perp (B \mid A)$.

(CR6) states that conditional revision should bring about no change for conditionals that are already in line with the revising conditional, and (CR7) guarantees that no conditional change contrary to the revising conditional is caused by conditional revision.

In [10] it is shown that any revision operator satisfying (CR0) – (CR7) also fulfills the principles (C1) – (C4) of [4], and representation theorems are proved for (CR5) – (CR7), i.e. for the postulates crucial for the idea of conditional preservation. We will cite these representation results here, phrased for ordinal conditional functions:

Theorem 1 ([10]). *The conditional revision operator $\star$ satisfies* (CR5) *iff for each ordinal conditional function κ and for each conditional $(B \mid A)$ it holds that:*

$$\kappa(\omega) \leq \kappa(\omega') \quad \text{iff} \quad (\kappa \star (B \mid A))(\omega) \leq (\kappa \star (B \mid A))(\omega') \tag{5}$$

for all worlds $\omega, \omega' \in Mod(AB)$ $(Mod(A\overline{B}), Mod(\overline{A})$, respectively).

Theorem 2 ([10]). *Suppose $\star$ is a conditional revision operator satisfying (CR5). Let κ be an ordinal conditional function, and let $(B \mid A)$ be a conditional.*

1. *$\star$ satisfies* (CR6) *iff for all $\omega \in Mod(AB)$, $\omega' \in Mod(A\overline{B})$, $\kappa(\omega) < \kappa(\omega')$ implies $(\kappa \star (B \mid A))(\omega) < (\kappa \star (B \mid A))(\omega')$.*
2. *$\star$ satisfies* (CR7) *iff for all $\omega \in Mod(AB)$, $\omega' \in Mod(A\overline{B})$, $(\kappa \star (B \mid A))(\omega') < (\kappa \star (B \mid A))(\omega)$ implies $\kappa(\omega') < \kappa(\omega)$.*

Example 2. A physician has to make a diagnosis. The patient he is facing obviously feels ill, and at first glance, the physician supposes that the patient is suffering from disease D causing mainly two symptoms, S_1 (major symptom) and S_2 (minor symptom). To obtain certainty about these symptoms, further examinations will be necessary the next days. The following table reflects the ordinal conditional function κ representing the epistemic state of the physician under these conditions:

D S_1 S_2	κ	D S_1 S_2	κ
0 0 0	5	1 0 0	4
0 0 1	2	1 0 1	2
0 1 0	1	1 1 0	0
0 1 1	4	1 1 1	3

So in this epistemic state, $DS_1\overline{S_2}$ is considered to be the most plausible world, i.e. $Bel(\kappa) \models D, S_1, \overline{S_2}$. Moreover, we have $\kappa \models (S_1 \mid D), (D \mid S_1)$ and $(\overline{S_2} \mid D)$.

In the evening, the physician finds an article in a medical journal, pointing out that symptom D_2 has recently proved to be of major importance for disease D. So the physician wants to revise κ to incorporate the new conditional belief $(S_2 \mid D)$.

In the next section, we will formalize a *principle of conditional preservation* for ranking functions in a very intuitive way. Nevertheless, this principle will prove to be both theoretically meaningful and practically relevant: Presupposing two very fundamental properties of revision, all postulates (CR5) – (CR7) can be derived from it, and it will provide a simple method to adequately realize revision by conditionals.

5 The principle of conditional preservation

Minimality of change is a crucial paradigm for belief revision, and a "principle of conditional preservation" is to realize this idea of minimality when conditionals are involved in change. Minimizing absolutely the changes in conditional beliefs, as in [2], is an important proposal to this aim, but it does not always lead to intuitive results (cf. [4]). The idea we will develop here rather aims at *preserving the conditional structure of knowledge* within an epistemic state.

Suppose κ is an OCF representing an epistemic state. Let $\kappa^* = \kappa \star (B \mid A)$ be the result of a revision of κ by a new conditional belief $(B \mid A)$. If this revision is expected to be as close to κ as possible, then certainly any change should clearly be caused by the revising conditional. Looking at the representation (1) of a conditional as a generalized indicator function in section 2, we see that a conditional induces an equivalence relation on worlds via

$$\omega_1 \equiv_{(B|A)} \omega_2 \quad \text{iff} \quad (B \mid A)(\omega_1) = (B \mid A)(\omega_2)$$

On all worlds within $Mod(AB)$, $Mod(A\overline{B})$ and $Mod(\overline{A})$, respectively, $(B \mid A)$ acts in the same way, so all these worlds are indistinguishable with respect to $(B \mid A)$. Therefore the plausibility of these worlds should be changed in the same way, i.e. their *relative changes* should be equal:

Principle of conditional preservation:

$$\kappa^*(\omega_1) - \kappa(\omega_1) = \kappa^*(\omega_2) - \kappa(\omega_2) \quad \text{whenever} \quad \omega_1 \equiv_{(B|A)} \omega_2 \tag{6}$$

Indeed, this is a very elementary postulate being based only on the intuitive and nearly indisputable representation (1) of a conditional. (6) is equivalent to

$$\kappa^*(\omega_1) - \kappa^*(\omega_2) = \kappa(\omega_1) - \kappa(\omega_2) \quad \text{whenever} \quad \omega_1 \equiv_{(B|A)} \omega_2 \qquad (7)$$

so the plausibility structure within each of the equivalence classes $Mod(AB)$, $Mod(A\overline{B})$ and $Mod(\overline{A})$ is maintained. (7) is a quantified version of (5), taking the relative distances provided by κ into account. It will soon be shown that essentially, any revision fulfilling (6) also satisfies (CR5) – (CR7) (cf. theorem 4 below). But first we present an easy characterization for revisions following the principle of conditional preservation:

Theorem 3. *The revision $\kappa \star (B \mid A)$ satisfies the principle of conditional preservation (6) iff there exist integers α_0, α^+ and α^- such that*

$$\kappa \star (B \mid A)(\omega) = \begin{cases} \kappa(\omega) + \alpha^+ & \text{if} \quad \omega \models AB \\ \kappa(\omega) + \alpha^- & \text{if} \quad \omega \models A\overline{B} \\ \kappa(\omega) + \alpha_0 & \text{if} \quad \omega \models \overline{A} \end{cases} \qquad (8)$$

Proof. If $\kappa \star (B \mid A) = \kappa^*$ has the form (8), then $\kappa^* - \kappa$ is constant on each of the equivalence classes $Mod(AB), Mod(A\overline{B})$ and $Mod(\overline{A})$. So (6) is satisfied.

Conversely, suppose $\kappa^* = \kappa \star (B \mid A)$ is a revision of κ satisfying the principle of conditional preservation (6). Choose $\omega_0 \in Mod(\overline{A}), \omega^+ \in Mod(AB)$ and $\omega^- \in Mod(A\overline{B})$ arbitrarily. Then $\alpha^+ := \kappa^*(\omega^+) - \kappa(\omega^+)$, $\alpha^- := \kappa^*(\omega^-) - \kappa(\omega^-)$, $\alpha_0 := \kappa^*(\omega_0) - \kappa(\omega_0)$, are integers yielding the desired representation (8). □

Now we establish the announced relationship between the principle of conditional preservation and the postulates (CR5) – (CR7):

Theorem 4. *If the revision $\kappa \star (B \mid A)$ follows the postulates (CR1) and (CR2) and satisfies the principle of conditional preservation (6), then $\kappa \star (B \mid A)$ is in accordance with (CR5) – (CR7).*

Proof. Let the revision $\kappa^* = \kappa \star (B \mid A)$ follow the postulates (CR1) and (CR2) and fulfill the principle of conditional preservation (6). We will use the representation theorems 1 and 2 for the proof.

Let ω, ω' be worlds in the same set $Mod(AB), Mod(A\overline{B})$ or $Mod(\overline{A})$. Then $\omega \equiv_{(B|A)} \omega'$ and therefore $\kappa^*(\omega) - \kappa(\omega) = \kappa^*(\omega') - \kappa(\omega')$ or, equivalently, $\kappa^*(\omega) - \kappa^*(\omega') = \kappa(\omega) - \kappa(\omega')$. This shows $\kappa^*(\omega) \leq \kappa^*(\omega')$ iff $\kappa(\omega) \leq \kappa(\omega')$. Thus by theorem 1, postulate (CR5) is satisfied.

Now that (CR5) is proven, we can make use of theorem 2.

If $\kappa \models (B \mid A)$, then by (CR2), $\kappa^* = \kappa$, and (CR6) and (CR7) are trivially fulfilled. So assume now $\kappa \not\models (B \mid A)$. Then $\kappa(A\overline{B}) \leq \kappa(AB)$, by lemma 3. According to theorem 3, there exist integers α_0, α^+ and α^- such that κ^* is of the form (8). By (CR1), κ^* is a successful revision, i.e. $\kappa^* \models (B \mid A)$, so again by lemma 3, $\kappa(AB) + \alpha^+ = \kappa^*(AB) < \kappa^*(A\overline{B}) = \kappa(A\overline{B}) + \alpha^-$. This implies $\alpha^+ < \alpha^-$. Let $\omega \models AB, \omega' \models A\overline{B}$.

To prove (CR6), assume $\kappa(\omega) < \kappa(\omega')$. Then $\kappa(\omega) + \alpha^+ < \kappa(\omega') + \alpha^+ < \kappa(\omega') + \alpha^-$, thus $\kappa^*(\omega) < \kappa^*(\omega')$, which was to be shown.

Similarly, to prove (CR7), presuppose $\kappa^*(\omega') < \kappa^*(\omega)$. Then $\kappa(\omega') + \alpha^- < \kappa(\omega) + \alpha^+$, and therefore $\kappa(\omega') < \kappa(\omega) + \alpha^+ - \alpha^- \leq \kappa(\omega)$.

Theorem 3 provides a simple method to realize revisions following the principle of conditional preservation. Intuitively, the involved integers $\alpha_0, \alpha^+, \alpha^-$ should be chosen appropriately to keep the amount of change minimal. Obviously, the following revision operator $\star_0$ also intends to meet this requirement:

For an ordinal conditional function κ and a conditional $(B \mid A)$, we define $\kappa \star_0 (B \mid A)$ by setting

$$\kappa \star_0 (B \mid A)(\omega) = \begin{cases} \kappa(\omega) - \kappa(B \mid A) & \text{if } \omega \models AB \\ \kappa(\omega) + \beta^- + 1 & \text{if } \omega \models A\overline{B} \\ \kappa(\omega) & \text{if } \omega \models \overline{A} \end{cases} \qquad (9)$$

where $\beta^- = -1$, if $\kappa \models (B \mid A)$, and $= 0$, else. Thus we obtain a revision of the form (8) with $\alpha^+ = -\kappa(B \mid A), \alpha^- = \beta^- + 1$, and $\alpha_0 = 0$.

$\star_0$ not only satisfies the principle of conditional preservation, but also all of the postulates (CR0) – (CR4), therefore, by theorem 4, it is a revision operator obeying (CR0) – (CR7).

Example 3 (continued). We make use of the revision operator $\star_0$ introduced by (9) to calculate a revised ordinal conditional function $\kappa \star_0 (S_2 \mid D)$ in the diagnosis example 2 above. Here we have $\kappa(DS_2) = 2, \kappa(D\overline{S_2}) = 0$, so $\kappa(S_2 \mid D) = 2 = -\alpha^+$, and $\kappa \not\models (S_2 \mid D)$, therefore $\beta^- = 0$ and $\alpha^- = 1$. Following (9), we obtain

$D\ S_1\ S_2$	κ	$\kappa \star_0 (S_2 \mid D)$
0 0 0	5	5
0 0 1	2	2
0 1 0	1	1
0 1 1	4	4
1 0 0	4	5
1 0 1	2	0
1 1 0	0	1
1 1 1	3	1

So after any revision following (8) resp. (9), again a thorough description of an epistemic state is available which is concisely represented by an ordinal conditional function and may undergo further changes.

6 Summary and outlook

In this paper, we develop a formal description of the *principle of conditional preservation* for ordinal conditional functions; both notions are of crucial meaning for iterated belief revision (cf. [11], [4]). We base this principle in an elementary way on the representation of conditionals as generalized indicator functions, providing a simple and intuitive motivation for it. Despite its amazing simplicity, the formalization (6) presented here proves to be quite powerful: It essentially implies the postulates (CR5) – (CR7) describing the idea of conditional preservation in [10], and thus generalizes the postulates (C1) – (C4) introduced in [4] as a first approach to handle conditionals coherently under propositional revision. In this paper, we deal more generally with revision by conditional beliefs throughout.

Two problems are still left open and will be pursued in forthcoming papers: The characterization theorem 3 involves integers $\alpha_0, \alpha^+, \alpha^-$ which describe completely the revised ranking function. These integers, however, may be chosen rather arbitrarily. Which integers are most adequate to keep the amount of total change minimal?

Here we only deal with revision by one single conditional. In many applications, several conditionals have to be taken into regard simultaneously. The generalization of our principle of conditional preservation to be applicable to sets of conditionals is a topic of our ongoing research.

References

1. C.E. Alchourrón, P. Gärdenfors, and P. Makinson. On the logic of theory change: Partial meet contraction and revision functions. *Journal of Symbolic Logic*, 50(2):510–530, 1985.
2. C. Boutilier and M. Goldszmidt. Revision by conditional beliefs. In *Proceedings 11th National Conference on Artificial Intelligence (AAAI'93)*, pages 649–654, Washington, DC., 1993.
3. P.G. Calabrese. Deduction and inference using conditional logic and probability. In I.R. Goodman, M.M. Gupta, H.T. Nguyen, and G.S. Rogers, editors, *Conditional Logic in Expert Systems*, pages 71–100. Elsevier, North Holland, 1991.
4. A. Darwiche and J. Pearl. On the logic of iterated belief revision. *Artificial Intelligence*, 89:1–29, 1997.
5. B. DeFinetti. *Theory of Probability*, volume 1,2. John Wiley and Sons, New York, 1974.
6. D. Dubois and H. Prade. Conditioning, non-monotonic logic and non-standard uncertainty models. In I.R. Goodman, M.M. Gupta, H.T. Nguyen, and G.S. Rogers, editors, *Conditional Logic in Expert Systems*, pages 115–158. Elsevier, North Holland, 1991.
7. P. Gärdenfors. *Knowledge in Flux: Modeling the Dynamics of Epistemic States*. MIT Press, Cambridge, Mass., 1988.
8. M. Goldszmidt and J. Pearl. Qualitative probabilities for default reasoning, belief revision, and causal modeling. *Artificial Intelligence*, 84:57–112, 1996.
9. H. Katsuno and A. Mendelzon. Propositional knowledge base revision and minimal change. *Artificial Intelligence*, 52:263–294, 1991.
10. G. Kern-Isberner. Postulates for conditional belief revision. In *Proceedings IJCAI-99*. Morgan Kaufmann, 1999. (to appear).
11. W. Spohn. Ordinal conditional functions: a dynamic theory of epistemic states. In W.L. Harper and B. Skyrms, editors, *Causation in Decision, Belief Change, and Statistics, II*, pages 105–134. Kluwer Academic Publishers, 1988.

A Theory of
First-Order Counterfactual Reasoning

Michael Thielscher

Dresden University of Technology
mit@inf.tu-dresden.de

Abstract. A new theory of evaluating counterfactual statements is presented based on the established predicate calculus formalism of the Fluent Calculus for reasoning about actions. The assertion of a counterfactual antecedent is axiomatized as the performance of a special action. An existing solution to the Ramification Problem in the Fluent Calculus, where indirect effects of actions are accounted for via causal propagation, allows to deduce the immediate consequences of a counterfactual condition. We show that our theory generalizes Pearl etal.'s characterization, based on causal models, of propositional counterfactuals.

1 Introduction

A counterfactual sentence is a conditional statement whose antecedent is known to be false as in, "Had the gun not been loaded, President Kennedy would have survived the assassination attempt." If counterfactuals are read as material implications, then they are trivially true due to the presupposed falsehood of the condition. Nonetheless it can be both interesting and important to use a more sophisticated way of deciding acceptability of a counterfactual query like, e.g., "Had the hit-and-run driver immediately called the ambulance, the injured would not have died." Counterfactuals can also teach us lessons which can be useful in the future when similar situations are encountered, as in "Had you put your weight on the downhill ski, you would not have fallen" [11]. Other applications of processing counterfactuals are fault diagnosis, determination of liability, and policy analysis [1].[1]

The first formal theory of reasoning about counterfactuals was developed in [8], where a counterfactual sentence was accepted iff its consequent holds in all hypothetical 'worlds' which are 'closest' to the factual one but satisfy the counterfactual antecedent. A first method for processing actual counterfactual queries on the basis of a concrete concept of worlds and closeness was proposed in [5]. Generally, the value of a theory of processing counterfactuals depends crucially on how well the expected consequences of a counterfactual condition are determined. It has been observed, among others, by the authors of [1,4] that

---

[1] It is worth mentioning that the importance of theories of counterfactual reasoning for the field of AI is documented by the fact that Judea Pearl receives this year's *IJCAI award for research excellence* also for his pioneering work on causality.

knowledge of causality is required to this end. Accordingly, their method is based on so-called *causal models*. A distinguishing feature of this approach is that it deals with probabilities of statements and the way these probabilities change under counterfactual conditions. On the other hand, causal models are essentially propositional. This does not allow for processing counterfactuals which involve disjunctions or quantifications as in, "Had you worn a safety helmet, or had you taken any other route, you would not have been hurt by a roof tile." Further restrictions of the causal models approach to counterfactual reasoning are entailed by the requirement that the value of each dependent variable is uniquely determined by the exogenous variables (see Section 4 on the implications of this).

The Fluent Calculus is a general strategy for axiomatizing knowledge of actions and effects using full classical predicate logic [14]. Based on a Situation Calculus-style branching time structure, the plain Fluent Calculus allows for processing a particular kind of counterfactual statements, namely, where the counterfactual antecedent asserts a sequence of actions different from the one that has actually taken place as in, "Had the assassin shot at the vice president, the president would have survived the shot:" Consider the generic predicate $Holds(f, s)$ denoting that fluent[2] f holds in situation s, and the generic function $Do(a, s)$ denoting the situation reached by performing action a in situation s. Then it is a simple exercise to axiomatize, by means of the Fluent Calculus, knowledge of the effect of $Shoot(p)$, denoting the action of shooting p, in such a way that the following is entailed:[3]

$$Holds(Alive(President), S_0) \wedge Holds(Alive(Vice), S_0)$$
$$\wedge S_1 = Do(Shoot(President), S_0) \wedge S_1' = Do(Shoot(Vice), S_0)$$
$$\supset \neg Holds(Alive(President), S_1) \wedge Holds(Alive(President), S_1')$$

The obvious reason for this to work without further consideration is that the two statements $\neg Holds(Alive(President), S_1)$ and $Holds(Alive(President), S_1')$ do not mutually contradict due to the differing situation argument.

Counterfactual assertions about situations instead of action sequences cannot be processed in such a straightforward manner. If to an axiom like,

$$Holds(Loaded(Gun), S_0) \wedge \tag{1}$$
$$\neg Holds(Alive(President), Do(Shoot\text{-}with(Gun, President), S_0))$$

the counterfactual condition $\neg Holds(Loaded(Gun), S_0)$ is added, then a plain inconsistency is produced. Our theory for processing counterfactual statements with the Fluent Calculus solves this problem by associating a new situation term with a counterfactual antecedent that modifies facts about a situation. The step from an actual one to a situation thus modified is modeled by performing an action which has the very modification as effect. Suppose, e.g., the 'action'

---

[2] A *fluent* represents an atomic property of the world which is situation-dependent, that is, whose truth value may be changed by actions.

[3] A word on the notation: Predicate and function symbols, including constants, start with a capital letter whereas variables are in lower case, sometimes with sub- or superscripts. Free variables in formulas are assumed universally quantified.

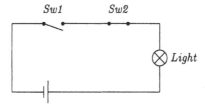

Fig. 1. An electric circuit consisting of a battery, two switches, and a light bulb, which is on if and only if both switches are closed.

Modify(Not-Loaded(x)) has the sole effect of x becoming unloaded, then a suitable axiomatization in the Fluent Calculus yielded,

$$(1) \land S_0' = Do(Modify(Not\text{-}Loaded(Gun)), S_0)$$
$$\supset Holds(Alive(President), Do(Shoot\text{-}with(Gun, President), S_0')) \qquad (2)$$

which corresponds to the counterfactual statement above, "Had the gun not been loaded, the president would have survived the assassination attempt."

A counterfactual antecedent may have consequences which are more immediate, that is, which do not refer to another action like in (2). Consider, for example, the simple electric circuit depicted in Fig. 1, taken from [9]. A counterfactual statement whose truth is obvious from the wiring (and the assumption that battery, light bulb, and the wires are not broken) is, "Were *Sw1* closed in the situation depicted, light would be on." This conclusion is rooted in this so-called *state constraint*:

$$Holds(On(Light), s) \equiv Holds(Closed(Sw1), s) \land Holds(Closed(Sw2), s) \quad (3)$$

However, it is less obvious how to formalize the grounds by which is rejected the counterfactual statement, "Were *Sw1* closed in the situation depicted, *Sw2* would be open," suggested by the implication, $Holds(Closed(Sw1), s) \land \neg Holds(On(Light), s) \supset \neg Holds(Closed(Sw2), s)$, which is a logical consequence of axiom (3). This question is closely related to the Ramification Problem [6], that is, the problem of determining the indirect effects of actions. If someone closes *Sw1* in our circuit, does the light come on as an indirect effect or does *Sw2* jump open to restore state consistency?

A standard extension of the Fluent Calculus addresses the Ramification Problem, and in particular meets the challenge illustrated with the circuit, by means of directed causal relations where indirect effects are obtained via causal propagation [12]. This solution, together with our proposal of modeling counterfactual antecedents as actions, furnishes a ready method for accommodating the immediate consequences caused by a counterfactual assertion.

Our proposal for evaluating counterfactuals with the Fluent Calculus and the approach of [1, 4] thus have in common the property of being grounded on causality. On the other hand, two features are not shared even if we confine ourselves to the restricted case of propositional counterfactuals. First, our theory is not restricted by the so-called *reversibility* property [4]. Second, a rather unique feature of our account of counterfactuals is that a counterfactual antecedent may be rejected as being unacceptable in the current state of affairs. Both non-reversibility and the possibility of rejecting counterfactual conditions will be discussed in detail below. We will also show formally how, despite these differences, the approach of [1, 4], if restricted to causal models without probabilities, is embedded in our proposal.

2 The Fluent Calculus

In what follows we give a concise introduction to the axiomatization strategy of the Fluent Calculus. The reader who is unfamiliar with it might want to consult the (electronically available) gentle introduction [13]. A distinguished purpose of the Fluent Calculus, which roots in the logic programming formalism of [7], is to address not only the representational but also the inferential aspect [2] of the classical Frame Problem. We use a many-sorted second order language with equality, which includes sorts for fluents, actions, situations, and states. Fluents are reified propositions. That is to say, we use terms like $On(Sw1)$ to denote fluents, where On is a unary function symbol. States are fluents connected via the binary function symbol "$\circ$", written in infix notation, which is assumed to be both associative and commutative and to admit a unit element, denoted by $\emptyset$. Associativity allows us to omit parentheses in nested applications of $\circ$. A function $State(s)$ relates a situation s to the state of the world in that situation, as in the following partial description of the initial state in our circuit example:

$$\exists z\, [\, State(S_0) = Closed(Sw2) \circ z \wedge \forall z'.\ z \neq Closed(Sw1) \circ z'\,] \qquad (4)$$

Put in words, of state $State(S_0)$ it is known that $Closed(Sw2)$ is true and possibly some other fluents z hold, too—with the restriction that z does not include $Closed(Sw1)$, of which we know that it is false.

Fundamental for any Fluent Calculus axiomatization is the axiom set $EUNA$ (the *extended unique names-assumption*) [12]. This set comprises the axioms AC1 (i.e., associativity, commutativity, and unit element) and axioms which entail inequality of two state terms whenever these are not AC1-unifiable. In addition, we have the following foundational axiom, where f is of sort fluent,

$$\forall s, f, z.\ State(s) \neq f \circ f \circ z \qquad (5)$$

by which double occurrences of fluents are prohibited in any state which is associated with a situation. Finally, we need the $Holds$ predicate introduced in Section 1, though it is not part of the signature but a mere abbreviation of an equality sentence: $Holds(f, s) \overset{\text{def}}{=} \exists z.\ State(s) = f \circ z$.

So-called state update axioms specify the entire relation between the states at two consecutive situations. Regarding our circuit example, let the only direct effect of an action called $Toggle(x)$ be that switch x changes its position from open to closed or vice versa. Ignoring indirect effects for the moment, this is a suitable pair of state update axioms:

$$\neg Holds(Closed(x), s) \supset State(Do(Toggle(x), s)) = State(s) \circ Closed(x)$$
$$Holds(Closed(x), s) \supset State(Do(Toggle(x), s)) \circ Closed(x) = State(s)$$

That is, if $Toggle(x)$ is performed in a situation s where x is not closed, then the new state equals the old state plus $Closed(x)$. Conversely, if x happens to be closed, then the new state plus $Closed(x)$ equals the old state. In other words, in the first case $Closed(x)$ is the only positive direct effect, while it is the only negative direct effect in the second case.

A crucial extension of the basic Fluent Calculus introduced so far copes with the Ramification Problem. Recall, for instance, state constraint (3). It gives rise, among others, to the indirect effect that light turns on if $Sw1$ gets closed whenever $Sw2$ is already closed. Such indirect effects are accounted for by the successive application of directed causal relations [12]. An example for such a relation, which holds for the circuit, is $Closed(Sw1)$ <u>causes</u> $On(Light)$, <u>if</u> $Closed(Sw2)$ while the following should *not* be formalized and added to the axiomatization: $Closed(Sw1)$ <u>causes</u> $\neg Closed(Sw2)$, <u>if</u> $\neg On(Light)$.

It cannot be gathered from a mere state constraint which of its logical consequences correspond to 'real' indirect effects. Yet with the aid of additional domain knowledge about potential causal influence it is possible to automatically extract suitable causal relationships from state constraints [12]:[4]

Consider a given binary relation $\mathcal{I}$ among the underlying fluents.[5] For all state constraints C, all prime implicates $L_1 \vee \ldots \vee L_m$ of C, all $i = 1, \ldots, m$, and for all $j = 1, \ldots, m$, $j \neq i$: If $(atom(L_i), atom(L_j)) \in \mathcal{I}$,[6] then this is a valid causal relationship:

$$\neg L_i \text{ \underline{causes} } L_j, \text{ \underline{if} } \bigwedge_{\substack{k = 1, \ldots, m \\ k \neq i; k \neq j}} \neg L_k$$

For example, $\mathcal{I} = \{(Closed(Sw1), On(Light)), (Closed(Sw2), On(Light))\}$ is the appropriate influence relation for our electric circuit. If $\mathcal{I}$ is used for the generation of causal relationships from state constraint (3), then this is the result of the above algorithm:

$$
\begin{aligned}
& Closed(Sw1) \text{ \underline{causes} } On(Light), \text{ \underline{if} } Closed(Sw2) \\
& Closed(Sw2) \text{ \underline{causes} } On(Light), \text{ \underline{if} } Closed(Sw1) \\
& \neg Closed(Sw1) \text{ \underline{causes} } \neg On(Light) \\
& \neg Closed(Sw2) \text{ \underline{causes} } \neg On(Light)
\end{aligned}
\tag{6}
$$

---

[4] The following procedure assumes state constraints to have a format where each occurrence of $Holds(\varphi, s)$ is replaced by the simple atomic expression φ. For the sake of simplicity, we confine ourselves to constraints with the universally quantified s being the only variable. A generalization can be found in [12].

[5] If $(F, G) \in \mathcal{I}$, then this indicates that fluent F may have direct causal influence on fluent G.

[6] By $atom(L)$ we denote the atom of a literal L.

The axiomatization of each single causal relationship in the Fluent Calculus is based on a predicate $Causes(z, e, z', e')$, which shall be true if, according to the causal relationship, in the current state z the occurred effects e give rise to an additional, indirect effect resulting in the updated state z' and the updated current effects e'. Let $\mathcal{R}$ be a set of causal relationships, then by $\Pi[\mathcal{R}]$ we denote the corresponding set of Fluent Calculus axioms defining $Causes$ in this way.

In order to account for possible indirect effects, the state update axioms from above are refined as follows:

$\neg Holds(Closed(x), s) \supset$
$\quad [\, z = State(s) \circ Closed(x) \supset Ramify(z, Closed(x), State(Do(Toggle(x), s))) \,]$

$Holds(Closed(x), s) \supset$
$\quad [\, z \circ Closed(x) = State(s) \supset Ramify(z, -Closed(x), State(Do(Toggle(x), s))) \,]$

where the term $-F$ represents the occurrence of a negative effect and where $Ramify(z, e, z^*)$ means that state z^* is reachable from z, e by the successive application of (zero or more) causal relationships. Following [12], $Ramify$ is defined via a standard second-order axiom characterizing the reflexive and transitive closure of $Causes$.

To summarize, let $\Sigma_{Circuit}$ be the union of the two state update axioms just mentioned, state constraint (3), $\Pi[(6)]$, the second-order definition of $Ramify$, foundational axiom (5), and the appropriate axioms $EUNA$. This Fluent Calculus theory we will use in the next two sections to illustrate various features of our approach to counterfactual reasoning.

3 Axiomatizing Counterfactuals

We now propose a theory for evaluating counterfactual queries whose antecedent changes what is known about one or more situations. Our theory is implicitly defined by an axiomatization strategy—based on the Fluent Calculus—for counterfactual statements. Consider, for example, the atomic counterfactual condition, for some x and s, "If $Closed(x)$ were true in situation s, ...". By making this assertion one wishes to talk about a situation which is like s except for $Closed(x)$ being true *and* except for all consequences caused by that modification. Generally, our theory allows to process counterfactual statements of the form, "If Φ, then Ψ," where the antecedent Φ expresses modifications—of one or more situations—which can be modeled as actions, and Ψ is a statement about what holds in these (and possibly other) situations. A unified treatment of modifications according to Φ is provided by the following generic state update axiom, which defines the action $Modify(p, n)$ where p and n are finite collections of fluents which shall become true and false, resp., as requested by the counterfactual antecedent. All further consequences of this update are obtained as indirect effects via ramification. Hence, $Modify(p, n)$ is suitably defined by,[7]

$Poss(Modify(p, n), s) \supset$
$[z \circ n = State(s) \circ p \supset Ramify(z, p \circ \neg n, State(Do(Modify(p, n), s)))]$

where the generic predicate $Poss(a, s)$ means that action a is possible in situation s. The state update axiom is accompanied by this action precondition axiom:[8] $Poss(Modify(p, n), s) \supset \overline{Holds}(p, s) \wedge Holds(n, s)$.

To enhance readability, we introduce the following notation: The expression

$$s \lhd f_1 \wedge \ldots \wedge f_m \wedge \neg f_{m+1} \wedge \ldots \wedge \neg f_n$$

denotes the situation $Do(Modify(f_1 \circ \ldots \circ f_m, f_{m+1} \circ \ldots \circ f_n), s)$. For example, the term $S_0 \lhd Closed(Sw1)$ shall denote $Do(Modify(Closed(Sw1), \emptyset), S_0)$.[9]

The axiomatization of a counterfactual conclusion Ψ refers to the modified situation(s) produced by the counterfactual antecedent, as in the following proposition, which asserts the correctness of, "If $Sw1$ were closed in the situation depicted in Fig. 1, then light would be on:"

Proposition 1. *The formulas $\Sigma_{Circuit} \cup \{(4)\}$ entail,*

$$Poss(S_0 \lhd Closed(Sw1)) \supset Holds(On(Light), S_0 \lhd Closed(Sw1))$$

A counterfactual statement may also involve the performance of actions in the hypothetical situation(s) as in, "If $Sw1$ were closed in the situation depicted in Fig. 1, then light would be off after toggling $Sw1$:"

Proposition 2. *The formulas $\Sigma_{Circuit} \cup \{(4)\}$ entail,*

$$Poss(S_0 \lhd Closed(Sw1)) \supset$$
$$\neg Holds(On(Light), Do(Toggle(Sw1), S_0 \lhd Closed(Sw1)))$$

As opposed to existing, propositional accounts of counterfactual reasoning, our theory allows evaluating counterfactual antecedents which exploit the full expressive power of first-order logic and, for instance, include disjunctions of modifications and modifications of more than one situation as in, "If either $Sw2$ would have been open in the initial situation as depicted in Fig. 1, or if $Sw1$ were open now that we have toggled it, then light would be off now:"

Proposition 3. *The formulas $\Sigma_{Circuit} \cup \{(4)\}$ entail,*

$$Poss(S_0 \lhd \neg Closed(Sw2)) \wedge Poss(Do(Toggle(Sw1), S_0) \lhd \neg Closed(Sw1))$$
$$\wedge [S_1 = Do(Toggle(Sw1), S_0 \lhd \neg Closed(Sw2))$$
$$\vee S_1 = Do(Toggle(Sw1), S_0) \lhd \neg Closed(Sw1)] \supset \neg Holds(On(Light), S_1)$$

---

[7] Below, $-(f_1 \circ \ldots \circ f_m)$ means $-f_1 \circ \ldots \circ -f_m$.

[8] Below, $\overline{Holds}(f_1 \circ \ldots \circ f_m, s)$ means $\neg Holds(f_1, s) \wedge \ldots \wedge \neg Holds(f_m, s)$. The usefulness of preconditions for the *Modify* action will become clear in Section 5, where we consider the rejection of counterfactual antecedents.

[9] With a slight abuse of notation, $Poss(Modify(f_1 \circ \ldots \circ f_m, f_{m+1} \circ \ldots \circ f_n), s)$ shall similarly be written as $Poss(s \lhd f_1 \wedge \ldots \wedge f_m \wedge \neg f_{m+1} \wedge \ldots \wedge \neg f_n)$.

144

4 Non-reversibility

The part of our theory where the immediate consequences of a counterfactual antecedent are determined via causal propagation, and the approach of [1,4] based on causal models, have in common the notion of causality. Nonetheless and even if we cut down the expressiveness of our theory to propositional counterfactuals, there are some important properties which are not shared, one of which is *reversibility* [4]. Informally, reversibility means that if counterfactually asserting that a fluent F has a value x results in a value y for some fluent G, and on the other hand asserting G to have value y results in F achieving value x, then F and G will have the respective values x and y anyway, that is, without any counterfactual assertion.

No comparable property is implied by a set of causal relationships in our approach. This allows to process counterfactuals of the following kind, which cannot be dealt with in the approach of [1,4]. Suppose the two switches in our main example are tightly mechanically coupled so that it cannot be the case that one is open and the other one is closed [12]. Then both of these counterfactuals are obviously true: "If $Sw1$ were in a different position than it actually is, $Sw2$, too, would assume a different position," and "If $Sw2$ were in a different position than it actually is, $Sw1$, too, would assume a different position." Yet this does not imply, contrary to what reversibility would amount to, that both $Sw1$ and $Sw2$ actually do occupy different positions than they do.

In order to evaluate these counterfactuals, let $\Sigma'_{Circuit}$ be $\Sigma_{Circuit}$ augmented by the state constraint $Holds(Closed(Sw1),s) \equiv Holds(Closed(Sw2),s)$, along with the causal relationships (or rather the Fluent Calculus axiomatization thereof) which are determined by the constraint if the influence relation is extended by $(Closed(Sw1), Closed(Sw2))$ and $(Closed(Sw2), Closed(Sw1))$.

Proposition 4. *The formulas* $\Sigma'_{Circuit}$ *entail,*

$$Poss(S_0 \lhd Closed(Sw1)) \wedge S'_0 = S_0 \lhd Closed(Sw1)$$
$$\vee\ Poss(S_0 \lhd \neg Closed(Sw1)) \wedge S'_0 = S_0 \lhd \neg Closed(Sw1)$$
$$\supset\ [\,Holds(Closed(Sw2), S'_0) \equiv \neg Holds(Closed(Sw2), S_0)\,]$$

To the same class as this example belong counterfactuals which talk about properties that by their very definition are mutually dependent as in, "If the president were alive, he would not be dead—and vice versa." The reversibility property of the causal models approach to counterfactual reasoning prohibits processing this kind of counterfactual statements.

5 Rejecting Counterfactual Antecedents

Causal propagation of indirect effects is in general not guaranteed to produce a unique result, nor to produce any result at all [12]. In the context of the Ramification Problem, the lack of a resulting state is known as an instance of the Qualification Problem: Rather than giving rise to indirect effects of an action, a state constraint implies an implicit precondition.[10] For our theory of counterfactuals this property of causal relationships implies the rather unique feature that counterfactual antecedents may be rejected if desired.

Consider, for example, this counterfactual sentence due to [11]: "If there had been another car coming over the hill when you passed the car, there would have been a head-on collision." But suppose you as the driver knew that you are on a on-way street, then it seems most appropriate to reject the counterfactual assertion by saying, "But there could not have been another car coming because we are on a one-way." This answer is indeed obtained in our approach by a straightforward formalization of the underlying scenario. Consider, to this end, the state constraint, $Holds(Oncoming\text{-}car, s) \wedge Holds(Passing, s) \supset Holds(Collision, s)$. Both fluents $Oncoming\text{-}car$ and $Passing$ may influence $Collision$. Hence, these two causal relationships are determined by the constraint:

$$Oncoming\text{-}car \underline{causes} Collision, \underline{if} Passing$$
$$Passing \underline{causes} Collision, \underline{if} Oncoming\text{-}car$$

Next we add the knowledge that on a one-way road there are no oncoming cars, formalized by $Holds(One\text{-}way, s) \supset \neg Holds(Oncoming\text{-}car, s)$. Changing the status of a road may causally affect the flow of oncoming traffic but not the other way round, which means that the only causal relationship triggered by the new constraint is, $One\text{-}way \underline{causes} \neg Oncoming\text{-}car$. Let $\Sigma_{Collide}$ denote the complete Fluent Calculus axiomatization of this scenario along the line of $\Sigma_{Circuit}$ in Section 2, then we have the following result:

Proposition 5. $\Sigma_{Collide} \cup \{Holds(Passing, S_0) \wedge \neg Holds(Oncoming\text{-}car, S_0) \wedge \neg Holds(Collision, S_0) \wedge Holds(One\text{-}way, S_0)\}$ entails,

$$\neg Poss(S_0 \lhd Oncoming\text{-}car)$$

The reason is that no available causal relationship allows to restore consistency wrt. the state constraint, $Holds(One\text{-}way, s) \supset \neg Holds(Oncoming\text{-}car, s)$. Proposition 5 is to be interpreted as a rejection of the counterfactual antecedent, "If there had been another car coming over the hill, ... " as 'unrealistic' in the state of affairs. In this way, counterfactual antecedents are only accepted if a world can be constructed around them which is consistent with the state constraints and which does not require 'acausal' modifications.

In the approach of [1, 4], 'acausal' modifications are also not permitted, but the realization of counterfactual conditions involves annulling some of what corresponds to our state constraints, namely, those which normally determine the values of the fluents being altered by the counterfactual antecedent. Any antecedent is thus accepted.

---

[10] A standard example is the constraint which says that in certain cultures you cannot be married to two persons. This axiom gives rise to the (implicit) precondition that you cannot marry if you are already married. The constraint should not imply the indirect effect of automatically becoming divorced [10].

146

The possibility of a counterfactual being rejected, desired as it could be in general, may not always be accepted a reaction. Consider the counterfactual, "If light were on in the situation depicted in Fig. 1, the room would not be pitch dark." As it stands, our axiomatization would reject the condition of this counterfactual on the grounds that the light could not possibly be on because the controlling switches are not in the right position. Insisting upon the counterfactual condition in question, and coming to the conclusion that the counterfactual statement holds, would require to deny the background knowledge of the relation between the switches and the light bulb. Making explicit the desire to deny this relation, the counterfactual statement can be evaluated without rejection if state constraint (3) is replaced by,

$$\neg Holds(Denied(Switch\text{-}Light\text{-}Relation), s) \supset$$
$$[\, Holds(On(Light), s) \equiv Holds(Closed(Sw1), s) \wedge Holds(Closed(Sw2), s)\,]$$

The generic fluent $Denied(x)$ shall be used in general whenever there is desire to weaken a state constraint in this fashion. Situations which do not result from counterfactual reasoning are supposed to not deny any underlying relation among fluents. This is expressed by these three axioms:

$$Factual(S_0)$$
$$Factual(s) \wedge \forall p, n.\, a \neq Modify(p, n) \supset Factual(Do(a, s))$$
$$Factual(s) \supset \neg Holds(Denied(x), s)$$

Let $\Sigma^*_{Circuit}$ be $\Sigma_{Circuit}$ thus modified. Then the above counterfactual antecedent is acceptable if the denial of the dependence of the light is made explicit:

Proposition 6. $\Sigma^*_{Circuit} \cup \{(4)\}$ *is consistent with,*

$$Poss(S_0 \lhd On(Light) \wedge Denied(Switch\text{-}Light\text{-}Relation))$$

6 Axiomatizing Causal Model-Counterfactuals

In concentrating on the crucial connection between reasoning about counterfactuals and causal reasoning, the approach of [1, 4] based on causal models has a strong relation to the proposal of the present paper. Despite the conceptual difference between probabilistic, propositional causal models and the second-order Fluent Calculus with the full expressive power of logic, and despite the further differences discussed in the preceding two sections, counterfactual reasoning in causal models, in the non-probabilistic case, can be embedded into our theory. For the sake of simplicity and clarity, we assume all variables in causal models to be binary. The following definitions follow [4].

Definition 1. *A causal model is a triple* $M = \langle \mathcal{U}, \mathcal{V}, \mathcal{F} \rangle$ *where* $\mathcal{U}$ *and* $\mathcal{V} = \{V_1, \ldots, V_n\}$ *are disjoint sets of propositional variables (exogenous and endogenous, resp.), and* $\mathcal{F}$ *is a set of propositional formulas* $\{F_1, \ldots, F_n\}$ *such that (i)* F_i *contains atoms from* $\mathcal{U}$ *and* $\mathcal{V} \setminus \{V_i\}$. *(The set of variables from* $\mathcal{V}$ *that occur in* F_i *is denoted by* PA_i *(the parents of* V_i*).)*

(ii) For each interpretation for the variables in $\mathcal{U}$ there is a unique model of $(V_1 \equiv F_1) \wedge \ldots \wedge (V_n \equiv F_n)$.

As an example, consider the causal model $M_{Circuit}$ consisting of $\mathcal{U} = \{U_1, U_2\}$; $\mathcal{V} = \{Sw1, Sw2, Light\}$; and $F_{Sw1} = U_1$, $F_{Sw2} = U_2$, and $F_{Light} = Sw1 \wedge Sw2$, which models the electric circuit of Fig. 1 using two additional, exogenous variables U_1 and U_2 that determine the positions of the two switches.

Definition 2. *Let $M = \langle \mathcal{U}, \mathcal{V}, \mathcal{F} \rangle$ be a causal model, $\mathcal{X} \subseteq \mathcal{V}$, and $\iota_{\mathcal{X}}$ a particular interpretation for the variables in $\mathcal{X}$. A submodel of M is the causal model $M_{\iota_{\mathcal{X}}} = \langle \mathcal{U}, \mathcal{V}, \mathcal{F}_{\iota_{\mathcal{X}}} \rangle$ with $\mathcal{F}_{\iota_{\mathcal{X}}} = \{F_i \in \mathcal{F} : V_i \notin \mathcal{X}\} \cup \{X \equiv \iota_{\mathcal{X}}(X) : X \in \mathcal{X}\}$, provided $M_{\iota_{\mathcal{X}}}$ is a causal model according to Def. 1.*

For $Y \in \mathcal{V}$, let $Y_{\iota_{\mathcal{X}}}(\iota_{\mathcal{U}})$ denote the truth-value for Y in the (unique) model of $F_{\iota_{\mathcal{X}}}$ with interpretation $\iota_{\mathcal{U}}$ for $\mathcal{U}$. Then $Y_{\iota_{\mathcal{X}}}(\iota_{\mathcal{U}})$ is the evaluation of the counterfactual sentence, "If $\mathcal{X}$ had been $\iota_{\mathcal{X}}$, then Y," in the setting $\iota_{\mathcal{U}}$.

E.g., a submodel of $M_{Circuit}$ is given by, $\mathcal{F}_{\{Sw1 = True\}} = \{F_{Sw1} = True, F_{Sw2} = U_2, F_{Light} = Sw1 \wedge Sw2\}$. Consider $\iota_{\mathcal{U}} = \{U_1 = False, U_2 = True\}$, which characterizes the situation depicted in Fig. 1. Then $Light_{\{Sw1 = True\}}(\iota_{\mathcal{U}}) = True$, which confirms the counterfactual, "If $Sw1$ were closed, light would be on."

We will now present a correct Fluent Calculus axiomatization of causal models and the evaluation of counterfactuals. The (propositional) fluents are the variables of the model. The definitions $\{F_1, \ldots, F_n\}$ of the endogenous variables are directly translated into state constraints, each of which can possibly be denied, that is, $\neg Holds(Denied(definition\text{-}of\text{-}V_i), s) \supset HOLDS(F_i, s)$.[11] The state constraints determine a collection of causal relationships on the basis of the influence relation $\mathcal{I} = \{(V, V_i) : V \in PA_i\}$. For a causal model M, let Σ_M denote the Fluent Calculus axiomatization which consists of the foundational axioms, including a suitable set $EUNA$, along with the state constraints and the axiomatizations of the causal relationships determined by M as just described.

Theorem 1. *Let $M = \langle \mathcal{U}, \mathcal{V}, \mathcal{F} \rangle$ be a causal model wit Fluent Calculus axiomatization Σ_M. Consider a subset $\mathcal{X} \subseteq \mathcal{V}$ along with a particular realization $\iota_{\mathcal{X}}$ such that $M_{\iota_{\mathcal{X}}}$ is a submodel, a variable $Y \in \mathcal{V}$, and a particular realization $\iota_{\mathcal{U}}$ for $\mathcal{U}$. Let $\Sigma = \Sigma_M \cup \{\bigwedge_{U \in \mathcal{U}}[Holds(U, S_0) \equiv \iota_{\mathcal{U}}(U)]\}$ and let $S_0' = S_0 \triangleleft \bigwedge_{X \in \mathcal{X}}[X \equiv \iota_{\mathcal{X}}(X)] \wedge \bigwedge_{X \in \mathcal{X}} Denied(definition\text{-}of\text{-}X)$. Then,*

1. $\Sigma \cup \{Poss(S_0')\}$ is consistent.
2. $\Sigma \models Poss(S_0') \supset Holds(Y, S_0')$ iff $Y_{\iota_{\mathcal{X}}}(\iota_{\mathcal{U}}) = True$.
3. $\Sigma \models Poss(S_0') \supset \neg Holds(Y, S_0')$ iff $Y_{\iota_{\mathcal{X}}}(\iota_{\mathcal{U}}) = False$.

Proof. The fact that $M_{\iota_{\mathcal{X}}}$ admits a unique model $\iota_{\mathcal{V}}$ under $\iota_{\mathcal{U}}$ implies that there is a unique state which complies with $\iota_{\mathcal{U}}$ and $\iota_{\mathcal{X}}$ and which satisfies the state constraints. From the construction of the underlying causal relationships it follows that this state is reachable by ramification, which proves claim 1. The construction of the causal relationships also implies that no relationship can

---

[11] $HOLDS(F, s)$ is F but with each atom A replaced by $Holds(A, s)$.

be applied by which is modified any fluent representing a variable from $\mathcal{U}$ or from $\mathcal{X}$. Hence, the aforementioned state is the only one which can be consistently assigned to $State(S_0')$. This proves claims 2 and 3, since this state agrees with ι_ν on all variables.

7 Discussion

The author of [5] argues against pushing too far the connection between counterfactual and causal reasoning, on two grounds. First, a counterfactual statement may stress that antecedent and conclusion are *not* causally linked, as in, "Even if I were free tonight, I still would not have dinner with you." This is perfectly compatible with our theory, by which the example counterfactual would be confirmed *because* of the lack of a causal connection. Second, a counterfactual statement may reverse the direction of causality to serve as explanation as in, "If John had Koplic spots, he would have measles." In order to accommodate such explanatory counterfactuals, which amounts to saying which of the *causes* of a denied supposition require modification, an extension of our theory is needed which allows to carefully add appropriate explanatory 'causal' relationships which only apply when performing a *Modify* action.

References

1. A. Balke and J. Pearl. Counterfactuals and policy analysis. In P. Besnard and S. Hanks, ed.'s, *Proc. of UAI*, pp. 11–18. Morgan Kaufmann, 1995.
2. W. Bibel. A deductive solution for plan generation. *New Gener. Comput.*, 4:115–132, 1986.
3. W. Bibel. Let's plan it deductively! *Artif. Intell.*, 103(1–2):183–208, 1998.
4. D. Galles and J. Pearl. An axiomatic characterization od causal counterfactuals. In *Foundations of Science*. Kluwer Academic, 1998.
5. M. L. Ginsberg. Counterfactuals. *Artif. Intell.*, 30:35–79, 1986.
6. M. L. Ginsberg and D. E. Smith. Reasoning about action I: A possible worlds approach. *Artif. Intell.*, 35:165–195, 1988.
7. S. Hölldobler and J. Schneeberger. A new deductive approach to planning. *New Gener. Comput.*, 8:225–244, 1990.
8. D. Lewis. *Counterfactuals*. Harvard University Press, 1973.
9. V. Lifschitz. Frames in the space of situations. *Artif. Intell.*, 46:365–376, 1990.
10. F. Lin and R. Reiter. State constraints revisited. *J. of Logic and Comput.*, 4(5):655–678, 1994.
11. J. McCarthy and T. Costello. Useful counterfactuals and approximate theories. In C. Ortiz, ed., *Prospectes for a Commonsense Theory of Causation*, AAAI Spring Symposia, pp. 44–51, Stanford, 1998. AAAI Press.
12. M. Thielscher. Ramification and causality. *Artif. Intell.*, 89(1–2):317–364, 1997.
13. M. Thielscher. Introduction to the Fluent Calculus. *Electr. Transact. on Artif. Intell.*, 1998. (Submitted). URL: http://www.ep.liu.se./ea/cis/1998/014/.
14. M. Thielscher. From Situation Calculus to Fluent Calculus: State update axioms as a solution to the inferential frame problem. *Artif. Intell.*, 1999. (To appear).

Logic-Based Choice of Projective Terms*

Ralf Klabunde

University of Heidelberg
Centre for Computational Linguistics
Karlstr. 2, D–69117 Heidelberg
klabunde@novell1.gs.uni-heidelberg.de

Abstract. Lexical choice is the problem of determining what words to use for the concepts in the domain representation. In this paper, we offer a model of the choice of projective terms such as *vor, davor, vorn* etc. that is based on abductive reasoning. After a motivation for the need to treat these lexical items as candidates for lexical choice, we outline the semantic and pragmatic conditions for their use. We present an abductive proof method that allows us to collect these conditions. Sets of collected conditions can be mapped unambiguously to corresponding projective items.

1 Introduction

One of the challenging problems for a natural language generation system is the choice of adequate lexical items which will appear in the final output text. Usually lexical choice is confined to open–class or content words referring to concepts or discourse referents in a knowledge base. Closed–class and/or function words have not received much attention as a genuine problem of lexical choice yet. Only content words are assumed to be candidates for a choice process while function words are selected from the lexicon and realized by grammatical decisions during the formulation process. For example, the concept house might be realized as *house, building, red house, hovel* and so on, depending on the available concepts and attributes in the knowledge base and especially the wanted effect of the use of a specific item for the listener. As a rule of thumb in lexical choice a concept should be expressed with the most specific word available, because that would convey the most information (cf. Stede, 1995).[1] Contrary to the choice of content words the usage of function words like conjunctions, prepositions etc. is usually governed by grammatical decisions and not subject to a proper choice process.

This widely accepted view implies that primarily nouns, adjectives, and verbs belong to the range of a function mapping concepts to lexical items, but prepositions and adverbs in general do not. However, such a principal distinction is

---

* This research was supported by a grant from the Deutsche Forschungsgemeinschaft (DFG) which is gratefully acknowledged.
[1] However, such an account can fail to account for basic–level preferences so that the utterance can carry unwanted implicatures (Reiter, 1990).

not appropriate because prepositions and adverbs as closed–class items very often express a content as well. This holds especially for spatial prepositions and adverbs referring to spatial relations between objects.

Spatial prepositions and adverbs can be divided into two classes, topological and projective[2] ones. While topological items like *in, drinnen, an, daran* etc. refer to relations between objects that can be established by means of topological concepts as, e.g., contact or inclusion, projective terms like *hinter, hinten* or *dahinter* require a spatial frame of reference providing a system of axes and directions in order to refer to a spatial relation. Using the common terminology, we refer to the object providing the axis–based region the reference object RO and the object that is located within this region the primary object PO. The system of axes as frame of reference has a zero point, the so–called origo, which can be one of the interlocutors (which corresponds to a deictic frame of reference), the reference object (which corresponds to an intrinsic frame of reference), or a third entity. Each axis has two directions so that in 3D–space six directions appear which we will call front, back, left, right, top, and bottom (cf. Eschenbach et al., in press; Herskovits, 1986).

The choice of projective prepositions and adverbs differs from the choice of, e.g., nouns because it is affected by several tasks. From the seven component tasks Cahill & Reape (1998) assume for applied NLG systems, at least the three tasks of salience determination, referring expression generation, and lexicalization affect the choice of projective prepositions and adverbs.

Salience of discourse referents affects primarily pronominalization because salient or focussed referents are usually realized as pronouns. Referents which are very salient in discourse can even be ommitted so that ellipses become appropriate linguistic means to realize certain information. This holds also true for the use of projective adverbs because these adverbs can only be used to express a spatial relation with an implicit reference object. However, in order to be cooperative, a speaker should be safe that the listener is able to identify the reference object. Hence, only identifiable and salient discourse referents should be used as reference objects. Additionally, the choice of referring expressions is a task for projective terms as well. Projective prepositions are primarily used with a reference object that is already identifiable for the listener. Therefore, it must be realized as a definite noun phrase in German. Obviously lexicalization affects the choice as well because one and the same spatial relation can be realized by a number of prepositions and adverbs.[3]

The relevance of these tasks indicates that in addition to object–specific applicability conditions additional pragmatic constraints must be taken into account as well. For example, if only spatial features are taken into consideration,

---

[2] These spatial terms are also called dimensional prepositions or adverbs.

[3] In this paper we do not draw a distinction between lexicalization as the conversion of concepts to lexical items and lexical choice as the decision between lexical alternatives (Cahill, 1998). Since we assume a pipelined NLG–architecture so that the semantic input for the formulator allows an unambiguous mapping from conceptual conditions to lexical items, lexical choice and lexicalization can be unified to one process.

we would not be able to choose between, e.g., the preposition *hinter* and the adverb *dahinter* because both items are able to express the same spatial relation. The difference is grounded in the accessibility state of the discourse referent for the reference object and the influence of the meaning of the locative anaphor *da*.

We should mention that approaches to the computation of spatial relations in geometric abstractions of a real–world environment (Gapp, 1994, 1997; Wazinski, 1992) do not concern lexical choice in its actual sense because these models are restricted to geometrical aspects solely. The model of lexical choice we are proposing could rather be based on such an approach. For example, Gapp's work concerns the computation of spatial relations, not their expression in a natural language, although he does not draw this distinction properly. However, the existence of preposition–free languages (like, e.g., Finnish) as well as the differing lexical means to express spatial relations in closely related languages like German and English (Carroll, 1997) clearly show that both tasks should be kept separate. Consequently the constraints we are giving hold for German only. For example, the semantic differences expressed by *hinter*, *dahinter*, and *hinten* are not lexically encoded in English; these three lexical items can be translated as *behind*.

We propose a decompositional approach to the choice of dimensional prepositions and adverbs where features of the reference object and its kind of textual embedding are collected in order to receive an unambiguous mapping. Feature collection is realized as abductive reasoning.

2 Semantics and Pragmatics of Projective Prepositions and Adverbs

Before we introduce our abductive approach, we will explicate the conditions for the use of projective prepositions and adverbs. These conditions emerge from an investigation of village descriptions (for details, see Klabunde, in press).

First of all, we have to draw a distinction between semantic and pragmatic conditions. The main difference between the semantics of projective prepositions and the semantics of the adverbs *vorn*, *hinten*, *unten*, and *oben* is the localization of the primary object in a region that is external and internal, respectively, to the reference object. While prepositions like *vor* locate a primary object with respect to a region that is outside the reference object, the corresponding adverb *vorn* locates the primary object in an internal region provided by a reference object. However, a preferred interpretation with an internal relation does not hold for the adverbs *rechts* and *links* referring to horizontal axis-based regions. Furthermore, the class of projective adverbs is more heterogenous than the class of projective prepositions. Next to the already mentioned adverbs the adverbs *davor*, *rechts davon* etc. have different meanings based on an external relation again, and *linkerhand* and *rechterhand* possess a different pragmatics. The following simple examples demonstrate the essential meaning differences.

(1) *Das Stadttor ist hinter der Kirche* (the city gate is behind the church)

(2) *Das Stadttor ist hinten*

(3) *Das Stadttor ist dahinter*

(4) *Das Stadttor ist rechts von der Kirche*
    (the city gate is to the right of the church)

(5) *Das Stadttor ist rechts*

(6) *Das Stadttor ist rechts davon*

(7) *Das Stadttor ist rechterhand*

In (1) and (3) the city gate is located in an external back–region of the church, provided by a specific frame of reference. Due to the spatial features of churches with their intrinsic fronts, the origo might be the church, but the speaker or listener as well. In (2) however the city gate is more likely located in an inner region, e.g., a square or inner room, depending on the context. Such a distinction is not given by (4) – (7). In all cases the right direction holds primarily for an external relation, although (5) might be interpreted with an internal relation as well.

In addition to this distinction between internal and external regions there are salience– and pragmatic–based constraints. First of all, all prepositions and adverbs are used with identifiable reference objects. In German identifiability is usually expressed by definiteness. Localizations like *hinter einem Haus steht ein Brunnen* (there is a fountain behind a house) with a house as unidentifiable reference object are principally possible, but they do not occur in coherent descriptions. The reason is that the reference objects function as "anchors" for the single localizations at what has been said before. Identifiability of the reference object holds for the use of adverbs as well, although that reference object is omitted when the adverbs are used. In addition to being identifiable, the reference object must also be focused in the discourse model, because it can only be omitted if the listener is able to pick up the right referent.

The meaning of the so–called pronominal adverbs like *darüber* or *rechts davon* differs from the meaning of the corresponding prepositions by means of the use of the locative anaphor *da*. This anaphor refers to an identifiable and focussed region, not to an object. This region is the place the reference object occupies.

Finally, the adverbs *rechterhand* and *linkerhand* are identical in their meaning with *rechts* and *links* or the morphologically complex prepositions *rechts von* and *links von*, but their use is confined to a humanlike origo, i.e., a fictiver observer wandering through the object constellation, or the speaker or listener.

These linguistically motivated constraints are exemplary summarized in Table 1. The remaining prepositions and adverbs are associated with the same corresponding conditions except that the spatial directions are different. If the conditions in a row are met, we are able to map these conditions onto the corresponding preposition or adverb. Table 2 gives the literals we are using to represent these conditions. Localizations are represented by means of two literals. A literal p(R,RO,PV) states that the reference object RO is standing in relation p to a region R, determined by a point of view PV. A functor loc(PO,R) locates the primary object PO in region R.

lemma	identifiability of RO	focused RO	spatial direction	relation external	relation internal	additional constraints
vor	+	−	front	+	−	
davor	+	+	front	+	−	RO's place provides the region
vorn	+	+	front	−	+	
rechts von	+	−	right	+	−	
rechts	+	+	right	+	+	
rechts davon	+	+	right	+	−	RO's place provides the region
rechterhand	+	+	right	+	−	humanlike origo

Table 1. Constraints for the use of some projective prepositions and adverbs

constraint	literal
reference object	`ro(X)`
primary object	`po(X)`
identifiability	`id(X)`
focussed referent	`foc(X)`
spatial localizations	`front(V,RO,PV)` ∧ `loc(PO,V)` `back(B,RO,PV)` ∧ `loc(PO,B)` etc.
internal relation	`container(RO)`
place of the RO	`place_of(RO,P)`
humanlike origo	`human(PV)`

Table 2. Used literals

3 Planning by Abduction

How can the conditions given in Table 1 be satisfied during content planning? We propose an abductive approach to content planning which guarantees by means of the proof method that conditions for the use of specific prepositions and adverbs are satisfied.

Abductive reasoning is reasoning about the best explanation for a given observation. To make precise what counts as a good explanation, one introduces a preference criterion by which alternative explanations can be compared. A preferred explanation for an observation might be the least specific one, the most specific one, the one with the lowest proof costs, etc. Abductive explanation is classically characterized as follows (cf. Maier & Pirri, 1996): a knowledge base K, the usual consequence relation $\models$, and an observation E to be explained, such that $K \not\models E$, are given. A statement H is taken as the best explanation of E in K iff:

1. $K \cup \{H\} \models E$; and

2. H is "better" than any other statement in the set $\{H' \mid K \cup \{H'\} \models E\}$, according to the preference criterion.

We use a generalized version of what Stickel (1990) calls *predicate specific abduction*, where only elements from a distinguished set of literals may be assumed. What counts as the best explanation will be based on the (preferably minimal) number of assumptions made.

Abduction has been used in natural language processing for interpretation tasks such as metonymy resolution, understanding vague expressions, or plan recognition (cf. Hobbs et al., 1993). Recently, abductive reasoning has also been proposed for use in generation, partly for planning (Lascarides & Oberlander, 1992; Thomason et al., 1996), and as a framework for both the interpretation and the generation of discourse (Thomason & Hobbs, 1997). The basic idea behind these approaches is to find the best way to obtain a communicative goal state by modifying the discourse model with applicable operators. The plan is the set of hypotheses discovered by an abductive proof of the proposition that the goal state has been achieved. These approaches are primarily used in integrative architectures of NLG systems.

For our purposes it is helpful to view abductive proofs as essentially *relational*. An abductive proof determines the relation between a knowledge base, an observation, a specification of what assumptions can be made, and proved and assumed literals that jointly provide an explanation for the initial observation. The prototype we have implemented in Prolog make this relation available explicitly, and great care was taken to ensure that queries such as (8), where not all arguments are instantiated, are handled correctly by generating a manageable subset of all possible solutions.

(8) ?- abduce(Goal, Assumable, Proved, Assumed).

In the above query, Goal is the observation to be proved by the abductive meta-interpreter, Assumable is a set of literals that may be assumed, and Proved and Assumed are multisets of literals that were used or assumed, respectively, during the abductive proof of Goal. During generation the meta-interpreter is invoked with the goal instantiated, while the set of assumable literals is specified. An exemplary query for the choice of projective terms is given in (9), although it is possible to start with other instantiations. This query is interpretable as: given a spatial relation, which further conditions can be proven so that we are able to determine an appropriate preposition or adverb? In addition to the spatial relation only the proved literals will be handed down to the lexical chooser because the semantic and pragmatic constraints for a specific lexical item should be safe knowledge. Hence, the instantiation of Pr is what we are interested in for lexical choice.

(9) ?- abduce([front(v,a,_), loc(b,v)], [id(a), foc(a),
         container(a)], Pr, As).

This query gives the spatial relation between two objects and asks for constraints that allow an unambiguous mapping of the spatial relation to a preposition or adverb. The literals front(v,a,_), loc(b,v) are to be interpreted as "object a provides a region v determined by some origo that is of no interest for the lexical choice. Object b is located in region v." From the fact that queries like (9) are accepted it is clear that the abduction scheme we use is somewhat more general than predicate specific abduction: we supply information as to what literals may be assumed, whereas predicate specific abduction would only specify the functors and arities of those literals.

Since we do not use weighted abduction (Hobbs et al., 1993), we are not faced with the problem of giving costs and weights an appropriate probabilistic interpretation. On the other hand, what should we use as a preference criterion? A sequence of several criteria is used. First, proofs are preferred for the number of provable literals used, the more the better. In the cases we consider, there seems to be a loose correspondence between this criterion and the Gricean maxims of relevance and quality. Second, proofs are preferred compared to other proofs if they involve less assumptions. The number of assumptions made is determined by the cardinality of the set that is the reduction of the multiset found during an abductive proof. Third, everything else being equal, we prefer proofs with the highest amount of assumption re-use. This is determined by the difference between the cardinalities of the multiset of assumptions and of its corresponding set. The relevant idea—an assumption becomes more plausible if it is used to explain more than one thing—is essentially the same as the one behind the factoring rule of Stickel (1990).

Proving the conditions for the use of specific lexical items is managed by a rule that calls for proofs in the following order:

1. prove whether the entities given in the query can function as reference and primary object. This proof involves the check whether both objects are neighboured and whether the reference object is more salient than the primary object. If the entity RO in the literal p(R,RO,PV) can only be assumed and not be proven as an object, we have to prove that it is the place an object occupies and that this object is the reference object. If this proof is successful we receive a condition for the choice of a pronominal adverb.

2. prove that the reference object is identifiable for the listener. The identifiability–condition is satisfied if the referent belongs to the mutual beliefs.[4] If only the speaker/system knows a certain referent but not the listener that referent is unidentifiable.

3. prove whether the reference object is focused. If foc(X) is provable, we receive a condition for the choice of an adverb. An object's discourse referent is focused if it is identifiable and had been mentioned in the preceding proposition.[5] Otherwise it is not in the focus of attention.

---

[4] This is a slight oversimplification because identifiability of a discourse referent does not only imply that both speaker and listener know this referent. It is also possible that the speaker assumes the listener is able to infer a certain referent or that he is able to single out a specific referent by perceptual means (cf. Lambrecht, 1994).

[5] This is achieved by a predicate active/3 that is used to drag along, test, and update the discourse state. active/3 must be resolved exactly once.

4. try to prove that the reference object is a container object that establishes an internal relation. If this can be successfuly proven we receive a condition for the choice of adverbs like *vorn, hinten, rechts* etc.

5. if the spatial relation is `right` or `left` try to prove that the origo is provided by a human. A successful proof results in the choice of the adverbs *rechterhand* and *linkerhand*, respectively.

Together with the spatial relation the proven literals represent the constraints for the lexical choice. For example, in order to choose the lemma *dahinter*, the following literals with corresponding instantiations of the variables must be proven:

(10) {ro(Z), po(Y), back(H,X,_), loc(Y,H), place_of(X,Z), id(Z), foc(Z)}

Once the proof is complete lexical choice is a simple process. Together with the literals in the goal to be proven the proven literals are matching conceptual conditions represented in the corresponding lemmas.

4 An Outline of Two Example Proofs

We sketch two example proofs that illustrate the used techniques. The lemma we want to choose should express that an object **b** is located behind an object **a** as seen from some point of view. Additionally, we know that **a** belongs to the mutual beliefs and that both objects can function as reference and primary object, respectively.

facts: church(a), city_gate(b), in_mutual_beliefs(a), etc.
assumable: ro(a), po(b), id(a), foc(a)
to prove: back(r,a,_), loc(b,r)

According to the query and rule given above the proven literals are ro(a), po(b), id(a). The assumed literal is foc(a). Recalling the conditions in Table 1 again, the localization together with the proven literals allows the choice of the preposition *hinter*.

Now suppose we know that **a** is not an object but the place an object **c** occupies and **c** is in the mutual beliefs and identifiable. The same query results in the proven literals place_of(a,c), id(c), foc(c) with no assumed literals. In this case the conditions for the choice of the adverb *dahinter* are met. An alternative explanation would result in the proven literal po(b) and the assumed literals ro(a), id(a), and foc(a). Since our preference criterion for the best explanation involves the minimization of the number of assumptions, this explanation will not be chosen as the best one.

5 Concluding Remarks

Abductive approaches have pros and cons. One disadvantage of abductive reasoning is its computational complexity which makes it an inappropriate method for real–time generation. However, translating this logic–based approach into a systemic framework (Patten, 1988) should be straightforward. Choosers connected to systemic networks can check whether the aforementioned conditions are met in the environment. Hence, this disadvantage of abductive approaches might be avoidable in the systemic paradigm. On the other hand, if abduction is used in a small domain like the one presented in this paper we receive the results in acceptable time. Additionally, the usefulness of abductive reasoning is not confined to content planning only; the same abductive method can be used for syntactic realization as well (cf. Klabunde & Jansche, 1998). This implies that the substantial advantage of an abductive approach is its provision of a common and uniform framework for lexical choice and syntactic realization. To summarize, we have shown that expressing spatial relations in a natural language is a task for lexical choice. This task can be efficiently realized in an abductive framework.

References

1. Cahill, L. (1998) Lexicalisation in applied NLG systems. Brighton/ Edinburgh; 1st deliverable of the RAGS project. Available at http:\\www.itri.brighton.ac.uk\projects\rags\.
2. Cahill, L. & Reape, M. (1998) Component tasks in applied NLG systems. Brighton/Edinburgh; 2nd deliverable of the RAGS project. Available at http:\\www.itri.brighton.ac.uk\projects\rags\.
3. Carroll, M. (1997) Changing place in English and German: language–specific preferences in the conceptualization of spatial relations. In: J. Nuyts & E. Pederson (eds.) *Language and Conceptualization*. Cambridge: Cambridge University Press; 137–161.
4. Eschenbach, C.; Habel, C. & Leßmöllmann, A. (in press) Multiple frames of reference in interpreting complex projective terms. To appear in: P. Olivier (ed.) *Spatial Language. Cognitive and Computational Aspects*. Dordrecht: Kluwer.
5. Gapp, K.-P. (1994) From Vision to Language: A Cognitive Approach to the Computation of Spatial Relations in 3D Space. In: *Proceedings of the 1st European Conference on Cognitive Science in Industry*. Luxembourg; 339–358.
6. Gapp, K.-P. (1997) *Objektlokalisation*. Wiesbaden: Deutscher Universitätsverlag.
7. Herskovits, A. (1986) *Language and Spatial Cognition: An Interdisciplinary Study of the Prepositions in English*. Cambridge: Cambridge University Press.
8. Hobbs, J.R., M.E. Stickel, D.E. Appelt & P. Martin (1993) Interpretation as abduction. *Artificial Intelligence* 63, 69–142.
9. Lambrecht, K. (1994) *Information Structure and Sentence Form*. Cambridge: Cambridge University Press.
10. Lascarides, A. & Oberlander, J. (1992) Abducing temporal discourse. In: R. Dale, E. Hovy, D. Rösner, and O. Stock (eds),*Aspects of Automated Natural Language Generation*, Berlin: Springer, 167–182.
11. Klabunde, R. (in press) Semantik und Pragmatik dimensionaler Adverbien. To appear in: C. Habel & C. von Stutterheim (eds.) *Sprache und Raum*. Tübingen: Niemeyer.
12. Klabunde, R. & Jansche, M. (1998) Abductive Reasoning for Syntactic Realization. *Proceedings of the 9th International Workshop on Natural Language Generation*. Niagara–on–the–Lake, Ontario, Canada; 108–117.

13. Mayer, M.C., & Pirri, F. (1996) A study on the logic of abduction. In: W. Wahlster (ed), *Proceedings of the 12th European Conference on Artificial Intelligence (ECAI 96)*, Chichester: Wiley, 18–22.
14. Patten, T. (1988) *Systemic text generation as problem solving.* Cambridge: Cambridge University Press.
15. Reiter, E. (1990) A New Model for Lexical Choice for Open–Class Words. *Proceedings of the 5th International Natural Language Generation Workshop.* Dawson, PA.; 23–30.
16. Stede, M. (1995) Lexicalization in Natural Language Generation: A Survey. *Artificial Intelligence Review,* **8**; 309–336.
17. Stickel, M.E. (1990) Rationale and methods for abductive reasoning in natural language interpretation. In: R. Studer (ed), *Natural Language and Logic,* Berlin: Springer, 233–252.
18. Thomason, R.H., J.R. Hobbs, and J.D. Moore (1996) Communicative goals. In: K. Jokinen, M. Maybury, M. Zock, and I. Zukerman (eds), *Proceedings of the ECAI 96 Workshop Gaps and Bridges: New Directions in Planning and Natural Language Generation,* Budapest, 7–12.
19. Thomason, R.H., and J.R. Hobbs. 1997. Interrelating interpretation and generation in an abductive framework. Paper presented at the AAAI Fall 1997 Symposium on Communicative Action in Humans and Machines. Cambridge, MA.
20. Wazinski, P. (1992) Generating Spatial Descriptions for Cross–modal References. In: *3rd Conference on Applied Natural Language Processing.* Trento; 56–63.

Knowledge Based Automatic Composition and Variation of Melodies for Minuets in Early Classical Style

Mathis Löthe

Universität Stuttgart
Institut für Informatik
Breitwiesenstr. 20-22
D-70565 Stuttgart
Germany
email: loethems@informatik.uni-stuttgart.de

Abstract. Composition of piano minuets in early classical style takes an intermediate position between restricted and free computerized composition problems. One of its distinct features is that microstructure (i.e. relations between notes), macrostructure (i.e. relations between larger parts) and their interdependences have to be dealt with. In this paper, we give an overview of the process of composition which consists of the subtasks planning of macrostructure, construction of melody and rhythm, variation of melodic motives, and addition of bass and middle voices. We then present a knowledge-based approach with strategies and results for note by note construction of melody and methods for melody variation.

1 Overview

This paper describes a knowledge based approach to composition of minuets for piano in early classical style (that is the period of the young W.A. Mozart) by a computer. The goal is, to approximate the style of this period as closely as possible.

The result of the composition process is music in staff notation. The interpretation of the music, the sound synthesis, is not dealt with here.

The era of classical style started around 1750. Influenced by the philosophical ideals of the era of enlightenment, composers strived for simplicity and rationality. The classical ideal is to have simple melody and harmony and a regular form with strong structural relationships between parts.

A minuet is a genre of dance music in triple meter (usually 3/4) and was known at the french court since 1650. Minuets need a regular form and well perceivable relations between the parts such allowing the dancers to dance the elaborate and complicated figures of such a formal court dance.

By having those structural properties, the genre of minuets soon became a model genre for classical composition itself. Simple and strict minuets were written as dance music, for piano and composition teaching. Artful minuets were composed as movements of sonatas and symphonies.

Fig. 1. Beginning of Minuet KV. 2 W.A.Mozart 1762

The composition of minuets can be discussed at two levels. On the level of *microstructure*, relations between notes in horizontal direction (melodic intervals) and in vertical (synchronous) direction (harmony) are discussed. *Macrostructure* deals with the large structural parts in horizontal direction. Decisions in the microstructure of a part depend on the role the part plays in the macrostructure of the piece. Therefore, musical parts can not just be concatenated together, they have to be purpose-built to fit into their role.

The process of classical minuet composition can be divided into several subtasks:

– planning the macrostructure of the piece, that is the sequence of parts and their harmonic relations.
– planning the harmonic relations inside a part.
– constructing the melodic parts from scratch by composition.
– constructing the melodic parts by variation of existing parts.
– constructing the bass (and middle voices if wanted).

Here, the subtasks are written down in a linear order. Though it might be possible to compose by strictly following this order, it has to be noted, that human composers usually mix the subtasks. E.g. a composer might decide to change his plan of the macrostructure (or to start thinking about macrostructure at all), after he has composed the first four bars of the melody. Research about the importance of the composition strategy in the field of chorale harmonization can be found in [WPA99].

The strategies for the different subtasks need different kinds of knowledge. To achieve our goal of composing minuets which sound like a typical example of the period's style, it is important to represent musical knowledge adequately in the system. Section 2 describes possibilities to acquire this knowledge.

Related work on computerized composition and its common features with minuet composition are described in section 3. Section 4 describes the base for experiments on different subtasks of minuet composition. The experiments have focussed on melody construction. The final summary presents some thoughts about other fields of AI, which might make use of strategies developed for minuet composition.

2 Sources for Musical Knowledge

A system for automatic composition of minuets can be seen as a knowledge based system. The author of such a system has to acquire the domain knowledge, has to formalize it and finally has to provide an algorithm which the computer can use to find a solution.

Often, it is helpful to look at the ways human experts codify their knowledge, e.g. their literature. Explicit knowledge about composition is needed for teaching composition students or for scientific works – e.g. for comparisons between styles or for classification of borderline cases of works. Humans usually learn some subtasks, such as planning of macrostructure, harmonic relations, and voice leading by intellect or by formal rules. Other aspects like melody composition are usually learned by analogy after a profound study of many works in the desired style. [Cop91] simulates the latter in his pattern-matching based composition system.

For the field of classical minuets, the following knowledge sources are considered:

1. Period literature[1]. [Rie52] has an explicit (sometimes chatting) style of explaining the composition technique by examples and rules (described by words). [Koc82] consists of textual descriptions and partially formal rules about melody, voice leading and dissonance treatment. Both authors describe the specific genre of minuets because of its model position for classical composition in general.
   The usage of period literature ensures a correct view of the style. Modern musicological literature with an historical approach like [Bud83] serves the same purpose.
2. Ahistoric theories such as theory of harmony and Schenkerian analysis look for properties universal to all tonal music.
   Theory of harmony classifies harmonic chords and their relations. Schenkerian analysis as used by [Ebc86] explains the structure of a piece by taking a simple "*Ursatz*" and adding details step by step.
   Ahistoric theories are usually more formalized than period literature knowledge. So they can be helpful to get insight in the structural form of tonal music in general in order to prepare implementation. But they can not give information about classical style and the genre of minuets.
3. To ask musicologists, composition teachers and composers which are familiar with the style about explicit rules and composing methods. This gives access to explicit knowledge which might not have been written down.
4. Inductive analysis of works: Classical minuets are searched for evidence in favour or against a hypothetical rule. Composition examples in period literature are especially interesting, because they have been composed with the purpose to show, how it is done.
5. Computer experiments: The composition algorithm is run at first with a ruleset including a hypothetic rule and then with another ruleset without that particular rule. The results are compared (or judged by an expert).

---

[1] In musicology, *period* or *contemporary* literature (zeitgenössische Literatur) is literature written in the same time period as the musical period or style it describes.

6. Cognitive introspection. Ask someone, who has just composed a minuet, how and why he made certain decisions.

The last three knowledge sources should give access to implicit knowledge, which is not available in literature.

3 Related Work

In the past, approaches to automated composition have focussed either on composition styles and problems with firm restrictions or on those with very few restrictions.

One example for work on tonal composition with strong restrictions is the system for four-part-harmonization of Bach chorales by [Ebc86]. The system is given the chorale melody as input and returns a four part harmonization. The macrostructure is with the chorale melody. Ebcioglu uses Schenkerian analysis to determine the melody's microstructural properties and then constructs the harmonization.

A system, which creates music from scratch (without music material as input) is the automatic 16th century counterpoint of [Sch89]. Schottstaedt uses period literature – the *Doctor gradus ad parnassum* by J.J. Fux [2] – as knowledge base. The musical knowledge consists of rules, which look at small groups of consecutive notes. In our terms, Schottstaedt only considers microstructure.

In both styles, the solution has to fulfill strict microstructural requirements, with the latter being a bit less restricted. Both systems represent their musical knowledge and have a single strategy to make use of it.

Other composition tasks have a larger degree of freedom for the solution e.g. approaches to algorithmic composition[3]. Here, the emphasis is more on creativity and there are few rules to judge the result. Algorithmic composition systems such as Common Music [Tau94] have the purpose to assist human composers in their work. Parts are generated by an algorithm and put together to a larger piece. Some systems even offer graphical editors which allow direct manipulation of the piece's macrostructure.

An approach to deal with macrostructure in classical style can be found in [Ber95], who construct a piano sonata in the style of Mozart out of "building blocks". The macrostructure of such a piano sonata is more complicated than that of a short minuet. Berggren identifies elements in microstructure called *formulas* and uses them to identify different classes of building blocks, eg. different kinds of accompaniment texture. His system does not deal with the adaptions in microstructure which are necessary to fit blocks together.

---

[2] The *doctor gradus at parnassum* is written at the beginning of the 18th century. In a very strict sense, it is not period literature, because it is written much later and musicologists have found slight influence from baroque harmony. Nevertheless Fux had the firm intention to describe Palestrina's ancient practice.

[3] an atonal, experimental style of music

Compared with those approaches, minuet composition is in the middle between free and restricted styles and has to integrate different subtasks with their strategies.

4 Subtasks of Classical Minuet Composition

The composition strategy is based on a representation of a musical score in a *score chart*. It stores notes with pitch and duration (the same information classical composers wrote down). The score chart also has to store meta information about the composition in form of *structural elements*.

4.1 Rule-Based Note by Note Composition of Melodies

Figure 2 shows a random sequence of notes, which follows the basic parameters of a minuet. It has a fixed key, 3/4 meter, the pitches and the durations stay inside the usual bounds for minuets and there are no syncopes. It is still far away from a classical minuet, because it is perceived as a sequence of unrelated notes instead of a melody. The main reason is, that the intervals between notes are much too large. In [Rie52] there are also examples for "melodies" perceived as unrelated.

Fig. 2. Random sequence of notes in G-major

Two basic types of melodies can be observed in minuets. The first type is a *scale-based* melody, which consists mostly of seconds, some thirds and repetitions and a few larger intervals. This type is a kind of successor to the old style counterpoint melodies as in [Sch89]. Relatedness is established by the preference for small intervals.

The other type of melody is based on broken chords, which consist of steps to the next chord note (up or down). Because this type of melody puts more emphasis on harmony, chord based melodies have become quite popular with early classical style. With chord based melodies, relatedness can be established by different means. One is, to join two chord based sections with a second. Another is, to have sections which are similar to each other e.g. by variation (see section 4.2). Then the interval at the joint is rather unimportant because the perceptual connection is established by similarity, not by using a small interval.

Using this approach, simple rules for interval based composition can be formulated. Melody construction has to consider the prerequisites given by the role the part to be constructed plays in macrostructure. This gives every position a *structural environment* which limits the choice of notes. So some rules have a reference to the structural environment in the chart among their preconditions. One example for a parameter is the harmonic base – a sequence of harmonies – given before starting the note by note composition.

During melody construction, further structural elements can be added to the structural environment. Some rules start a new structural element e.g. the transition rule (No. 5) described below. A transition is a structural element, because it influences several consecutive notes. In contrast to other structural elements, the position of a transition can not be given as a prerequisite beforehand, because the possibility to use a transition depends on beginning of the melody. The existence of this structural element is *"found"* during the process of note-by-note composition. This method of finding structures can be seen as a first small step towards bottom-up planning.

Some rules have been implemented and tests with different sets of rules were made. Here are some examples of rules for scale-based melodies for the beginning of a minuet. Earlier rules and experiments can be found in [Löt99].

1. **Start note** The melody starts on the first degree of the key. [4]
2. **Standard scale-based intervals:**
   In a scale-based melody, repetitions and seconds are allowed.
3. **Thirds in scale-based melodies**
   Thirds can be used in scale-based melodies if they don't follow immediately.
4. **Consider harmonic base:**
   A harmonic base for a time interval can be specified. Melody notes have to be a member of the harmonic chord valid at the starting time of the note.
5. **Start of transitions:**
   if There is an unaccented beat
   without change of harmonic base and
   there is not an unfinished transition in the chart
   then a second is allowed.
   A structural element of type "transition" is started and
   it is noted in the chart.
   The duration of the transition note must not be longer than the duration of the note before.
6. **Closing of transition**
   In a transition, the next interval has to be a second in the same direction as the opening of a transition.
   This ends the transition.
7. **Leading notes:** (scale based melodies)
   If the preceding note has degree 7, only a minor second up is allowed.
8. **Ambitus:** The pitch has to be in the a range of two octaves.

While a few rules about treatment of large intervals and leading notes are written down in [Koc82], most of those rules have been found by analyzing examples.

Here are two rules used for composition of chord based melodies.

8. **Links between chords**
   if the harmonic base changes
   then only seconds up and down are allowed

---

[4] This is the common way to start a minuet ([Rie52]).

9. **Chordsteps**
   if the harmonic base stays the same
   then repetitions, and steps to the next chord note up and down are allowed

To avoid a stylistically unsuitable distribution of melodic and rhythmic weight, it is useful to limit the use of repetitions. These rules have been found by judging results of experiments.

10. **repetition limit** There must not be more than n consecutive notes of equal pitch. (possible limits: $n = 2, n = 3$)

11. **no repetition split on downbeat**
    There must not be two notes with the same pitch on beat 1.

12. **no weak-strong repetion** In a repetition, the following note must have same or less weight than the previous.

Classical composition sometimes use chromatic transition. For the experiments shown in figure 3, the same preconditions and closing rule as for normal (diatonic) transitions have been used.

13. **Start of chromatic transition**
    if a transition is possible (see rule 5)
    then a semitone (augmented unsion) is allowed,
        and a structural element of type "chromatic transition"
        (a subclass of transiction) is started

Minuets are dance music and therefore have their obligatory meter (3/4) and their characteristic rhythm. [Rie52] defines the following kind of bars:

- Perfect [5] minuet bars consist of 3 quarters. It is allowed to split up to two of the quarters into eighths.
- Imperfect [6] minuet bars consist of a half and a quarter or vice versa. The quarter may be split.
- Dead [7] bars consist of a dotted half.
- Running [8] bars contain 6 eighths.

Those kind of bars are utilized the following way.

- A minuet normally consists of perfect bars.
- One bar out of a two bar block may be imperfect.
- As final note of cadences, dead bars are allowed.
- Running notes are allowed only to prepare the final cadence of the whole piece, i.e. the 3rd and 2nd last bar.

Riepels duration rules are overly strict. He does not allow a combination of a dotted quarter [9] with an eighth though it can be frequently observed in original compositions. Running bars (consisting of 6 eighths) are also found in minuets at earlier positions. Nevertheless, the rules describe a characteristic rhythm for minuets.

---

[5] Original: *vollkommene*
[6] Original: *unvollkommene*
[7] Original: *todte* (historic spelling)
[8] Original: *laufende*
[9] Riepel calls this combination *only fit for a limping dancemaster*

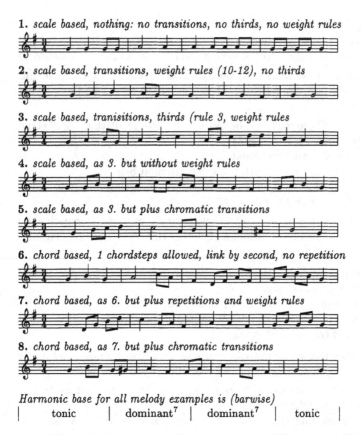

1. *scale based, nothing: no transitions, no thirds, no weight rules*

2. *scale based, transitions, weight rules (10-12), no thirds*

3. *scale based, tranisitions, thirds (rule 3, weight rules*

4. *scale based, as 3. but without weight rules*

5. *scale based, as 3. but plus chromatic transitions*

6. *chord based, 1 chordsteps allowed, link by second, no repetition*

7. *chord based, as 6. but plus repetitions and weight rules*

8. *chord based, as 7. but plus chromatic transitions*

Harmonic base for all melody examples is (barwise)

| tonic | dominant[7] | dominant[7] | tonic |

Fig. 3. Scale-based melodies contructed by different sets of interval rules

For experiments, a rule interpreter has been implemented. Its input are an empty melody, the ruleset and the parameters (such as harmonic base, desired length, end note ...). Its output is a set of alternatives for continuing the melody. Each alternative is described by the interval, the duration and the structural elements involved.

This constitutes a search problem as described in [RN95]. The state is the current state of the score chart (melody + structural elements), the successor function is given with the rule interpreter. The goal predicate is true, if the melody has the right length. Extra conditions can be specified. The fail predicate is true, if the melody exceeds the desired length. The start state is a score chart which contains all structural information given as parameters but no notes.

Different search strategies have been implemented:

- A standard backtracking strategy.
- A random backtracking strategy. which chooses by random the branch to be searched next.
- A repetetive-attempt random search. The next state is also selected by random. In case of a fail, no backtracking is done, but the strategy starts its next attempt with start state again.

With the current state of the ruleset, the repetetive-attempt random search, is the quickest.

Figure 3 shows 4 bar motives constructed with different subsets of the interval rules. Example 1 forbids too much and the melody consist mostly of repetitions. With its use of transitions Example 2 is a smooth scale based melody, while occasional thirds can make melodies more interesting as in example 3. Example 4 has a bad distribution of weight, because the weight rules are not used. The chromatic transition in example 5 doesn't fit into classical style but rather sounds like 19th century danse music. The chord-based melodies with repetitions (No. 7) and without (No 6) are quite good, the effect of the chromatic transition (No. 8) is the same as with the scale based melody (No. 5).

In general, the 4 bar motives generated by note by note composition lack the inner relationships typical for classical style. The purpose of the 4 bar examples is to allow a good judgement of the properties of the ruleset. For use in a minuet, composition by variation should start atleast at the level of 2 bar motives.

4.2 Variation of melodic motives

Classical style requires relations and similarities between structural parts. A frequently observed technique is the variation of melodic motives. Variation means creating something which is similar, but not equal. That means keeping the most, but modifying some of the properties of the orginal. [10]

Some operators of variation like *adding* or *removing a suspension* or *filling a skip with a scale* can easily be expressed by representing a melody as a sequence of notes. Other techniques of variation can more easily be expressed, when the motive is represented in a different formalism. E.g. it is easier to describe inversion as *inverting all interval directions while keeping their size* than to express it in terms of pitches.

A general abstract model of variation consists of the following steps:

1. Select formalism (set of attributes) to describe the motive to be varied.
2. Compute attribute values for the description.
3. Modify some of the attribute values in the description.
   Keep others.
4. Retransform the description into notation.

From a theoretical point of view, if two arbitrary motives are given, it is possible to construct a description formalism, which allows to see one motive as a variation of the other. In spite of the existence of such a description formalism, those two motives might be perceived as totally unrelated.

It is therefore necessary to select the attribute set and the attribute modification in a way, that they are musically meaningful. It must be possible to perceive both, the similarity of motives and the difference. The selected description formalism should be musically justified e.g. by the fact, that its terms are also used in period literature.

---

[10] Theory about variation can be found in [Lob44]

168

Notation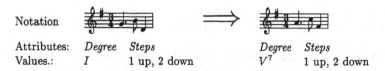

Attributes:	Degree	Steps		Degree	Steps
Values.:	I	1 up, 2 down		V^7	1 up, 2 down

Fig. 4. Example: chord analogon operator

Composition of scale-based melodies focusses on intervals between notes, so a description as relative notes (intervals + duration) is appropriate for them. For chord based melodies, which have been composed by thinking in chordsteps, a description by chordsteps (+ durations) has to be considered. The *chord-analogon* operator which is shown in figure 4 uses chordsteps as description formalism. The motive is described as a pattern of chordsteps. The variation applies the same step pattern to a different chord. This operator can also be seen between bar 1 and bar 2 of the Mozart minuet in figure 1.

4.3 Structure Plans

The classical principles of simplicity, balance and symmetry of formal parts require a regular form (usually two 2-bar units form a 4-bar unit, two 4-bar units form an 8-bar unit and so on) for a piece. The representation of such a macrostructural plan seems to be easy at first: Compose one motive of two bars, use a variation operator of your choice, or compose from scratch again, put the parts together and you have four bars which are related to each other. Go on this way by either variating or composing from scratch and you have a large regularly built piece of music. But a piece composed this way will not fit into classical style.

Classical style requires additional relations between the formal parts, for example harmonic relations and different strengths in the cadence of a part. Fortunately, modern and period literature discusses those aspects thoroughly as *tonal order* [11] of a piece.

Concatenating melodic parts in the score chart is easy to realize. The difference between a meaningful macrostructure for a minuet and an arbitrary sequence of motives are the extra constraints of the tonal order. The note-by-note composition subtask and the variation operators have to consider them.

There is a prototypical form of minuet, which consists of four 4 bar phrases [Bud83].

- Bars 1-4, end with a weak cadence on the tonic[12].
- Bars 5-8 modulate to the key of the dominant and close the first half of the piece with a strong cadence on the dominant[13].
- Bars 9-12 often have the character of a middle part. So they can have a different character and basic parameters and more distant harmony. The phrase usually close on a weak cadence on the dominant[14].

---

[11] Original: Tonordnung
[12] Original: Grundabsatz
[13] Original: Quintkadenz
[14] Original: Quintabsatz

– Bars 13-16 close with a strong cadence in the tonic[15]. They are often a reprise of the Bars 1-4 (but with a strong cadence) or a reprise of bars 5-8 (but on the tonic).

Bars 1-8 and 9-16 are often enclosed in repeat signs. Minuets can have a trio, which is a second minuet after which the main minuet is repeated (usually without its inner repetitions).

The macrostructure of most minuets follows this scheme with slight variations. E.g. the minuet in figure 1 has a strong cadence instead of a weak one in bar 4. Other minuets might have a weak cadence on bar 8 instead of a strong one. Both variations of the scheme can not be used in one minuet, because the important closing of first half (bar 8) of the minuet should be stronger (that means at least equal) than the intermediate closing at bar 4. For human composers, musical knowledge is organized by such musical background reasons and it therefore is easy to remember.

4.4 Evaluation of Music

Many AI-Systems have taken advantage of the fact, that it is often easier to criticize something, than to improve it (or to do it better from the beginning). Strategies like "Generate-and-test" or genetic algorithms make use of this fact.

Stepwise front-to-end construction algorithms can not foresee the effect of a decision on the rest of the piece. When evaluating a complete composition, different properties can be detected and weighted against each other. This allows the use of elements, which are not forbidden but must not be used too often and whose usage has to correspond to structure, e.g. appogiaturas.

5 Summary

Composition of classical minuets takes a middle position between restricted and free composition problems. The composition process is split into subtasks using different strategies and different kinds of musical knowledge. Different structural levels must be considered, while decisions can be either made bottom-up or top-down.

Knowledge is one of the most important components of AI. Systems, which integrate knowledge from different sources and in different formats, require methods to administrate the knowledge in order to detect inconsistencies. One strategy will be the documentation of the knowledge source, which makes manual corrections easier. Another help can be meta knowledge – the background reasons and principles humans use to organize their knowledge – which allows the detection of errors and conflicts in the computational knowledge of the strategies.

---

[15] Original: Grundkadenz

170

References

[Ber95] Ulf Berggren. *Ars combinatoria - Algorithmic Construction of Sonata Movements by Means of Building Blocks Derived from W.A.Mozart's Piano Sonatas.* Dissertation, Uppsala University, 1995. 174 S.

[Bud83] Wolfgang Budday. *Grundlagen musikalischer Formen der Wiener Klassik: an Hand der zeitgenössischen Theorie von Joseph Riepel u. Heinrich Christoph Koch dargest. an Menuetten u. Sonatensätzen (1750 - 1790).* Bärenreiter Verlag, Kassel, 1983.

[Cop91] David Cope. *Computers and Musical Style.* Computer Music and Digital Audio Series. Oxford University Press, 1991.

[Ebc86] Kemal Ebcioglu. An expert system for harmonization of chorales in the style of J.S.Bach. Technical Report 86-09, Department of Computer Science, University of Buffalo - State University of New York, 1986.

[Koc82] Heinrich Christoph Koch. *Versuch einer Anleitung zur Composition, 3 Bände entstanden 1782-1793.* Reprint 1969 Georg Olms Verlag Hildesheim, 1782.

[Lob44] Johann Christian Lobe. *Compositions-Lehre : oder umfassende Theorie von der thematischen Arbeit und den modernen Instrumentalformen.* Reprint 1988 Olms, Hildesheim, 1844.

[Löt99] Mathis Löthe. Computerized composition of minuets in early classical style. In Geraint Wiggins, editor, *Proceedings of the AISB'99 Symposium on Musical Creativity*, pages 124–129. Edinburgh College of Art & Division of Informatics, University of Edinburgh, 1999.

[Rie52] Joseph Riepel. *Sämtliche Schriften zur Musiktheorie (entstanden 1752-1782), Ed. Thomas Emmerich.* Reprint 1996 Böhlau Verlag, Wien, Köln, Weimar, 1752.

[RN95] Stuart J. Russell and Peter Norvig. *Artificial intelligence : a modern approach.* Prentice Hall series in artificial intelligence. Prentice Hall, Englewood Cliffs, NJ, 1995. XXVIII, 932 S.

[Sch89] William Schottstaedt. Automatic counterpoint. In Max V. Mathews and John R. Pierce, editors, *Current Directions in Computer Music Research*, System Development Foundation Benchmark. MIT Press, 1989.

[Tau94] Heinrich Taube. Common music. Technical report, Zentrum für Kunst und Medientechnologie, Karlsruhe, Germany, 1994. available at ccrma-ftp.stanford.edu:/pub/Lisp/cm.tar.Z.

[WPA99] Geraint Wiggins and Somnuk Phon-Amnuasiuk. The four-part harmonization problem: A comparison between genetic algorithms and a rule-based system. In Geraint Wiggins, editor, *Proceedings of the AISB'99 Symposium on Musical Creativity*, pages 28–34. Edinburgh College of Art & Division of Informatics, University of Edinburgh, 1999.

Inferring Flow of Control in Program Synthesis by Example

Stefan Schrödl and Stefan Edelkamp

Institut für Informatik
Albert-Ludwigs-Universität, Am Flughafen 17, D-79110 Freiburg, Germany,
e-mail:{schroedl,edelkamp}@informatik.uni-freiburg.de

Abstract. We present a supervised, interactive learning technique that infers control structures of computer programs from user-demonstrated traces. A two-stage process is applied: first, a minimal deterministic finite automaton (DFA) M labeled by the instructions of the program is learned from a set of example traces and membership queries to the user. It accepts all prefixes of traces of the target program. The number of queries is bounded by $O(k \cdot |M|)$, with k being the total number of instructions in the initial example traces. In the second step we parse this automaton into a high-level programming language in $O(|M|^2)$ steps, replacing jumps by conditional control structures.

1 Introduction

1.1 Program Synthesis from Examples

The ultimate goal of program synthesis from examples is to teach the computer to infer general programs by specifying a set of desired input/output data pairs. Unfortunately, the class of total recursive functions is not identifiable in the limit [8]. For tractable and efficient learning algorithms either the class has to be restricted or more information has to be provided by a cooperative teacher.

Two orthogonal strains of research can be identified [6]. Until the late 1970s, the focus was on inferring functional (e.g., Lisp) programs based on traces. Since the early 1980s the attention shifted towards model-based and logic approaches.

All functional program synthesis mechanism are based on two phases: *trace generation* from input/output examples, and *trace generalization* into a recursive program. Biermann's *function merging mechanism* [4] takes a one-parameter Lisp function whose only predicate is *atom*, and decomposes the output in an algorithmic way into a set of nested basic functions. Subsequently, they are merged into a minimal set that preserves the original computations by introducing discriminant predicates. These mechanisms perform well on predicates that involve structural manipulation of their parameters, such as list concatenation or reversal. However, their drawbacks are two-fold. The functional mapping between input and output terms cannot be determined in this straightforward way for less restrictive applications; on the other hand, manually feeding the inference algorithm with example traces can be a tedious and error-prone task. Secondly, the merging algorithms require exponential time in general.

The second direction of research (frequently called *Inductive Logic Programming*) is at the intersection between empirical learning and logic programming. A pioneering work was Shapiro's *Model Inference System* [13] as a mechanism for synthesizing Prolog programs from positive and negative facts. The system explores the search space of clauses using a configurable strategy. The subsumption relation assists in specializing incorrect clauses implying wrong examples, and in adding new clauses for uncovered ones. The critical issues are the undecidability of subsumption in the general case, the large number of required examples, and the huge size of the search space.

1.2 Programming in the Graphical User Interface

The last decades have seen a revolutionary change in human-computer interfaces. Instead of merely typing cryptic commands into a console, the user is given the illusion of moving around objects on a "desktop" he already knows from his everyday-life experience. Users can refer to an action by *simply performing* the action, something they already know how to do. Therefore, they can more easily handle *end user programming* tools.

Many spreadsheet programs and telecommunication programs have built-in *macro recorders*. Similarly to a tape recorder, the user presses the "record" button, performs a series of keystrokes or mouse clicks, presses "stop", and then invokes "play" to replay the entire sequence. Frequently, the macro itself is internally represented in a higher programming language (such as Excel macros in Visual Basic).

Moreover, the current trends in software development tools show that even programming can profit from graphical support. "Visual computing" aims at relieving conventional programming from the need of mapping a visual representation of objects being moved about the screen into a completely different textual representation of those actions. In an ideal general-purpose programming scenario, we could think of a domain-independent graphical representation for standard data structures, such as arrays, lists, trees, etc. which can be visually manipulated by the user.

Cypher gives an overview of current approaches [5]. Lieberman's *Tinker* system permits a beginning programmer to write Lisp programs by providing concrete examples of input data, and typing Lisp expressions or providing mouse input that directs the system how to handle each example. The user may present multiple examples that serve to incrementally illustrate different cases in conditional procedures. The system subsequently prompts the user for a distinguishing test. However, no learning of program structures takes place.

Based on these observations, we argue that program synthesis from traces could regain some attraction. The burden of trace generation can be greatly alleviated by a graphical user interface and thus becomes feasible.

In this paper, we propose an efficient interactive learning algorithm which solves the complexity problem of the merging algorithm in functional program synthesis. Contrary to the latter approach, we focus on imperative programming languages. They also reflect more closely the iterative nature of interaction with graphical user interfaces. The flow of control in imperative languages is constituted by conditional branches and loops; their lack in most current macro recorders is an apparent limitation.

2 Editing a First Example Trace

Figure 1 shows our prototypical graphical support. The user generates a first example trace by performing a sequence of mouse selections, mouse drags, menu selections, and key strokes.

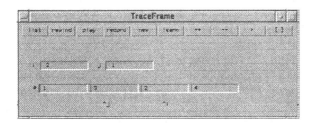

Fig. 1. Trace Frame.

Throughout the paper, we will exemplify the inference mechanism with the well-known *bubble-sort* algorithm. The user might start with the sample array a = [2, 1] of length n = 2. A variable i is introduced to hold the number of remaining iterations, and is initialized to one (int i=n-1). Then he states that the end is not yet reached (i>0). Subsequently he initializes another variable j to zero, meant as an index for traversing the array (int j=0). Now the array element with index 0 is compared to its successor (a[j]>a[j+1]). Since the comparison 1>2 fails (F) he swaps the elements (swap(j,j+1)). For ease of exposition, we assume that the swap-procedure has already been programmed to interchange two values in the array. The user increases j (j++) and then observes that the array has been traversed up to position i (j<i; F) in which case i is decremented (i--). The next iteration starts. But since i now has reached the left border (i>0; F) the sorting is accomplished and the procedure stops (return). In summary, the example generated by the end user is given as follows: i=0;i>0;T;j=0;a[j]<a[j+1];F; swap(j,j+1);j++;j<i;F;i--;i>0;F;return.

3 The *ID*-Algorithm

Grammar inference is defined as the process of learning an unknown grammar given a finite set of labeled examples. An important, widely used subset of formal languages are *regular grammars*, which can be generated and recognized by deterministic finite automata (DFA). However, given a finite set of positive examples and a finite, possibly empty set of negative examples, the problem of learning a minimum state DFA equivalent to the target is *NP*-hard [9]. Hence, the learner's task has to be simplified by imposing certain desired criteria on the examples (like structural completeness, characteristic samples), or by providing the learner with access to sources of additional information, like a knowledgeable teacher (oracle) who responds to queries generated by the learner.

Our algorithm is based on Angluin's *ID*-algorithm which is briefly recalled in this section. It may be skipped in a first reading.

Let Σ be the set of symbols, Σ^* be the set of strings, and λ be the empty string. Furthermore, let $M = (Q, \delta, \Sigma, q_0, F)$ be a DFA according to the usual quintuple definition and $L(M)$ be the language accepted by M. A state q in M is *alive* if it can be reached by some string α and left with some string β such that $\alpha\beta \in L(M)$. In a minimal DFA there is only one state d_0 that is not alive. A set of strings P is said to be *live-complete* w.r.t. M if for every live state q in M there exists a string $\alpha \in P$ such that $\delta(q_0, \alpha) = q$. Therefore, $P' = P \cup \{d_0\}$ represents all states in M. In order to find a string representation of the state reached on reading an input b from the state represented by α we define a function $f : P' \times \Sigma \to \Sigma^* \cup \{d_0\}$ by $f(d_0, b) = d_0$ and $f(\alpha, b) = \alpha b$. The *transition set* T' denotes the set of all elements of P', together with all elements $f(\alpha, b)$ for all $(\alpha, b) \in P \times \Sigma$. Analogously to P we define $T = T' - \{d_0\}$.

Input: a live complete set P and a teacher to answer membership queries
Output: a description of the canonical DFA M for the target regular grammar

$i = 0; v_i = \lambda; V = \{\lambda\}, T = P \cup \{f(\alpha, b) | (\alpha, b) \in P \times \Sigma\}; T' = T \cup \{d_0\}, E_0(d_0) = \emptyset$;
for each $\alpha \in T$
    if $(\alpha \in L)$ $E_0(\alpha) = \{\lambda\}$ **else** $E_0(\alpha) = \emptyset$;
while $(\exists \alpha, \beta \in P'$ and $b \in \Sigma$ such that $E_i(\alpha) == E_i(\beta)$ but $E_i(f(\alpha, b)) \neq E_i(f(\beta, b)))$
    let $\gamma \in E_i(f(\alpha, b)) \oplus E_i(f(\beta, b))$
    let $v_{i+1} = b\gamma$
    let $V = V \cup \{v_{i+1}\}$ and $i = i + 1$
    for each $\alpha \in T$
        if $(\alpha v_i \in L)$ $E_i(\alpha) = E_{i-1}(\alpha) \cup \{v_i\}$; **else** $E_i(\alpha) = E_{i-1}(\alpha)$;
Extract the automaton M for L from the sets E_i and T (see text)

Fig. 2. Angluin's *ID*-algorithm.

The goal of the *ID* algorithm (Figure 2) is to construct a partition of T' that places all the equivalent elements in one state [2]. The equivalence relation is the Nerode relation such that the resulting DFA will be minimal [1]. The algorithm starts with an initial partition of one accepting and one non-accepting state and refines it successively. In each step i of *ID* a string v_i is drawn such that for any two states q and q' there exists a $j \leq i$ with $\delta(q, v_j) \in F$ and $\delta(q', v_j) \notin F$ or vice versa. Thus, we define the i-th partition E_i as follows: $E_i(d_0) = \emptyset$ and $E_i(\alpha) = \{v_j | j \leq i, \alpha v_j \in L(M)\}$. Then for every two strings $\alpha, \beta \in T$ with $\delta(q_0, \alpha) = \delta(q_0, \beta)$ we have $E_j(\alpha) = E_j(\beta)$ for all $j \leq i$. For each i the algorithm searches for a separating pair α, β and a symbol b such that $E_i(\alpha) = E_i(\beta)$ but $E_i(f(\alpha, b)) \neq E_i(f(\beta, b))$. Let γ be any string that is either in $E_i(f(\alpha, b))$ and not in $E_i(f(\beta, b))$ or vice versa. Then we define $v_{i+1} = b\gamma$ and construct the $(i+1)$-th partition as follows. For each $\alpha \in T$ we query the string αv_{i+1}. If $\alpha v_{i+1} \in L(M)$ we set $E_{i+1} = E_i \cup \{v_{i+1}\}$; otherwise, we let $E_{i+1} = E_i$ unchanged.

We iterate until no separating pair α, β exists and extract M from the sets E_i and the transition set T as follows. The states of M are the sets $E_i(\alpha)$, for $\alpha \in T$. The initial state of M is $E_i(\lambda)$. The accepting states of M are the sets $E_i(\alpha)$, where $\alpha \in T$ and $\lambda \in E_i(\alpha)$. If $E_i(\alpha) = \emptyset$ then we add self loops on the state $E_i(\alpha)$ for all $b \in \Sigma$; else we set the transition $\delta(E_i(\alpha), b) = E_i(f(\alpha, b))$ for all $\alpha \in P$ and $b \in \Sigma$.

Angluin proved that ID asks no more than $n \cdot |\Sigma| \cdot |P|$ queries, where n is the number of states in M: the algorithm iterates through the $while$-loop at most n times, since each time at least one set E_i (corresponding to a state) is partitioned into two subsets. It asks $|T|$ questions, where T contains no more than $|\Sigma| \cdot |P|$ elements.

4 Customizing *ID* for Program Traces

4.1 Naive Approach

A simple strategy to apply the *ID*-algorithm to the problem of program inference from traces goes as follows. The alphabet Σ consists of all program lines occurring in the examples. More precisely, we partition Σ into $\Gamma \cup \Lambda \cup \Delta \cup \{\text{return}\}$, where Γ is the set of non-branching instructions (e.g. assignments), Λ is the set of (boolean) tests (e.g. numerical comparisons), $\Delta = \{\text{T}, \text{F}\}$ is the set of boolean values, and **return** signals the end of the procedure. The language L to be learned is regular and consists of all prefixes of valid execution traces. Programs are represented as finite state machines, where transitions are labeled with the respective instructions. Let $Pr(\alpha)$ be the set of all prefixes to α. The live-complete set P for the *ID*-algorithm can now be fixed as $P = Pr(S) \cup \{\lambda\}$, with S being the example trace.

For the initial examples in P, the user is free to choose any data, such as the array $[2, 1]$ in our case. As a heuristic guideline, the first examples are supposed both not to be overly lengthy (in order to reduce the number of subsequent questions), but at the same time cover all states of the automaton (in order to specify P). However, this requirement is not compulsory: in the version *IID* of the algorithm [11], the initial set of examples need not to be *live-complete*; the user is allowed to incrementally refine the automaton structure by presenting additional (positive or negative) examples later on.

Using this scheme, the number of queries (2158) asked for our bubble-sort case is clearly inacceptably high. Fortunately, the majority of them can immediately be answered by the system itself.

4.2 Pruning

We make the following general assumptions to hold for all execution traces α in Σ^*.

1. If $\alpha a \in L$ for some $a \in \Sigma$ then also $\alpha \in L$. In words: every prefix of a word in L is itself in L.
2. If $\alpha ab \in L$ and $\alpha ac \in L$ where $a \in \Gamma \cup \Delta$ and $b, c \in \Sigma$, then $b = c$. In words: There is only one instruction that follows a non-branching instruction or a boolean.
3. If $\alpha ab \in L$ and $a \in \Lambda$ then $b \in \Delta$. In words: A test is only followed by a boolean denoting its outcome.
4. We have $\alpha ab \notin L$ for $a = $ return and all $b \in \Sigma$. In words: No instruction may follow the end statement.

If condition 3. or 4. is violated, the trace is malformed and is hence rejected.

According to condition 1., we can efficiently store both the example traces and the query traces confirmed by the user in a *trie* data structure [10]. The bold path in Figure 3 corresponds to the first example trace of Section 2. Given a query string $\alpha b \gamma$, we tentatively insert it into the trie. If it is already contained, the answer is "yes". If the new trie forks at a non-branching instruction, condition 2. is violated and thus the answer is "no". Otherwise, the user is prompted. Unless his response is positive, the query string is removed.

For example at branch (3) in Figure 3 the system asks: int i=n-1;i>0;T; int j=0; a[j]>a[j+1]; T; swap(j,j+1); j++; j<i; T; int i=n-1; $\in L$? The user will answer "no".

In the further course of the session, the system will eventually "guess" all possible instructions as $b\gamma$ until the correct one a[j]>a[j+1] is found. As a further simplification, we can allow the user to edit the question and to immediately type in the right continuation.

4.3 Selection of Example Data

Ideally, the system should present its queries by animating a sequence of instructions for a suitable instantiation of the variables. Given only the raw code fragments, it might be difficult for the user to find the correct continuation.

This raises the question of how to select data which is consistent with a given trace, i.e., how to find an assignment to the variables that makes one choice point true and another one which makes it false. Two options are conceivable: the user could be asked to give a pool of examples independently of (prior or alternating to) the learning process, from which the system can choose some appropriate one. Alternatively, he can provide a specification to generate random data. E.g.,

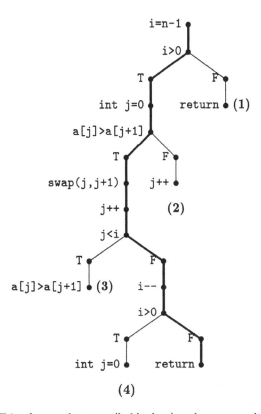

Fig. 3. Trie of example traces (bold edges) and query results (thin).

the bubble-sort algorithm should sort every permutation of the array elements, which we w.l.o.g. fix to be $[1, 2, \ldots, n]$. For instance, the array $a = [3, 1, 2]$ of length $n = 3$ leads to the following instantiation for question (3): int i=2; 2=i>0; T; int j=0; 2=a[0]>a[1]=1; T; swap(0,1); j++; 1=j<i=2; T; i=2; $\in L$? The user responds by replacing i=2 by the next step which compares $3 = a[1] > a[2] = 2$, i.e., the test a[j]>a[j+1].

Figure 4 depicts the finite state machine for the bubble-sort program inferred by the *ID*-algorithm. All states are accepting, and all omitted transitions lead to the dead state d_0.

4.4 Query Complexity

Every affirmatively answered membership question and every edited answer string inserts at least one node into the trie. Incrementally extending the trie in this way contributes to reduce the number of user questions. The total number is bounded by the size of the final trie minus that of the the initial one. In our bubble-sort example, this bound corresponds to the number of thin edges in Figure 3. Actually, the user is asked four instead of 2158 times.

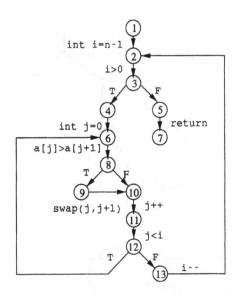

Fig. 4. DFA for bubble-sort.

Due to the restrictions on well-formed traces, we can specify a tighter upper bound on the number of user questions, compared to that of Angluin. If $\alpha \in T$ violates one of conditions 2. - 4., then $\alpha v_i \notin L$ for all distinguishing strings v_i. We count the number of the remaining valid elements $\tilde{T}$.

We assume that all given examples in P are complete traces, i.e. end with a **return** statement. Therefore, extensions $f(\alpha, b) \in T$ to $\alpha \in P$ are only available at proper prefixes of elements in the example set. However, if α ends with a non-branching instruction, restriction 2. constrains $f(\alpha, b)$ to be in the set P. In case α ends with a test instruction, condition 3. leaves us with two choices T and F for b. With k denoting the total number of tests in the example set, we have that $|\tilde{T}|$ is bounded by $k + |P|$. Finally, we conclude that the total number of membership queries is bounded by $n \cdot (k + |P|) = O(n \cdot |P|)$.

5 Transforming Automata into Structured Programs

It is straightforward to write down any generated automaton as a program using some form of jumps (e.g., *goto*-statements).

For more complex algorithms such flow charts quickly become confusing. In most current high-level programming languages, jump statements are either strongly discouraged (e.g., in C), or do not exist at all (e.g., in Pascal). Instead, high-level constructs are available for conditional branching and looping.

Therefore, we do not regard the automaton generated by the *ID*-algorithm as the final output, but rather apply a transformation in order to replace jumps by control structures.

Our algorithm transforms the automaton graph step by step by repeatedly collapsing a subgraph into a new edge, for which we keep track of extra information: its type (e.g., simple, test, sequence, *while*-loop, etc.), possibly its subcomponents, and the set of its successors.

Two adjacent edges labeled with arbitrary instructions other than tests or booleans can be merged into a *sequence* if they are the only inward or outward edges of the enclosed node. Connected tests can be merged into (compound) *conditions* containing boolean operators depending on the role of their T- and F-edges in the obvious way. For example, if test t_1 is connected to test t_2 via its T-edge, and the F-edges of both t_1 and t_2 point to the same node v, then a compound condition $t_1 \wedge t_2$ is formed whose T-edge leads to the same node pointed to by the T-edge of t_2, and whose F-successor-node is v.

The more interesting cases are the instances where control structures are inferred: a (simple or compound) condition whose T-edge leads to a non-test edge with successor node v, and whose F-successor-node is also v, can be merged into an *if-then*-statement pointing to v. Similarly, if the T- and the F-edge lead to different edges with the same successor node, then the resulting conditional statement additionally contains an else-part. A *while*-loop is a condition-edge c whose T-successor leads to an edge (i.e., the repeated block) which has, in turn, c as its successor. The resulting edge points to the destination of the F-successor-edge of c. If the two edges are interchanged, the condition in the generated *while*-statement is negated. In *do-while*-loops, the condition follows the edge for the repeated block.

First, the algorithm initializes the in-degrees of all nodes (in linear time). Then all n nodes are repeatedly checked for applicable transformation rules. If none is found, we are done; otherwise the automaton is altered accordingly, and the degree of affected nodes is adjusted. Both these operations require constant time. Since each transformations removes at least one node, at most n iterations are performed, giving an overall worst-case complexity of $O(n^2)$.

Note that, in principle, it is not always possible to transform jumps into control structures without reasoning about the semantics of a program or changing the set of variables (Fig. 5 sketches a critical loop structure). In these cases, the system should at least try to minimize the number of remaining *goto*s. Such graceful degradation is not covered by our algorithm and left as a topic for further research.

```
        int i=10, j=20;
11 :    i--;
12 :    if (j==0) return;
        j--;
        if (i > 0 && j>0) goto 11;
        goto 12;
```

Fig. 5. Without semantic information unfolding is impossible.

For our example, Figure 6 show the sequence of transformations applied to the original automaton of Figure 4. First, the edges $(6,8)$, $(8,9)$, $(8,10)$, and $(9,10)$ are collapsed into an *if-then*-statement (a). In the next step, the edge labeled j++ is appended to form a *sequence* edge (b). Now we create a *do-while*-loop, since the test edge $(11,12)$ appears after the repeated block (c). The two next steps summarizes it, together with the edges with respective labels int j=0 and i--, into a sequence (d). We create the outer *while*-loop (e), and then concatenate int i=n-1 and return to it, such that only one edge is left corresponding to the final program (f).

6   Conclusion and Discussion

We have presented a supervised, interactive learning algorithm which infers control structures of computer programs from example traces and queries.

First, a deterministic finite automaton is learned by a customized version of the *ID*-algorithm for regular language inference. By exploiting the syntactical form of programs and allowing the user to incrementally type in instructions, the number of questions is reduced from an infeasible to a moderate scale. An upper bound of $O(n \cdot |P|)$ membership queries is given. Secondly, the resulting automaton is rewritten in a high-level language with control structures using an $O(n^2)$ algorithm.

An early precursor of this work similar in spirit is presented by Gaines [7]. His approach infers a DFA by exhaustive and exponential search until an automaton is found that is consistent with the given traces.

Schlimmer and Hermens describe a note-taking system that reduces the user's typing effort by interactivly predicting continuations in a button-box interface [12]. An unsupervised, incremental machine learning component identifies the syntax of the input information. To avoid intractability the class of target languages is constrained to so-called k-reversible regular languages for which Angluin proposed an $O(n^3)$ inference algorithm [3]. However, for general proposed languages this class is too restrictive. It is not hard to find simple programs not covered by zero-reversible FSM's (as in the examples given in the paper). On the other hand, simply fixing k at a larger value sacrifices minimality of the generated automaton. Schlimmer and Hermens improve the system's accuracy by adding a decision tree to each state. However, prediction is not relevant to our approach since traces are deterministic: A new training example leads to a new FSM.

End users without programming knowledge can take benefit from inference of control structures. More powerful customization tools (e.g., macro recorders) are able to support them in solving more of the repetitive routine work which often needs elementary conditional branching and looping.

For the experienced programmer, the proposed inference mechanism might support the process of software development, mainly in view of integrity and incremental extensibility.

The final set of execution traces (as depicted by the resulting trie) uniquely determines the structure of the automaton. All source fragments in the generated

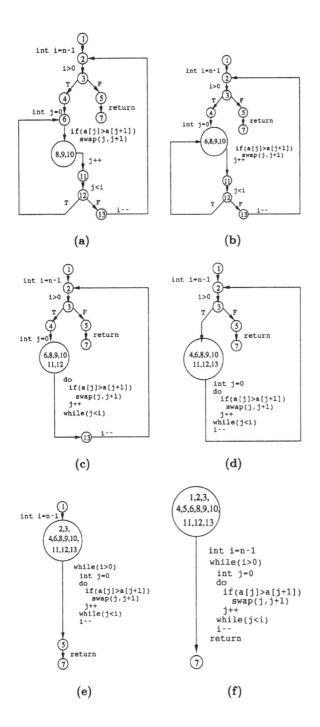

Fig. 6. Transformation of the DFA into a structured program.

program have been exercised in at least one example. Therefore, no untested code can arise. For recorded execution traces on concrete sample data, differences between the intended and the actual meaning of the program will occur by far more infrequently than bugs in programs developed without the control of explicit variable instantiations. In a way, both stages in the software development cycle, coding and testing, are performed more efficiently in parallel rather than in the usual alternating way.

A sequence of examples should start with simple examples and build to more complex and exceptional cases. Recursive and conditional procedures can be developed incrementally by starting with simple, "incorrect" definitions, and later adding more instances to handle more complicated and special purpose situations. Maintaining all used examples and only adding to this set ensures that previous examples are still covered and, that with growing complexity, no new bugs are introduced for cases which have already been successfully treated.

References

1. A. V. Aho, J. E. Hopcroft, and J. Ullman. *The design and analysis of computer algorithms*. Addison–Wesley, 1974.

2. D. Angluin. A note on the number of queries needed to identify regular languages. *Information and Control*, 51(1):76–87, 1981.

3. D. Angluin. Inferrence of reversible languages. *Journal of the Association of Computing Machinery*, 29:741–765, 1982.

4. A. W. Biermann. The inference of regular lisp programs from examples. *IEEE Trans. on Systems, Man, and Cybernetics*, 8(8):585–600, 1978.

5. A. Cypher, editor. *Watch What I Do: Programming by Demonstration*. MIT Press, 1993.

6. P. Flener. *Logic Program Synthesis from Incomplete Information*. Kluwer Academic Publishers, 1995.

7. B. Gaines. Behaviour/structure transformations under uncertainty. *International Journal of Man-Machine Studies*, 8(3):337–365, 1976.

8. E. M. Gold. Language identification in the limit. *Information and Control*, 10(5):447–474, 1967.

9. E. M. Gold. Complexity of automaton identification from given data. *Information and Control*, 37(3):302–320, 1978.

10. D. E. Knuth. *The Art of Computer Programming, Volume 3: Sorting and Searching*. Addison-Wesley Publishing Company, Reading, 1973.

11. R. Parekh, C. Nichitiu, and V. Honovar. A polynomial time incremental algorithm for regular grammar inference. Technical Report 97-03, Department of computer science, Iowa State University, 1997.

12. J. C. Schlimmer and L. A. Hermens. Software agents: Completing patterns and constructing user interfaces. *Journal of Artificial Intelligence Research*, 1:61–89, 1993.

13. E. Y. Shapiro. *Algorithmic Program Debuggging*. PhD thesis, Yale University, 1983. Published under the same title by MIT press.

Compilation Schemes: A Theoretical Tool for Assessing the Expressive Power of Planning Formalisms

Bernhard Nebel

Institut für Informatik, Albert-Ludwigs-Universität
Freiburg, Germany

Abstract. The recent approaches of extending the GRAPHPLAN algorithm to handle more expressive planning formalisms raise the question of what the formal meaning of "expressive power" is. We formalize the intuition that expressive power is a measure of how concisely planning domains and plans can be expressed in a particular formalism by introducing the notion of "compilation schemes" between planning formalisms. Using this notion, we analyze the expressive power of a large family of propositional planning formalisms and show, e.g., that Gazen and Knoblock's approach to compiling conditional effects away is optimal.

1 Introduction

The recent approaches of extending the GRAPHPLAN algorithm [3] to handle more expressive planning formalisms [1, 9, 10, 12] raise the question of what the formal meaning of *expressive power* is. In order to address the problem of measuring the relative expressive power of planning formalisms, we start with the intuition that a formalism $\mathcal{X}$ is *at least as expressive* as another formalism $\mathcal{Y}$ if *planning domains* and the corresponding *plans* in formalism $\mathcal{Y}$ can be *concisely expressed* in the formalism $\mathcal{X}$.

Bäckström [2] proposed to measure the expressiveness of planning formalisms using *ESP-reductions*, which are, roughly speaking, polynomial many-one reductions[1] on planning instances that *do not change the plan length*. Using this notion, he showed that all of the propositional variants of basic STRIPS not containing conditional effects or arbitrary logical formulae can be considered as expressively equivalent. However, taking our point of view, on one hand ESP-reductions are too restrictive. Firstly, plans must have identical size, while we might want to allow a moderate growth. Secondly, requiring that the transformation can be computed in polynomial time is overly restrictive. In fact, one

---

[1] We assume that the reader has a basic knowledge of complexity theory [8, 16], and is familiar with the notion of *polynomial many-one reductions* and the *complexity classes* P, NP, coNP, and PSPACE. All other notions will be introduced in the paper when needed.

formalism might be as expressive as another one, but the mapping between the formalisms might not recursive at all. On the other hand, ESP-reductions are too liberal because they allow for arbitrary transformations between planning instances. We, however, want to consider only transformations between planning domains, which are independent from the concrete initial state and goal specification.

Inspired by recent approaches in the area of *knowledge compilation* [5], we formalize the notion of *concisely expressible* using *compilation schemes*, which are solution-preserving mappings from planning domains in one formalism to planning domains in another formalism restricting the result to be of polynomial size. Furthermore, we distinguish between compilation schemes in whether they *preserve plan size exactly*, *linearly*, or *polynomially*.

Using the notion of *compilability*, we analyze a wide range of propositional planning formalisms, ranging from basic STRIPS to a planning formalism containing *conditional effects, arbitrary logical formulae*, and *partial state specifications*. As one of the results, we identify two equivalence classes of planning formalisms with respect to *polynomial-time* compilability preserving plan size exactly. This means that adding a language feature to a formalism without leaving the class does not increase the expressive power and should not affect the principal efficiency of the planning method. However, we also provide results that *separate* planning formalisms. Such separation results indicate that adding a particular language feature adds to the expressive power and to the difficulty of finding a plan. For example, we prove that conditional effects cannot be compiled away and that logical formulae cannot be compiled into conditional effects—provided the plans in the target formalism are allowed to grow only linearly. This answers, e.g., the question of whether Gazen and Knoblock's [9] approach to compile conditional effects away could be improved. If only linear plan growth is allowed, the size of the compiled domain structure is necessarily super-polynomial.

The rest of the paper is structured as follows. In Section 2, we introduce the range of propositional planning formalisms analyzed in this paper together with general terminology and definitions. Based on that, we introduce in Section 3 the notion of compilability between planning formalisms. In Section 4 we present polynomial-time compilation schemes between different formalisms that preserve the plan size, demonstrating that these formalisms are of identical expressiveness. For all of the remaining cases, we prove in Section 5 that there cannot be any compilation scheme preserving plan size linearly, even if there are no bounds on the computational resources of the compilation process. Finally, in Section 6 we summarize and discuss the results.

2 Propositional Planning Formalisms

First, we will define a very general propositional planning formalism, which appears to be as expressive as the propositional variant of ADL [17]. We will then consider various syntacticly restricted variants of this formalism.

Let Σ be a finite set of propositional **atoms**. Then, $\widehat{\Sigma}$ is defined to be the set consisting of the constants $\top$ (denoting truth) and $\bot$ (denoting falsity) as well as the **literals**, i.e., atoms and negated atoms, over Σ. The **language of propositional logic** over Σ is denoted by $\mathcal{L}_\Sigma$. Given a set of literals L, we define $\neg L$ to be the **element-wise negation** of L, i.e., $\neg L = \{p \mid \neg p \in L\} \cup \{\neg p \mid p \in L\}$.

A **state** s is a *truth-assignment* for the atoms in Σ, which is represented by the set of atoms that are true in this state. A **state specification** S is a subset of $\widehat{\Sigma}$, i.e., it is a *logical theory* consisting of literals only. It is called **consistent** iff it does not contain complementary literals or $\bot$. In general, a state specification describes many states, namely all those that satisfy S. Only in case that S is **complete**, i.e., for each $p \in \Sigma$ we have either $p \in S$ or $\neg p \in S$, S describes precisely one state. By abusing notation, we will refer to the **inconsistent** state specification by $\bot$, which is the **"illegal"** state specification.

Operators are pairs $o = \langle pre, post \rangle$. We use the notation $pre(o)$ and $post(o)$ to refer to the first and second part of an operator o, respectively. The **precondition** pre is a set of propositional formulae. The set $post$, which is the set of **postconditions**, consists of **conditional effects**, each having the form $\Gamma \Rightarrow L$, where the elements of $\Gamma \subseteq \mathcal{L}_\Sigma$ are called **effect conditions** and the elements of $L \subseteq \widehat{\Sigma}$ are called **effects**. Given a state specification S and an operator o, we define the set of **active effects** $A(S, post(o))$ and the set of **potentially active effects** $P(S, post(o))$ as follows:

$$A(S, post(o)) = \bigcup \{L \mid (\Gamma \Rightarrow L) \in post(o), S \models \Gamma\},$$

$$P(S, post(o)) = \bigcup \{L \mid (\Gamma \Rightarrow L) \in post(o), S \cup \Gamma \not\models \bot\}.$$

If for a given state specification S and an operator o, we have $A(S, post(o)) \neq P(S, post(o))$, this implies that the activated effects differ for different states described by the state specification. For this reason, we consider the application of the operator o as illegal in this situation.[2] Similarly to the rule that $A(S, post(o)) \neq P(S, post(o))$ leads to an illegal state specification, we require that if the precondition is not entailed by S or if the state specification is already inconsistent, the result of applying o to S results in $\bot$. This leads to the definition of the operator-result function R for operator o on the state specification S:

$$R(S, o) = \begin{cases} S - \neg A(S, post(o)) \cup A(S, post(o)) & \text{if } S \not\models \bot \text{ and} \\ & \quad S \models pre(o) \text{ and} \\ & \quad A(S, post(o)) \not\models \bot \text{ and} \\ & \quad A(S, post(o)) = P(S, post(o)) \\ \\ \bot & \text{otherwise} \end{cases}$$

A **planning instance** is now a tuple $\Pi = \langle \Xi, \mathbf{I}, \mathbf{G} \rangle$, where

- $\Xi = \langle \Sigma, \mathbf{O} \rangle$ is the **domain structure** consisting of a finite set of propositional atoms Σ and a finite set of operators $\mathbf{O}$,

---

[2] Of course, alternative semantics are possible.

- $\mathbf{I} \subseteq \widehat{\Sigma}$ is the **initial state specification**, and
- $\mathbf{G} \subseteq \widehat{\Sigma}$ is the **goal specification**.

When we talk about the **size of an instance**, symbolically $\|\Pi\|$, in the following, we mean the size of a (reasonable) encoding of the instance.

Let Δ be an element of $\mathbf{O}^*$, i.e., a finite sequence of operators, which is called **plan**. Then $\|\Delta\|$ denotes the size of the plan, i.e., the number of operators in Δ. We say that Δ is a c-**step plan** if $\|\Delta\| \leq c$. The result of applying Δ to a state specification S is recursively defined as follows:

$$Res(S, \langle \rangle) = S,$$
$$Res(S, \langle o_1, o_2, \ldots, o_n \rangle) = Res(R(S, o_1), \langle o_2, \ldots, o_n \rangle).$$

A sequence of operators Δ is said to be a **plan for** Π or a **solution of** Π iff (1) $Res(\mathbf{I}, \Delta) \not\models \perp$ and (2) $Res(\mathbf{I}, \Delta) \models \mathbf{G}$. Note that it can be easily verified [14] that any plan Δ for an instance Π is a *sound plan* in Lifschitz' [13] sense.

The propositional variant of basic STRIPS [6], which we will also call S in what follows, is a planning formalism that requires *complete state specifications*, *unconditional effects*, and *propositional atoms* as formulae. Less restrictive planning formalisms can have the following additional features:

Incomplete state specifications ($\mathcal{I}$): The state specifications may not be complete.

Conditional Effects ($\mathcal{C}$): Effects can be conditional.

Literals as formulae ($\mathcal{L}$): The formulae in preconditions and effect conditions can be literals.

Boolean formulae ($\mathcal{B}$): The formulae in preconditions and effect conditions can be arbitrary boolean formulae.

These extensions can also be combined. We will use combinations of letters to refer to such multiple extensions. For instance, $S_\mathcal{L}$ refers to the formalism S extended by literals in the precondition lists, $S_{\mathcal{IC}}$ refers to the formalism allowing for incomplete state specifications and conditional effects, but all formulae have to be atoms, and $S_{\mathcal{BIC}}$, finally, refers to the general planning formalism introduced above. Figure 1 displays the partial order on propositional planning formalisms defined in this way using a Hasse diagram. In the sequel we say that $\mathcal{X}$ **is a specialization** of $\mathcal{Y}$, written $\mathcal{X} \sqsubseteq \mathcal{Y}$, iff $\mathcal{X}$ is identical to $\mathcal{Y}$ or below $\mathcal{Y}$ in the Hasse diagram.

Comparing this set of planning formalisms with the one Bäckström [2] analyzed, one notices that despite small differences in the presentation of the planning formalisms S, $S_\mathcal{L}$, and $S_{\mathcal{LI}}$ are identical to CPS, PSN, and GT, respectively.

While one would expect that planning in S is much easier than planning in $S_{\mathcal{BIC}}$, it turns out that—taking a computational complexity perspective—this is not the case. When analyzing the computational complexity of planning in different formalisms, we consider as usual the problem of deciding whether there *exists a plan* for a given instance—the **plan existence problem** (PLANEX). We will use a prefix referring to the planning formalism if we consider the existence problem in a particular planning formalism.

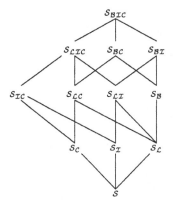

Fig. 1. Planning formalisms partially ordered by syntactic features

Theorem 1. $\mathcal{X}$-PLANEX *is* PSPACE-*complete for all* $\mathcal{X}$ *with* $\mathcal{S} \sqsubseteq \mathcal{X} \sqsubseteq \mathcal{S}_{\mathcal{BIC}}$.

Proof Sketch. PSPACE-hardness of $\mathcal{S}$-PLANEX follows from [4, Corollary 3.2]. Membership of $\mathcal{S}_{\mathcal{BIC}}$-PLANEX in PSPACE follows because each plan step can be verified by an NP-oracle, i.e., in PSPACE.[3] ∎

While this result may be interpreted as showing that we can easily solve the $\mathcal{S}_{\mathcal{BIC}}$ planning problem by transforming it to $\mathcal{S}$, it does not say anything about how easily we can transform *domain structures* and how long plans will be after the transformation.

3 Compilation Schemes

Basically, *compilation schemes* are solution-preserving mappings (which need not to be recursive) on domain structures.[4] Since in these mappings auxiliary atoms might be introduced, the compilation scheme may also influence the initial state and goal specifications. Furthermore, the initial state and goal specifications may also need some massaging, e.g., when compiling from partial to complete state specifications. The necessary *state-translation functions* should be very limited in computational power, however, so that they do not influence the principal expressive power.

A **compilation scheme from** $\mathcal{X}$ **to** $\mathcal{Y}$ is a tuple of functions $\mathbf{f} = \langle f_\xi, f_i, f_g, t_i, t_g \rangle$ that induces a function F from $\mathcal{X}$-instances $\Pi = \langle \Xi, \mathbf{I}, \mathbf{G} \rangle$ to $\mathcal{Y}$-instances $F(\Pi)$ as follows:

$$F(\Pi) = \langle f_\xi(\Xi), f_i(\Xi) \cup t_i(\Sigma, \mathbf{I}), f_g(\Xi) \cup t_g(\Sigma, \mathbf{G}) \rangle,$$

and satisfies the following conditions:

---

[3] Full proofs for all theorems can be found in the long version of the paper [14].

[4] This means that compilation schemes between planning formalisms are similar to knowledge compilation schemes [5], where the fixed part of a computational problem is the domain structure and the variable part consists of the initial state and goal specifications.

1. there exists a plan for Π iff there exists a plan for $F(\Pi)$;
2. the **state-translation functions** t_i and t_g are **modular**, i.e., for $\Sigma = \Sigma_1 \cup \Sigma_2$, $S \subseteq \widehat{\Sigma}$, and $S \not\models \bot$, the functions t_x (for $x = i, g$) satisfy

$$t_x(\Sigma, S) = t_x(\Sigma_1, S \cap \widehat{\Sigma_1}) \cup t_x(\Sigma_2, S \cap \widehat{\Sigma_2}),$$

and they are polynomial-time computable;
3. the size of the results of f_ξ, f_i, and f_g is polynomial in the size of the arguments.

Although there are no resource bounds on f_ξ, f_i, and f_g in the general case, we are also interested in *efficient compilation schemes*. We say that **f** is a **polynomial-time compilation scheme** if f_ξ, f_i, and f_g are polynomial-time computable functions.

In addition to that we measure the size of the corresponding plans in the target formalism. If a compilation scheme **f** has the property that for every plan Δ solving an instance Π there exists a plan Δ' solving $F(\Pi)$ such that $\|\Delta'\| \leq \|\Delta\| + k$ for some positive integer constant k, **f** is a **compilation scheme preserving plan size exactly** (modulo some additive constant). If $\|\Delta'\| \leq c \times \|\Delta\| + k$ for positive integer constants c and k, then **f** is a **compilation scheme preserving plan size linearly**, and if $\|\Delta'\| \leq p(\|\Delta\|, \|\Pi\|)$ for some polynomial p, then **f** is a **compilation scheme preserving plan size polynomially**. More generally, we say that a planning formalism $\mathcal{X}$ is **compilable** to formalism $\mathcal{Y}$ (in poly. time, preserving plan size exactly, linearly, or polynomially), if there exists a compilation scheme with the appropriate properties. We write $\mathcal{X} \preceq^x \mathcal{Y}$ in case $\mathcal{X}$ is compilable to $\mathcal{Y}$ or $\mathcal{X} \preceq_p^x \mathcal{Y}$ if the compilation can be done in polynomial time. The super-script x can be 1, c, or p depending on whether the scheme is exactly, linearly, or polynomially plan size preserving, respectively. As is easy to see, all the notions of compilability introduced above are reflexive and transitive.

Proposition 1. *The relations $\preceq^x$ and $\preceq_p^x$ are transitive and reflexive.*

Furthermore, it is obvious that when moving upwards in the diagram displayed in Figure 1, there is always a polynomial-time compilation scheme preserving plan size exactly. If π_i denotes the projection to the i-th argument and $\emptyset$ denotes the function that returns always the empty set, the generic compilation scheme for moving upwards in the partial order is $\mathbf{f} = \langle \pi_1, \emptyset, \emptyset, \pi_2, \pi_2 \rangle$.

Proposition 2. *If $\mathcal{X} \sqsubseteq \mathcal{Y}$, then $\mathcal{X} \preceq_p^1 \mathcal{Y}$.*

4 Compilability Preserving Plan Size Exactly

Proposition 2 leads to the question of whether there exist other compilation schemes than those implied by the specialization relation.

First, we will establish that $S_{\mathcal{LI}}$, $S_{\mathcal{L}}$, $S_{\mathcal{I}}$, and S are polynomial-time compilable into each other preserving plan size exactly. Having a closer look at

Bäckström's [2] equivalence proof for $S_{\mathcal{LI}}$ and S using an ESP-reduction, it turns out that the ESP-reduction he specified can be reformulated as a compilation scheme. From that the next theorem is immediate.

Theorem 2. $S_{\mathcal{LI}}$, $S_{\mathcal{I}}$, $S_{\mathcal{L}}$, and S are polynomial-time compilable to each other preserving plan size exactly.

One view on this result is that it does not matter from an expressivity point of view whether we allow for atoms only or permit also negative atoms and it does not matter whether we have complete or partial state specification—provided propositional formulae and conditional effects are not allowed. Interestingly, this view generalizes to the case where conditional effects are allowed.

In proving that, there are two additional complications. Firstly, one must compile conditional effects over partial state specifications to conditional effects over complete state specifications. This is a problem because the condition $A(S, post(o)) = P(S, post(o))$ in the definition of the function R must be tested. Secondly, when compiling a formalism with literals into a formalism that allows for atoms only, the condition $A(S, post(o)) \not\models \bot$ in the definition of R must be taken care of. Nevertheless, it is possible to solve these problems using a polynomial-time compilation preserving plan size exactly.

Theorem 3. $S_{\mathcal{LIC}}$, $S_{\mathcal{LC}}$, $S_{\mathcal{IC}}$, and $S_{\mathcal{C}}$ are polynomial-time compilable to each other preserving plan size exactly.

A summary of the results in this section is given in Figure 2. The two equivalence classes with respect to polynomial-time compilability preserving plan size exactly will be called $S_{\mathcal{LI}}$- and $S_{\mathcal{LIC}}$-class, in symbols $[S_{\mathcal{LI}}]$ and $[S_{\mathcal{LIC}}]$, naming them after their respective largest elements.

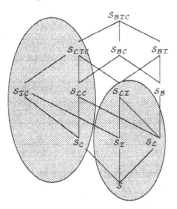

Fig. 2. Equivalence classes of planning formalisms created by polynomial-time compilation schemes preserving plan size exactly

5 The Limits of Compilation when Preserving Plan Size Linearly

In this section, we will show that there are no compilation schemes preserving plan size linearly other than those identified above and those implied by the specialization relation. First of all, we will prove that *conditional effects cannot be compiled away*. The deeper reason for this is that with conditional effects one can independently do a number of things in parallel, which is impossible in formalisms without conditional effects.

In order to illustrate this point, let us consider an example. We start with a set of n propositional atoms $\Sigma_n = \{p_1, \ldots, p_n\}$ and a disjoint copy of this set: $\Sigma_n^\# = \{p_i^\# \mid p_i \in \Sigma_n\}$. Furthermore, if $S \subseteq \widehat{\Sigma_n}$, then $S^\#$ shall denote the corresponding set of literals over $\widehat{\Sigma_n^\#}$. Consider now the following $\mathcal{S_{LIC}}$ domain structure:

$$\Sigma_{2n} = \Sigma_n \cup \Sigma_n^\#,$$

$$\mathbf{O}_{2n} = \left\{ \langle \top, \{p_i \Rightarrow p_i^\#, \neg p_i \Rightarrow \neg p_i^\# \mid p_i \in \Sigma_n\}\rangle \right\},$$

$$\Xi_{2n} = \langle \Sigma_{2n}, \mathbf{O}_{2n}\rangle.$$

From the construction it follows that for all pairs $(\mathbf{I}, \mathbf{G})$ such that $\mathbf{I}$ is a consistent and complete set over $\widehat{\Sigma_n}$ and $\mathbf{G} \subseteq \mathbf{I}^\#$, the instance $\Pi = \langle \Xi_{2n}, \mathbf{I}, \mathbf{G}\rangle$ has a one-step plan. Conversely, for all pairs $(\mathbf{I}, \mathbf{G})$ with $\mathbf{G} \cap \widehat{\Sigma^\#} \not\subseteq \mathbf{I}^\#$, there does not exist a solution. Constructing polynomially-sized $\mathcal{S_{BI}}$ domain structures with the same property turns out to be impossible, even if we allow for c-step plans.

Theorem 4. $\mathcal{S_{LIC}} \not\preceq^c \mathcal{S_{BI}}$.

Proof Sketch. Assume for contradiction that there exists a compilation scheme that compiles Ξ_{2n} to a $\mathcal{S_{BI}}$ structure Ξ'_{2n}. Since there are only polynomially many c-step plans, for a large enough n one plan Δ must be used for two different initial states in the target formalism. Since Δ is an unconditional plan, it adds and deletes the same atoms regardless of the initial state. One can then show that all goal atoms not added by Δ must be already present in both initial states, from which one concludes that identical goal states are produced, which means that $\mathbf{f}$ is not solution-preserving and, hence, not a compilation scheme, which is the desired contradiction. ∎

Next, we show that *general formulae cannot be compiled to conditional effects* by making use of results from circuit complexity. Assume that the atoms in Σ_n are numbered from 1 to n. Then a word w consisting of n bits could be encoded by the set of literals

$$\mathbf{I}_w = \{p_i \mid \text{if the } i\text{th bit of } w \text{ is } 1\} \cup$$
$$\{\neg p_i \mid \text{if the } i\text{th bit of } w \text{ is } 0\}.$$

We now say that the n-bit word w is **accepted with a one-step or c-step plan** by $\langle \Sigma_n \cup \{g\}, \mathbf{O}_n \rangle$ iff there exists a one-step or c-step plan, respectively, for the instance $\Pi_w = \langle \langle \Sigma_n \cup \{g\}, \mathbf{O}_n \rangle, \mathbf{I}_w \cup \{\neg g\}, \{g\} \rangle$.

A **uniform polynomially-sized family of domain structures** is an infinite sequence $\Xi = (\Xi_0, \Xi_1, \ldots)$, such that $|\Sigma_i| = i$ and the Ξ_i's can be generated by a $\log i$-space Turing machine. The language accepted by such a family with one-step (or c-step) plans is the set of words accepted using the domain structure Ξ_i for words of length i. Papadimitriou [16, p. 386] has pointed out that the class of languages accepted by *uniform polynomially-sized boolean expressions* is identical to the class of languages accepted by uniform polynomially-sized boolean circuits with logarithmic depth—the class NC^1. From that the next proposition follows immediately.

Proposition 3. *The class of languages accepted by uniform families of $\mathcal{S}_B$ domain structures using one-step plan acceptance is identical to NC^1.*

Theorem 5. $\mathcal{S}_B \not\preceq^c [\mathcal{S}_{\mathcal{LIC}}]$.

Proof Sketch. $\mathcal{S}_{\mathcal{LC}}$-plans with fixed goals and varying initial states can be represented by circuits with unbounded fan-in and constant depth, i.e., by AC^0-circuits. This means that c-step acceptance for families of polynomially-sized $\mathcal{S}_{\mathcal{LC}}$ structures is in AC^0. Assuming now that there exists a compilation scheme leads to the conclusion that non-uniform $\mathsf{AC}^0 \supseteq \mathsf{NC}^1$, which is impossible because of a result by Furst *et al.* [7], who showed that some languages in NC^1 are not in non-uniform AC^0. ∎

Finally, we show that *general formulae together with partial state specifications cannot be compiled away*, even if the plans grow polynomially. Let the **one-step plan existence problem** (1-PLANEX) be the PLANEX problem restricted to plans of size of 1. From the definition of the function R it is evident that $\mathcal{S}_{\mathcal{BIC}}$-1-PLANEX and $\mathcal{S}_{\mathcal{BI}}$-1-PLANEX are coNP-hard. Let p be some fixed polynomial, then the **polynomial step plan-existence problem** (p-PLANEX) is the PLANEX problem restricted to plans that have length bounded by $p(n)$, if n is the size of the planning instance. As is easy to see, this problem is in NP for all formalisms except $\mathcal{S}_{\mathcal{BIC}}$ and $\mathcal{S}_{\mathcal{BI}}$. Based on these observations and using a proof technique introduced by Kautz and Selman [11], we can prove the next result, which depends on the assumption that the *polynomial hierarchy does not collapse*. This assumption is stronger than $\mathsf{P} \neq \mathsf{NP}$, but is nevertheless commonly believed to be true.

Theorem 6. $\mathcal{S}_{\mathcal{BI}} \not\preceq^p \mathcal{S}_{\mathcal{BC}}$ *unless the polynomial hierarchy collapses.*

Proof Sketch. First, we construct a family of $\mathcal{S}_{\mathcal{BI}}$ structures such that for each formula φ of size n an initial state $\mathbf{I}_\varphi$ can be constructed such that $\langle \Xi_n, \mathbf{I}_\varphi, \{g\} \rangle$ has a one-step plan iff φ is unsatisfiable.

Given a set of n atoms, denoted by P_n, we define the set of clauses $\mathbf{A}_n$ to be the set containing all clauses with three literals that can be built using these atoms. The size of $\mathbf{A}_n$ is $O(n^3)$, i.e., polynomial in n. Let $\mathbf{D}_n$ be a set of new atoms $p_{\{l_1,l_2,l_3\}}$ corresponding one-to-one to the clauses in $\mathbf{A}_n$. Furthermore, let

$$\Phi_n = \bigwedge\Big\{ (l_1 \vee l_2 \vee l_3 \vee p_{\{l_1,l_2,l_3\}}) \mid \{l_1, l_2, l_3\} \in \mathbf{A}_n\Big\}.$$

We now construct the family of $\mathcal{S}_{\mathcal{BI}}$ domain structures as follows:

$$\Sigma_n = P_n \cup \mathbf{D}_n \cup \{g\},$$
$$\mathbf{O}_n = \{\langle\{\neg\Phi_n\}, \{\top \Rightarrow g\}\rangle\}.$$

Let $\mathbf{C}$ be a function that determines for all 3CNF formulae φ, which atoms in $\mathbf{D}_n$ correspond to the clauses in the formula , i.e., $\mathbf{C}(\varphi) = \{p_{\{l_1,l_2,l_3\}} \mid \{l_1, l_2, l_3\} \in \varphi\}$. Now, the initial state for any particular formula φ of size n can be computed by $\mathbf{I}_\varphi = \neg\mathbf{C}(\varphi) \cup (\mathbf{D}_n - \mathbf{C}(\varphi)) \cup \{\neg g\}$. From the construction, it follows that there exists a one-step plan for $\langle \Sigma_n, \mathbf{O}_n, \mathbf{I}_\varphi, \{g\}\rangle$ iff φ is unsatisfiable.

Let us now assume that there exists a compilation scheme $\mathbf{f}$ from $\mathcal{S}_{\mathcal{BI}}$ to $\mathcal{S}_{\mathcal{BC}}$ preserving plan size polynomially. From that we can construct an *nondeterministic, advice-taking Turing machine* with *polynomial advice* that decides unsatisfiability of a 3CNF formula in polynomial time (because $\mathcal{S}_{\mathcal{BC}}$-$p$-PLANEX is in NP). This means that we can decide a coNP-complete problem on a nondeterministic, polynomial advice-taking Turing machine in polynomial time. From that it follows that coNP $\subseteq$ NP/poly, which implies that the polynomial hierarchy collapses [19]. ∎

A summary of the results of this section is given in Table 1. The symbol $\sqsubseteq$

$\preceq^x$	$\mathcal{S}_{\mathcal{BIC}}$	$[\mathcal{S}_{\mathcal{LIC}}]$	$\mathcal{S}_{\mathcal{BC}}$	$\mathcal{S}_{\mathcal{BI}}$	$\mathcal{S}_{\mathcal{B}}$	$[\mathcal{S}_{\mathcal{LI}}]$
$\mathcal{S}_{\mathcal{BIC}}$	$=$	$\npreceq^p$ (6)	$\npreceq^p$ (6)	$\npreceq^c$ (4)	$\npreceq^p$ (6)	$\npreceq^p$ (6)
$[\mathcal{S}_{\mathcal{LIC}}]$	$\sqsubseteq$	$=$	$\sqsubseteq$	$\npreceq^c$ (4)	$\npreceq^c$ (4)	$\npreceq^c$ (4)
$\mathcal{S}_{\mathcal{BC}}$	$\sqsubseteq$	$\npreceq^c$ (5)	$=$	$\npreceq^c$ (4)	$\npreceq^c$ (4)	$\npreceq^c$ (5)
$\mathcal{S}_{\mathcal{BI}}$	$\sqsubseteq$	$\npreceq^p$ (6)	$\npreceq^p$ (6)	$=$	$\npreceq^p$ (6)	$\npreceq^p$ (6)
$\mathcal{S}_{\mathcal{B}}$	$\sqsubseteq$	$\npreceq^c$ (5)	$\sqsubseteq$	$\sqsubseteq$	$=$	$\npreceq^c$ (5)
$[\mathcal{S}_{\mathcal{LI}}]$	$\sqsubseteq$	$\sqsubseteq$	$\sqsubseteq$	$\sqsubseteq$	$\sqsubseteq$	$=$

Table 1. Separation Results

means that there exists a compilation scheme because the first formalism is a specialization of the second one. In all the other cases, we specify the separation and give the theorem number from which this result can be deduced, perhaps by applying Prop. 1 and 2. As can be seen, we have shown that for all pairs of formalisms for which we have not identified a polynomial-time compilation scheme preserving plan size exactly, a compilation scheme is impossible even if we allow for linear growth of the plan and unbounded computational resources.

There is of course the question whether compilation schemes preserving plan size polynomially are possible. In the long version of the paper [14] we show that between all pairs of formalisms for which such compilation schemes have not been ruled out, polynomial-time compilation schemes preserving plan size polynomially do exist.

6 Summary and Discussion

Motivated by the recent approaches to extend the GRAPHPLAN algorithm [3] to deal with *more expressive planning formalisms*, we asked what the term *expressive power* could mean in this context. One reasonable intuition seems to be that the term *expressive power* refers to how concisely domain structures and the corresponding plans can be expressed. Based on this intuition and inspired by recent work in the area of knowledge compilation [5], we introduced the notion of *compilability* in order to measure the relative expressiveness of planning formalisms.

Using this framework, we analyzed a large range of propositional planning formalisms. The most surprising result of this analysis is that we are able to come up with a complete classification. For each pair of formalisms, we were either able to construct a *polynomial-time compilation scheme* with the required size bound on the resulting plans or we could prove that such a compilation scheme is impossible—even if the computational resources for the compilation process are unbounded. In particular, we showed for the formalisms considered in this paper that incomplete state specifications and literals in preconditions can be compiled to basic STRIPS preserving plan size exactly, and that incomplete state specifications and literals in preconditions and effect conditions can be compiled away preserving plan size exactly, if we have already conditional effects. Moreover, we showed that there are no other compilation schemes preserving plan size linearly except those implied by the specialization relationship and those described above.

These results imply that Gazen and Knoblock's [9] approach to compiling conditional are optimal and demonstrate that adding general formulae and incomplete state specifications on top of that will increase expressive power—and the difficulty of finding a plan—even more.

One question one may ask is what happens if we consider formalisms with propositional formulae that are syntactically restricted, e.g. CNF or DNF formulae. However, we did not include an investigation of such formalisms in our analysis in order to keep it manageable (but see [15]). Another question might be whether *non-modular* state-translation functions could lead to more powerful compilation schemes. While this seems unlikely, it is, nevertheless, an interesting theoretical question.

Acknowledgments

The research reported in this paper was started and partly carried out while the author enjoyed being a visitor at the AI department of the University of New South Wales. Many thanks go to Norman Foo, Maurice Pagnucco, and Abhaya

Nayak and the rest of the AI department for the discussions and cappuccinos. I would also like to thank Birgitt Jenner and Jacobo Toran for some clarifications concerning circuit complexity.

References

1. C. R. Anderson, D. E. Smith, and D. S. Weld. Conditional effects in graphplan. In R. Simmons, M. Veloso, and S. Smith, eds., *Proc. AIPS-98*, p. 44–53. AAAI Press, Menlo Park, 1998.
2. C. Bäckström. Expressive equivalence of planning formalisms. *Artificial Intelligence*, 76(1–2):17–34, 1995.
3. A. L. Blum and M. L. Furst. Fast planning through planning graph analysis. *Artificial Intelligence*, 90(1–2):279–298, 1997.
4. T. Bylander. The computational complexity of propositional STRIPS planning. *Artificial Intelligence*, 69(1–2):165–204, 1994.
5. M. Cadoli and F. M. Donini. A survey on knowledge compilation. *AI Communications*, 10(3,4):137–150, 1997.
6. R. E. Fikes and N. Nilsson. STRIPS: A new approach to the application of theorem proving to problem solving. *Artificial Intelligence*, 2:189–208, 1971.
7. M. Furst, J. B. Saxe, and M. Sipser. Parity, circuits, and the polynomial-time hierarchy. *Mathematical Systems Theory*, 17(1):13–27, Apr. 1984.
8. M. R. Garey and D. S. Johnson. *Computers and Intractability—A Guide to the Theory of NP-Completeness*. Freeman, San Francisco, CA, 1979.
9. B. C. Gazen and C. Knoblock. Combining the expressiveness of UCPOP with the efficiency of Graphplan. In Steel and Alami [18], p. 221–233.
10. S. Kambhampati, E. Parker, and E. Lambrecht. Understanding and extending Graphplan. In Steel and Alami [18], p. 260–272.
11. H. A. Kautz and B. Selman. Forming concepts for fast inference. In *Proc. AAAI-92*, p. 786–793, San Jose, CA, July 1992. MIT Press.
12. J. Koehler, B. Nebel, J. Hoffmann, and Y. Dimopoulos. Extending planning graphs to an ADL subset. In Steel and Alami [18], p. 273–285.
13. V. Lifschitz. On the semantics of STRIPS. In M. P. Georgeff and A. Lansky, eds., *Reasoning about Actions and Plans: Proceedings of the 1986 Workshop*, p. 1–9, Timberline, OR, June 1986. Morgan Kaufmann.
14. B. Nebel. On the compilability and expressive power of propositional planning formalisms. Technical Report 101, Albert-Ludwigs-Universität, Institut für Informatik, Freiburg, Germany, June 1998.
15. B. Nebel. What is the expressive power of disjunctive preconditions? Technical Report 118, Albert-Ludwigs-Universität, Institut für Informatik, Freiburg, Germany, Mar. 1999.
16. C. H. Papadimitriou. *Computational Complexity*. Addison-Wesley, Reading, MA, 1994.
17. E. P. Pednault. ADL: Exploring the middle ground between STRIPS and the situation calculus. In R. Brachman, H. J. Levesque, and R. Reiter, eds., *Proc. KR-89*, p. 324–331, Toronto, ON, May 1989. Morgan Kaufmann.
18. S. Steel and R. Alami, editors. *Proc. ECP'97*, Toulouse, France, Sept. 1997. Springer-Verlag.
19. C. K. Yap. Some consequences of non-uniform conditions on uniform classes. *Theoretical Computer Science*, 26:287–300, 1983.

Generalized Cases: Representation and Steps Towards Efficient Similarity Assessment

Ralph Bergmann, Ivo Vollrath

University of Kaiserslautern
Department of Computer Science
PO-Box 3049, D-67653 Kaiserslautern, Germany
{bergmann, vollrath}@informatik.uni-kl.de

Abstract. For certain application areas of case-based reasoning, the traditional view of cases as points in the problem-solution space is not appropriate. Motivated by a concrete application in the area of electronic design reuse, we introduce the concept of a *generalized case* that represents experience that naturally covers a space rather than a point. Within a formal framework we introduce the semantics of generalized cases and derive a canonical similarity measure for them. Generalized cases can be represented in a very flexible way by using constraints. This representation asks for new means of similarity assessment. We argue that in principle fuzzy constraint satisfaction or non-linear programming can be applied for similarity computation. However, to avoid the computational complexity of these approaches, we propose an algorithm for an efficient estimation of similarity for generalized cases.

1 Introduction

In recent years, case-based reasoning (CBR) became very popular for a variety of application areas. While a lot of fielded industrial CBR applications have been developed already [4], there are also several new promising application fields that pose new important requirements on the CBR technology, triggering current research. In this paper, we present an extension of the traditional concept of a case that is motivated by requirements that emerged from the READEE project[1], which focuses on retrieval techniques for the reuse of designs in electronic engineering [12, 17, 5].

Typically, a case is a single point in the problem-solution space, i.e., a problem-solution pair that assigns a single solution to a single problem. During case-based problem solving, cases are retrieved from a case base using a similarity function, which compares the case description with the current query.

---

[1] READEE ['redɪ]: *R*euse *A*ssistant for *D*esigns in *E*lectronic *E*ngineering. Project at the University of Kaiserslautern, funded by "Stiftung Innovation Rheinland-Pfalz". Industrial cooperation partners are: Engineering office "Dr. Peter Conradi & Partner" and tec:inno GmbH, Germany. See http://wwwagr.informatik.uni-kl.de/~readee/

This paper introduces an extended view on cases which we call *generalized cases*. A generalized case covers not only a point of the problem-solution space but a whole subspace of it. A single generalized case immediately provides solutions to a set of closely related problems rather than to a single problem only. The solutions that a generalized case represents are very close to each other; basically they should be considered as (slight) variations of the same principle solution. In general, a single generalized case can be viewed as an implicit representation of a (possibly infinite) set of traditional "point cases".

Generalized cases naturally occur in certain CBR applications, such as for the recommendation of parameterized products within electronic commerce or brokerage services. The selection of reusable Intellectual Properties (IPs) of electronic designs is a very concrete example application. Such an IP is a design object whose major value comes from the skill of its producer [11], and a redesign of it would consume significant time. IPs usually span a design space because they are descriptions of flexible designs that have to be synthesized to hardware before they actually can be used.

The idea of generalizing cases was already implicitly present since the very beginning of CBR and instance-based learning research [10, 1, 16]. For the purpose of adaptation, recent CBR systems in the area of design and planning [9, 13, 3] use complex case representations that realize generalized cases.[2] In this paper we will present a general (and partially a more formal) view on the concept of generalized cases that also partially covers the above mentioned work.

First, Sect. 2 introduces the motivating application of IP reuse in some detail. Then, Sect. 3 presents the concept of a generalized case. Section 4 discusses representation of and reasoning with generalized cases while section 5 focuses on efficient similarity assessment. Finally, section 6 summarizes and outlines directions of future research.

2   Reuse of Electronic Design IPs

Increasingly, electronics companies use IPs ("Intellectual Properties") from third parties inside their complex electronic systems. An IP is a design object whose major value comes from the skill of its producer [11], and a redesign of it would consume significant time. However, a designer who wants to reuse designs from the past must have a lot of experience and knowledge about existing designs, in order to be able to find candidates that are suitable for reuse in his/her specific new situation. Currently, searching electronic IP databases can be an extremely time consuming task because, first, the public-domain documentation of IPs is very restricted and second there are currently no intelligent tools to support the designer in deciding whether a given IP from a database meets (or at least comes close to) the specification of his new application [12].

---

[2] The term "generalized case" was introduced in [3] but was restricted to planning cases. It has its origins in a concept for representing skeletal plans [2] and for acquiring them by machine learning algorithms.

Table 1. Selected parameters of the example IP.

parameter	description
frequency f	The clock frequency that can be applied to the IP.
area a	The chip area the synthesized IP will fit on.
width w	Number of bits per input/output word. Determines the accuracy of the DCT. Allowed values are 6, 7, ..., 16.
subword s	Number of bits calculated per clock tick. Changing this design space parameter may have a positive influence on one quality of the design while having a negative impact on another. Allowed values are 1, 2, 4, 8, and no_pipe.

IPs usually span a design space because they are descriptions of flexible designs that have to be synthesized to hardware before they actually can be used. The behavior of the final hardware depends on a number of *parameters* of the original design description. The valid value combinations for these parameters are constrained by different criteria for each IP.

Within the READEE project we developed a representative IP to study issues of representation, retrieval, and parameterization. This IP implements an algorithm for the *discrete cosine transform (DCT)* and its inverse operation *(IDCT)* which is needed as an important part of decoders and encoders of the widely known MPEG-2 video compression algorithm. While the complete electronics design in the VHDL language is described in [18], we now focus only on the variable parameters of the IP, which are shown in Table 1.

These parameters heavily depend on each other: increasing the accuracy w of the DCT/IDCT increases the chip area consumption a and decreases the maximum clock frequency f in a particular way, shown in Fig. 1. These relationships define a very specific design space that is spanned by the four mentioned parameters. The knowledge about this design space is very central for the selection of the IP during the process of design reuse. When case-based reasoning is applied for IP reuse, the question must be solved how to represent and how to reason with cases that encode design spaces rather than design points.

3 Generalized Cases

This section abstracts from the specific application described above and elaborates the general concept of a generalized case. At first, we basically apply an extensional view on generalized cases by considering them as sets of traditional cases. The definitions we provide in this section should be considered as a specification rather than a means for realizing representation mechanisms or the problem solving process. Representation issues will be addressed in Sect. 4.

3.1 Extensional Definition of Generalized Cases

Let $\mathbb{R}$ be the (possibly infinite) representation space for cases. In the IP application this representation space would include the parameters f, a, w, and s. One

$$f \leq \begin{cases} -0.66w + 115 & \text{if } s = 1 \\ -1.94w + 118 & \text{if } s = 2 \\ -1.74w + 88 & \text{if } s = 4 \\ -0.96w + 54 & \text{if } s = 8 \\ -2.76w + 57 & \text{if } s = \text{no} \end{cases}$$

$$a \geq \begin{cases} 1081w^2 + 2885w + 10064 & \text{if } s = 1 \\ 692w^2 + 2436w + 4367 & \text{if } s = 2 \\ 532w^2 + 1676w + 2794 & \text{if } s = 4 \\ 416w^2 + 1594w + 2413 & \text{if } s = 8 \\ 194w^2 + 2076w + 278 & \text{if } s = \text{no} \end{cases}$$

Fig. 1. Dependencies between the parameters of an example IP

could assume that $\mathbb{R}$ is subdivided into a problem space $\mathbb{P}$ and a solution space $\mathbb{S}$, as we assumed in the first approach to the formalization of generalized cases [5]. However, we drop this assumption here since it is less appropriate in design domains. A traditional case c or *point case* is a point in the representation space $\mathbb{R}$, i. e., $c \in \mathbb{R}$. A *generalized case*, can be defined as follows:

Definition 1. (Generalized Case) A *generalized case gc* is a subset of the representation space, i. e., $gc \subseteq \mathbb{R}$.

Hence, a generalized case stands for a set of point cases. However, a generalized case should not represent an arbitrary set. The idea is that a generalized case is an abbreviation for a set of closely related point cases that naturally occur as one entity in the real world (e. g. a single IP). Usually, these point cases all have some common problem and solution properties. They either have the same solution or their solutions don't differ drastically and can be represented in a compact form (e. g., the parametrized VHDL description of an IP).

3.2 Similarity Assessment

For retrieving generalized cases, the similarity between a query and a generalized case must be determined. As in traditional CBR, we assume that the query is a point in the representation space that may be only partially described. We further assume that a traditional similarity measure $\text{sim}(q, c)$ is given which assesses the similarity between a query q and a point case c. Such a similarity measure can be extended in a canonical way to assess the similarity $\text{sim}^*(q, gc)$ between a query q and a generalized case gc:

Definition 2. (Canonical Extension of Similarity Measures for Generalized Cases) The similarity measure $\text{sim}^*(q, gc) := \sup\{\text{sim}(q, c) \mid c \in gc\}$ is called the canonical extension of the similarity measure sim to generalized cases.

Applying sim^* ensures that those generalized cases are retrieved that contain the point cases which are most similar to the query. (For the rest of this paper, we will substitute max for sup. By doing this, we knowingly avoid some formal obstacles that will not be an issue in most technical application domains.)

3.3 Selecting Solutions from Generalized Cases

For reasoning with generalized cases, the traditional CBR cycle must be slightly modified. After retrieving a generalized case, there is not a single case or a single solution available, but a set of them. Hence, we have to select a single case depending on the current query. We continue the argument from the previous subsection and again assume the traditional similarity measure $sim(p, c)$ and derive from this a canonical selection approach as follows:

Definition 3. (Canonical Selection from Generalized Cases) The following set is called *canonical selection* from the generalized case gc for query q, based on the similarity measure sim:

$$Select^*(q, gc, sim) := \{c \in gc \,|\, sim(q, c) = sim^*(q, gc)\}$$

This simple definition selects those point cases from the generalized case with the highest similarity to the problem. These are the point cases that determine the value of the canonical similarity measure between the problem and the generalized case. For a canonically extended similarity measure sim^* and a canonical selection, the following property can be shown.

Lemma 1. Given a case base CB of point cases and a similarity measure sim, then for any case base CB^* of generalized cases such that $CB = \bigcup_{gc \in CB^*} gc$ holds: if c_{ret} is a case from CB which is most similar to a query q w.r.t. sim then there is a generalized case gc_{ret} from CB^* which is most similar to q w.r.t. sim^* such that $c_{ret} \in Select^*(q, gc_{ret}, sim)$.

This states that if we introduce generalized cases together with a sim^* and $Select^*$, the same cases are retrieved during problem solving, independent from the clustering of point cases into generalized cases. This provides a clear semantics of generalized cases defined in terms of traditional point cases and similarity measures. When building a CBR system that uses generalized cases, we can consider them as sets of point cases to which a traditional similarity measure is applied. For retrieving generalized cases we can apply the canonical extension of this similarity measure and a canonical selection. This ensures that the same cases are retrieved as if we had a CBR system which works with the point cases.

4 Representation of Generalized Cases

For building CBR systems that reason with generalized cases, efficient representations for generalized cases and efficient approaches for similarity assessment and selection must be developed. Obviously, a straight forward approach by applying the definitions from the previous section (e.g. computing the similarity for each point case covered by a generalized case and determining the maximum of the similarity values) can be quite inefficient if the generalized case covers a large subspace. This iterative approach would also not be able to cope with generalized cases that cover an infinite set of traditional cases. Hence, compact implicit representations for generalized cases must be developed.

In the following, we assume that cases are represented as attribute-value pairs, i.e., as a vector of n attributes encoding the representation space ($\mathbb{R} = \mathbb{R}_1 \times \cdots \times \mathbb{R}_n$). The four parameters shown in Table 1 are typically encoded as such attributes. Hence, a case can be encoded as $c = (c_1, \ldots, c_n)$ and a query as $q = (q_1, \ldots, q_n)$.

4.1 Representing Generalized Cases Using Constraints

Since a generalized case is a set of point cases, its representation must basically be able to encode a complex relation between the different attributes. For the IP shown in section 2, the relationship given in Fig. 1 must be represented. Constraints are a general and flexible way for representing such dependencies. Applying constraints in CBR is not new. Particularly in design tasks, similar ideas can be found, for example, in [9, 13]. In the following we suggest a representation for generalized cases that uses constraints for representing dependencies between attributes. The vocabulary [14] for representing generalized cases consists of the representation space $\mathbb{R} = \mathbb{R}_1 \times \cdots \times \mathbb{R}_n$ and of a set of variables $V = \{v_1, \ldots, v_n\}$, one for each attribute such that v_i holds values from $\mathbb{R}_i$. A generalized cases is represented by a set of constraints:

$$gc = \{C_1, \ldots, C_k\}$$

such that $V_i := \mathrm{Var}(C_i) \subseteq V.$[3] Such a generalized case represents the set of point cases whose attribute values fulfill all constraints.

Please note that, of course, a point case $(c_1, \ldots, c_n)$ can also be represented as a generalized case as follows: $gc = \{v_1 = c_1, \ldots, v_n = c_n\}$. To represent the IP from Sect. 2, linear ($v_i \leq a \cdot v_j + b$) and quadratic ($v_i \leq a \cdot v_j^2 + b \cdot v_j + c$) constraints are needed.

4.2 The General Problem of Similarity Assessment

The similarity assessment between a query $(q_1, \ldots, q_n)$ and a generalized case is as follows. Assume a given traditional similarity measure for point cases $\mathrm{sim}(q, c)$ that aggregates individual *local similarity measures* sim_i for the individual attributes by a function Θ that is monotone increasing in every argument[4]:

$$\mathrm{sim}((q_1, \ldots, q_n), (c_1, \ldots, c_n)) = \Theta(\mathrm{sim}_1(q_1, c_1), \ldots, \mathrm{sim}_n(q_n, c_n))$$

---

[3] $\mathrm{Var}(C)$ denotes the set of variables of the constraint C; V_i is an abbreviation for the set of variables that occur in C_i.

[4] Please note that this is one of the most common similarity measures applied in many CBR applications. Here, Θ is usually some variation of a weighted sum.

The computation of the canonical extension sim* given the similarity measures sim and the previously introduced constraint representation for generalized cases requires searching for values for the variables $v_1, \ldots, v_n$ such that every constraint $C_1, \ldots, C_n$ is satisfied and $\text{sim}((q_1, \ldots, q_n), (v_1, \ldots, v_n))$ is maximized.

4.3 Solution Selection by Nonlinear Programming and Fuzzy Constraints

For continuous domains, this similarity assessment problem can be formulated as a nonlinear programming (NLP) problem [6] or as a problem of nonlinear optimization [7]. An NLP problem is of the form:

$$\begin{aligned} \text{minimize} \quad & f(x), \quad x \in R^n, \\ \text{subject to} \quad & c_i(x) \geq 0, \quad i \in I, \\ & c_i(x) = 0, \quad i \in E \end{aligned}$$

In our context, the constraints c_i can be identified with the constraints $C_1, \ldots, C_k$ of *gc*. The *objective function* f corresponds to the similarity function sim. sim (and thus f) is not a linear function (this is true even if Θ is a linear function). Neither are the constraints c_i. So, in general, only numerical algorithms for solving such an NLP problem are applicable.

For discrete and mixed domains, our problem can be formulated as a fuzzy constraint satisfaction problem (FCSP) [8, 15, 20]. An FCSP is specified by a tuple (V, D, C). $V = \{v_j \mid j \in J\}$ is a finite set of variables ("attributes" in CBR terminology), $D = \{d_i \mid i \in I\}$ is a set of domains associated with the variables in V, and C is a set of constraints over these variables. The constraints in C are fuzzy relations with a *satisfaction index* indicating the degree of satisfaction for each variable binding. To model our problem of solution selection one could include in C all constraints from *gc* as *crisp* constraints. Additionally, the attribute values from the query can be used to derive an appropriate unary fuzzy constraint for each of these variables to be included in C. The satisfaction index si_i of each constraint is defined by the respective local similarity measure:

$$\text{si}_i(v_i) = \text{sim}_i(q_i, v_i)$$

The overall *solution function* of the FCSP is defined by the function Θ which also defines the overall similarity measure sim. The main difficulty in solving such an FCSP are the mixed (discrete and continuous) domains and the nonlinear continuous constraints occuring in some applications.

5 Efficient Similarity Estimation

The CBR problem of our application is not only to select the best solution from a given generalized case, but also to find the generalized case that contains the best overall solution out of a (possibly large) case base. The obvious approach would be to solve this retrieval problem by solving the case selection problem for each gc, and choosing the case that contains the best solution of all. Of course, this approach bears a very high computational effort. Because of this, we propose to employ a weaker but computationally more efficient retrieval approach that calculates an upper and lower bound for $\text{sim}^*(q, gc)$. There are two other reasons for using such an approach. First, the knowledge in many application domains (including IP reuse) is vague in important parts. In the presence of vague knowledge it is usually unreasonable to employ exact but computationally expensive algorithms. Second, the CBR systems are typically meant to be interactive tools. The solution selection phase (e. g. the parameter selection for a retrieved IP) needs to be supervised by an expert who might even call for comfortable interfaces to *explore* the design space of the generalized case.

5.1 Estimating Bounds on $\text{sim}^*(q, gc)$

In a first approach, bounds for the similarity $\text{sim}^*(q, gc)$ can easily be found by looking at the allowed ranges of individual attribute values only [5]. These bounds can provide valuable information during retrieval and they enable the filtering of cases for which a more detailed similarity assessment is not necessary any more. However, this first approach may be too rough if some attributes have large (or infinite) ranges, while they are still constrained depending on the values of other attributes. This is likely to be a problem if a generalized case not only contains unary constraints (like $3 \leq x_2 \leq 7$) but also constraints covering two or more variables (like $x_2 \geq ax_1 + b$). The following algorithm calculates a more accurate upper bound for $\text{sim}^*(q, gc)$ in such cases:

Input: $gc = \{C_1, \ldots, C_k\}$; q **Output:** upper bound for $\text{sim}^*(q, gc)$

1. Let $\mathbb{R}^{V_j} = \times\{\mathbb{R}_i \mid i \in V_j = \text{Var}(C_j)\}$. *(We will use the notation "$\circ$" for the (only informally defined) binary operator that merges two vectors $x_1 \in \mathbb{R}^{V_j}$ and $x_2 \in \mathbb{R}^{V \setminus V_j}$ into $x \in \mathbb{R}$ restoring the correct ordering of attributes.)*
2. For each constraint $C_j \in gc$ find a consistent variable assignment $x_{\max_j} \in \mathbb{R}^{V_j}$ that has the highest possible similarity for any given assignment $\bar{x} \in \mathbb{R}^{V \setminus V_j}$ of the remaining variables, i. e.:

$$\forall \bar{x} \in \mathbb{R}^{V \setminus V_j} : \forall x' \in \mathbb{R}^{V_j} : \text{sim}(q, \bar{x} \circ x_{\max_j}) \geq \text{sim}(q, \bar{x} \circ x')$$

($x_{\max_j}$ exists and can be defined regardless of $\bar{x}$ because of the monotony of Θ.)
3. $\mathbb{C}^{\text{ind}} := \left\{ C_j \mid \text{Var}(C_j) \cap \left(\bigcup_{i \neq j} \text{Var}(C_i) \right) = \emptyset \right\}$; $\mathbb{C}^{\text{dep}} := gc \setminus \mathbb{C}^{\text{ind}}$
   (We assume that every variable occurs in a constraint, i. e. $V = \bigcup_{C \in gc} \text{Var}(C)$.)
4. $\mathbb{V}^{\text{ind}} := \bigcup_{C \in \mathbb{C}^{\text{ind}}} \text{Var}(C)$; $\mathbb{V}^{\text{dep}} := \bigcup_{C \in \mathbb{C}^{\text{dep}}} \text{Var}(C)$
5. Let $\bar{x} \in \mathbb{R}^{V^{\text{ind}}}$ be the variable assignment according to all $x_{\max_j}$ for $C_j \in \mathbb{C}^{\text{ind}}$.
6. For each $C_j \in \mathbb{C}^{\text{dep}}$ calculate $\text{sim}_j = \text{sim}(q, \bar{x} \circ x_{\max_j} \circ x_{r_j})$ with x_{r_j} being an assignment for the remaining variables, such that their local similarities are maximized.
7. Return the minimum of all these sim_j (if $\mathbb{C}^{\text{dep}} = \emptyset$ return $\text{sim}(q, \bar{x})$).

A lower bound for $\text{sim}^*(q, gc)$ can be calculated by a very similar algorithm.

5.2 Calculating $x_{\max_j}$

The algorithm described above leaves open the task of calculating $x_{\max_j}$ (we will drop the index j now for $x_{\max}$ and C). We will now explain how an $x_{\max}$ can be found for piecewise differentiable numeric constraints over a continuous and ordered domain (this kind of constraints is quite typical for the design reuse applications we encountered). For the case that $q \in C$, $x_{\max}$ is the projection of q onto $\text{Var}(C)$. The case where $q \notin C$ is more difficult.

We assume that the similarity measures sim_i are compatible with the ordering of the elements in $\mathbb{R}_i$ for all $i = 1, \ldots, n$. For simplicity, we further assume that all constraints are closed, i.e., all points on the bounding surface Q_C of C are consistent with the constraint[5]. Given these assumptions it can be shown that a value for $x_{\max}$ according to the definition of the algorithm above can be found within Q_C. This allows us to search for the point of maximum similarity on the border of C without worrying about the inside region. In principle, this is just a matter of mathematical function analysis. One necessary (yet, not sufficient) condition for a maximum of similarity is that the derivation of the similarity function (if it is defined) sim is zero for all possible directions tangential to the surface of C. In fact, it is not necessary to check for *all* surface directions. It is sufficient to check the directions of m linear independent vectors, where m is equal to the dimension of the surface of C.

So, let $T_i(C, x)$ $(i = 1, \ldots, m; \ x \in Q_C)$ be linear independent vectors tangential to Q_C in the point x. To find a point $x_{\max}$ of maximum similarity to q it is necessary to find $x_{\max} \in Q_C$ for which the following condition holds for all $i = 1, \ldots, m$:

$$0 = \frac{\partial \, \text{sim}(q, x_{\max})}{\partial \, T_i(C, x_{\max})} = \langle T_i^*(C, x_{\max}) \, , \, grad \, \text{sim}(q, x_{\max}) \rangle$$
$$\Leftrightarrow 0 = \langle T_i(C, x_{\max}) \, , \, grad \, \text{sim}(q, x_{\max}) \rangle$$

With T^* denoting the normalization of T, and $\langle \cdot, \cdot \rangle$ being the scalar product. This formular says that the gradient of sim in $x_{\max}$ is perpendicular to the tangential vectors T_i.

The above condition can be used to find maxima and minima of piecewise differentiable constraints. In the application of IP reuse we expect linear and quadratic constraints to be sufficient for most applications. To illustrate how the above condition can be used to find $x_{\max}$ let us now consider a concrete example of a simple linear constraint between two variables. This kind of constraint occurs in many practical applications:

$$C = \{(x_1, x_2) \, | \, x_2 \geq ax_1 + b\}$$

where a and b are constants. This constraint can be interpreted as a set of values that is bounded by a hyperplane as shown in Fig. 2. The constraint is two-dimensional, so the constraint's surface is one-dimensional. Thus, we only

---

[5] This is only a theoretical – not a practical – restriction.

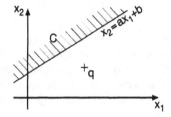 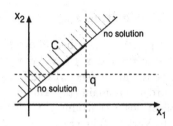

Fig. 2. A simple linear constraint C and a query q outside of that constraint.

Fig. 3. Most similar points of C to query q.

need one tangential vector $T(C, x)$ which we randomly choose by setting its first dimension to 1:

$$T(C, x) = \left(1, \frac{\partial\, ax_1 + b}{\partial x_1}\right) = (1, a)$$

Let us further assume a very simple (and common) definition of sim:

$$\mathrm{sim}(q, x) = w_1\mathrm{sim}_1(q_1, x_1) + w_2\mathrm{sim}_2(q_2, x_2)$$

with w_1 and w_2 being constant weights. This leads to

$$grad\,\mathrm{sim}(q, x) = \left(w_1\frac{\partial\,\mathrm{sim}_1(q_1, x_1)}{\partial\, x_1}\,,\, w_2\frac{\partial\,\mathrm{sim}_2(q_2, x_2)}{\partial\, x_2}\right)$$

To find the points of maximum similarity we have to find x such that:

$$0 = \langle T_i(C, x)\,,\, grad\,\mathrm{sim}(q, x)\rangle = w_1\frac{\partial\,\mathrm{sim}_1(q_1, x_1)}{\partial\, x_1} + aw_2\frac{\partial\,\mathrm{sim}_2(q_2, x_2)}{\partial\, x_2}$$

At this point we need the formulas for sim_i to calculate their derivations. Let

$$\mathrm{sim}_i(q_i, x_i) = 1 - \frac{|q_i - x_i|}{R_i}$$

where R_i is the difference between the maximum and the minimum allowed value of x_i (this is a quite common similarity measure). Deriving sim_i shows that for the special case where $w_1 = aw_2$ and $R_1 = R_2$, $\mathrm{sim}(q, \overline{x} \circ x_{\max})$ is maximal for all $x_{\max} = (x_1, x_2)$ that meet the following condition:

$$q_1 \geq x_1 \,\wedge\, q_2 \leq x_2 \,\wedge\, x_2 = ax_1 + b$$

This is illustrated in Fig. 3. For other cases ($w_1 \neq aw_2$) further treatment is necessary and easily possible. The outcome of this functional analysis are very simple decision criteria that can easily be implemented to calculate $x_{\max}$ for the special case of linear constraints. This functional analysis can also be done for more complex types of constraints, like quadratic constraints as they occur in the IP description of the example in Sect. 2.

6 Summary and Future Work

The concept of generalized cases presented in this paper is particularly appropriate if the knowledge from the real world domain which has to be captured in cases naturally covers a subspace rather than a point in the representation space. As we have demonstrated, this is true for the domain of IP reuse, but we observed that this is also true for all kinds of parameterizable products in electronic commerce applications [19].

Of course, the idea of generalizing cases (or examples) was implicitly present already since the very beginning of CBR and instance-based learning research [10, 1, 16]. However, these approaches are limited to hyperrectangular representations of generalized cases that cannot deal with dependencies between attributes.

We introduced the idea of using constraints as a flexible means for representing generalized cases. We also provided a general framework for similarity assessment of generalized cases as a canonical extension of a similarity measure on point cases (Salzberg [16] applied a special case of this approach based on the geometrical distance). Further, we proposed an efficient algorithm for estimating the canonical similarity between a query and a generalized case. With this algorithm, generalized cases like those occurring in the domain of IP reuse can be handled. A first version of this algorithm that is based on allowed attribute ranges [5] has recently been implemented as a component of the CBR tool CBR-Works[6]. The approach proposed in section 5.2 is currently under implementation. We expect significant improvements in accuracy of the similarity estimation, particularly in the IP reuse domain, since according to our experience, the constraints occurring in a generalized case for an IP are mostly independent.

Besides the similarity assessment of generalized cases, the problem of selecting the most similar point case from the retrieved generalized case must also be solved (see section 3.3). As we argued, non-linear programming or fuzzy constraint satisfaction techniques will be applicable for this task, but because of their computational complexity they should only be applied after case retrieval is finished and not for the similarity assessment itself. These issues will be further examined in the course of the READEE project.

References

1. Ray Bareiss. *Exemplar-Based Knowledge Acquisition: A unified Approach to Concept Representation, Classification and Learning.* Academic Press, 1989.
2. Ralph Bergmann. Knowledge acquisition by generating skeletal plans. In F. Schmalhofer, G. Strube, and Th. Wetter, editors, *Contemporary Knowledge Engineering and Cognition*, pages 125–133, Heidelberg, 1992. Springer.
3. Ralph Bergmann. *Effizientes Problemlösen durch flexible Wiederverwendung von Fällen auf verschiedenen Abstraktionsebenen.* PhD thesis, Universität Kaiserslautern, 1996. Available as DISKI 138, infix Verlag.

---

[6] CBR-Works was jointly developed by tec:inno GmbH and the University of Kaiserslautern during the funded projects INRECA, WiMo, and INRECA-II and is now marketed as a commercial product by tec:tnno.

4. Ralph Bergmann, Sean Breen, Mehmet Göker, Michel Manago, and Stefan Wess. *Developing Industrial Case Based Reasoning Applications: The INRECA Methodology.* Number 1612 in LNAI. Springer-Verlag, 1999.
5. Ralph Bergmann, Ivo Vollrath, and Thomas Wahlmann. Generalized cases and their application to electronic designs. In Erica Melis, editor, *Proceedings of the 7th German Workshop on CBR, Würzburg*, pages 6–19, 1999.
6. A. G. Buckley and J.-L. Goffin, editors. *Algorithms for Constrained Minimization of Smooth Nonlinear Functions.* Number 3016 in Mathematical Programming Study. The Mathematical Programming Society, Inc., North-Holland – Amsterdam, April 1982.
7. B. Cornet, V. H. Nguyen, and J. P. Vial, editors. *Nonlinear Analysis and Optimization.* Number 30 in Mathematical Programming Study. The Mathematical Programming Society, Inc., North-Holland – Amsterdam, Feb. 1987.
8. Didier Dubois, Hélène Fargier, and Henri Prade. The calculus of fuzzy restrictions as a basis for flexible constraint satisfaction. In *IEEE International Conference on Fuzzy Systems*, volume 2, pages 1131–1136, San Francisco, 1993.
9. Kefeng Hua, Ian Smith, and Boi Faltings. Integrated case-based building design. In Stefan Wess, Klaus-Dieter Althoff, and Michael M Richter, editors, *Topics in Case-Based Reasoning. Proc. of the First European Workshop on Case-Based Reasoning (EWCBR-93)*, Lecture Notes in Artificial Intelligence, 837, pages 436–445. Springer Verlag, 1993.
10. Janet L Kolodner. *Retrieval and Organizational Strategies in Conceptual Memory.* PhD thesis, Yale University, 1980.
11. Jeff Lewis. Intellectual property (IP) components. Artisan Components, Inc., [web page], http://www.artisan.com/ip.html, 1997. [Accessed 28 Oct 1998].
12. Peter Oehler, Ivo Vollrath, Peter Conradi, and Ralph Bergmann. Are you READEe for IPs? In Ralf Seepold, editor, *2nd GI/ITG/GMM-Workshop "Reuse Techniques for VLSI Design"*, FZI-Bericht, Karlsruhe, September 1998. Forschungszentrum Informatik.
13. Lisa Purvis and Pearl Pu. Adaptation using constraint satisfaction techniques. In Agnar Aamodt and Manuela Veloso, editors, *Case-Based Reasoning Research and Development, Proc. ICCBR-95*, Lecture Notes in Artificial Intelligence, 1010, pages 289–300. Springer Verlag, 1995.
14. Michael M. Richter. The knowledge contained in similarity measures. invited talk at the *International Conference on Case-Based Reasoning (ICCBR-95)*, 1995. http://wwwagr.informatik.uni-kl.de/~lsa/CBR/Richtericcbr95remarks.html.
15. Zs. Ruttkay. Fuzzy constraint satisfaction. In *IEEE International Conference on Fuzzy Systems*, volume 2, pages 1263–1268, Orlando, 1994. IEEE.
16. S Salzberg. A nearest hyperrectangle learning method. *Machine Learning*, 6:277–309, 1991.
17. Ivo Vollrath. Reuse of complex electronic designs: Requirements analysis for a CBR application. In Barry Smyth and Pádraig Cunningham, editors, *Advances in Case-Based Reasoning: 4th European Workshop, EWCBR-98, Proceedings*, volume 1488 of *Lecture Notes in Artificial Intelligence*, pages 136–147, Berlin, 1998. Springer.
18. Thomas Wahlmann. Implementierung einer skalierbaren diskreten Kosinustransformation in VHDL. Diploma thesis, University of Siegen, 1999.
19. W. Wilke. *Knowledge Management for Electronic Commerce.* PhD thesis, University of Kaiserslautern, 1998.
20. Jason H. Y. Wong, Ka-fai Ng, and Ho-fung Leung. A stochastic approach to solving fuzzy constraint satisfaction problems. In *Lecture Notes in Computer Science*, volume 1118, pages 568–569. Springer, 1996.

Be Busy and Unique ... or Be History — The Utility Criterion for Removing Units in Self-Organizing Networks

Bernd Fritzke[1]

Dresden University of Technology
Institute for Artificial Intelligence
Neural Computation Group
fritzke@inf.tu-dresden.de,
http://pikas.inf.tu-dresden.de/~fritzke

Abstract. A novel on-line criterion for identifying "useless" neurons of a self-organizing network is proposed and analyzed. The criterion is based on a utility measure. When it is used in the context of the growing neural gas model to guide deletions of units, the resulting method is able to closely track non-stationary distributions. Slow changes of the distribution are handled by adaptation of existing units. Rapid changes are handled by removal of "useless" neurons and subsequent insertions of new units in other places. Additionally, the utility measure can be used to prune existing networks to a specified size with near-minimal increase of quantization error.

1 Non-stationarity is difficult to handle ...

Non-stationary data distributions can be found in many technical, biological or economic processes. Self-organizing neural networks have rarely been considered for tracking those distributions since many of the models, e.g., the self-organizing map [Koh82], neural gas [MBS93] or the hypercubical map [BV95], use decaying adaptation parameters[1]. Once the adaptation strength has decayed, the network is "frozen" and thus unable to react to subsequent changes in the signal distribution. In figure 1 an example for the self-organizing map is shown.

2 ... even for incremental networks

Models with small constant parameters such as the incremental networks developed by the author [Fri94,Fri95a,Fri95b] are in a somewhat better position for handling non-stationary distributions. The non-decreasing adaptation rate enables the networks to follow slowly changing probability distributions such as, for example, a normal distribution with a slowly drifting mean. Rapid changes in the distribution, however, can in general not be handled properly as is illustrated for the growing neural gas (GNG) model [Fri95a] in figure 2.

---

[1] ... as does the classical k-means algorithm.

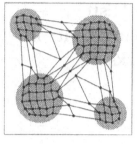 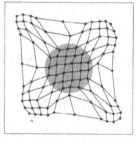 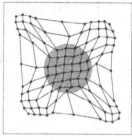

a) after 20000 signals   b) after 25000 signals   c) after 40000 signals

Fig. 1. A 10×10 self-organizing map (SOM) tries (and fails) to track a non-stationary signal distribution $p(\xi)$ which is uniform in the shaded areas. a) Initial state of $p(\xi)$ with the SOM network having adapted to 20000 input signals. At this point the probability distribution was changed. b) State of the network 5000 adaptation steps after the distribution had changed. A number of units has moved to the center c) After 40000 adaptation steps a fraction of the units have been adapted to the new regions of high probability density, but many have become dead units which are not adapted anymore. The decaying parameters were set according to parameter values as suggested by Ritter and Schulten [RMS91] and reached their minimum after 40000 input signals. Obviously, a change of the distribution at that time would have made it even more difficult for the SOM to adapt to the new situation.

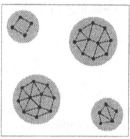 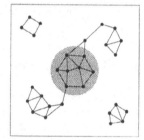 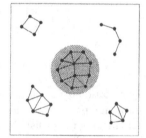

a) after 20000 signals   b) after 25000 signals   c) after 40000 signals

Fig. 2. A growing neural gas (GNG) network tries (and fails) to track a non-stationary signal distribution $p(\xi)$ which is uniform in the shaded areas. The maximum network size is set to 30. a) Initial state of $p(\xi)$ with the GNG network already grown to its maximum size. At this point the probability distribution was changed. b) State of the network 5000 adaptation steps after the distribution had changed. c) After 40000 adaptation steps some units have been adapted to the new regions of high probability density, but some have become dead units which are not adapted anymore. Parameter values (see [Fri95a] for details): $\lambda = 500$, $\varepsilon_b = 0.05$, $\varepsilon_n = 0.0006$, $\beta = 0.0005$, $a_{max} = 120$.

In the case of GNG it is easy to see why units may get stuck in former regions of high probability density and become so-called *dead units*. The network topology is updated – apart from the characteristic insertions – by two mechanisms.

"Competitive Hebbian Learning" [Mar93] is used to create new connections by always inserting a connection between the units s_1 and s_2 nearest and second-nearest to the current input signal ξ. When such a connection does already exist, its *age* parameter is set to zero. Moreover, the age of all edges adjacent to the winning unit s_1 is increased. If the age of an edge surpasses a maximum parameter a_{max}, the edge is removed and so are any units which become completely isolated from the network. This strategy leads to a connectivity which is a subgraph of the Delaunay triangulation and is optimally topology-preserving [Mar93].

One should note, that the edge ageing is a local process, since it happens only in the vicinity of the winning unit. Now consider the situation where a network has grown to a certain maximum size and is well adapted to a given probability distribution (figure 2a). When the probability density changes, some units are adapted rather quickly towards the new regions of high probability density (figure 2b). The edges connecting those units and the units remaining in the former regions of high probability density are quickly removed by the ageing process (figure 2c).

Thereafter, however, those units and connections which remain in the former regions of high probability density are not changed anymore. Those units will not be winner since they are too far from the current input signals. Consequently, the connections do not age since ageing is only performed for edges adjacent to the winner.

The result is a waste of network resources which may get even worse if the distribution changes several times. One could perhaps interpret the dead units as a kind of memory which may become useful again when the probability distribution takes on a previously held shape. In the following, however, we would like to discuss ways of using a network of limited size to track the distribution in each moment as faithfully as possible.

3   Objective function

As objective for the network we define the minimization of the expected quantization error

$$E = \int \|\xi - \mathbf{w}_{s(\xi)}\|^2 \, p(\xi) \, d\xi. \tag{1}$$

Here $s(\xi)$, also denoted as s_1, is the winning unit for the current input signal ξ, and $\mathbf{w}_{s(\xi)}$ is the *reference vector* associate with that unit. With $p(\xi)$ we denote the probability density of the input signals which is assumed to be unknown and non-stationary.

4   Local error measure

In a GNG network each unit c in the network has an error variable E_c which contains accumulated error information. For each input signal ξ the error variable of the winner s_1 for this signal is updated according to

$$E_{s_1} := E_{s_1} + \|\boldsymbol{\xi} - \mathbf{w}_{s_1}\|^2. \tag{2}$$

Moreover, after each adaptation step (small change of the reference vectors of s_1 and its topological neighbors, see [Fri95a]) the error variables of all units are subjected to exponential decay:

$$E_c := E_c - \beta E_c \quad \text{(for each unit } c\text{)} \tag{3}$$

whereby β is a suitable decay constant. The decay stresses the influence of more recently measured contributions to E_c. It is needed to account for the adaptation of the reference vectors and for the non-stationarity of the probability distribution. The parameter β is further analyzed in section 9.1.

The GNG algorithm always performs insertions of new units close to the unit q with maximum accumulated error.

5   Local utility measure

How can we assess the usefulness of a particular unit for the defined objective, namely minimizing the quantization error? A direct approach is to compute the change of this error caused by removal of the unit.

Let $\boldsymbol{\xi}$ be the current input signal and let the unit s_1 be the winner for $\boldsymbol{\xi}$. The quantization error for this input signal is

$$d_1 := \|\boldsymbol{\xi} - \mathbf{w}_{s_1}\|^2. \tag{4}$$

If we removed s_1, however, then $\boldsymbol{\xi}$ would be mapped to the second-nearest unit s_2 and the quantization error for $\boldsymbol{\xi}$ would be

$$d_2 := \|\boldsymbol{\xi} - \mathbf{w}_{s_2}\|^2 \tag{5}$$

which in general would be larger than d_1. The usefulness of s_1 for this particular input signal is simply the difference $d_2 - d_1$ of the two error values. To have an on-line criterion we associate with each unit c a local *utility* value U_c. For each input signal $\boldsymbol{\xi}$, the utility variable of the winner s_1 is updated according to

$$U_{s_1} := U_{s_1} + \|\boldsymbol{\xi} - \mathbf{w}_{s_2}\|^2 - \|\boldsymbol{\xi} - \mathbf{w}_{s_1}\|^2. \tag{6}$$

As is done for the local error values, we subject all utility variables to exponential decay after each adaptation step (with the same value for β):

$$U_c := U_c - \beta U_c \quad \text{(for each unit } c\text{)} \tag{7}$$

To maintain a certain level of utility, a unit must be "busy", i.e., it must be winner often enough to overcome the exponential decay (7) and it must also be "unique", i.e., have a sufficiently different reference vector from other units, since otherwise the increment in (6) becomes small.

This utility measure is an on-line variant of a measure recently proposed for batch vector quantization methods [Fri97].

6   Utility criterion for removal of units

How can the utility measure be used to remove neurons in a GNG network with fixed maximum size? We can not simply remove the unit with minimum value of U since there always will be one and this strategy would cause a constant change of the network structure even for perfectly stationary probability distributions. Let us recall that the stated objective is minimization of the overall quantization error (1). Therefore, removal with subsequent re-insertion at a different place should only be performed, if we can expect a decrease of this error. When we remove a unit, the GNG network with fixed maximum size immediately inserts a new unit near the unit q with maximum accumulated error, i.e., the unit such that

$$q := \arg \max_c E_c.$$

We can expect the newly inserted unit to reduce this error in the future, since it reduces the size of the Voronoi region of q. When the mean distance of the signals in q's Voronoi region is reduced by a certain factor, the (square) quantization error is reduced by an even larger factor. The analysis in section 9.2 indicates that a cautious estimate for the local error reduction is 75%, i.e. a reduction by a factor 4. An estimate of the increase in quantization error by removing a unit i is its current utility value U_i.

A suitable strategy is to remove a unit when its utility value falls below a certain fraction of the error variable E_q. Let i be the unit with minimum utility

$$i := \arg \min_c U_c.$$

Then i should be removed if

$$E_q/U_i > k \tag{8}$$

for some value of k. We denote (8) as the *utility criterion*. Large values of the parameter k require a large ratio of maximum error and minimum utility and will cause less frequent deletions of units. Smaller values will cause accordingly more and faster deletions which leads to faster tracking of non-stationary distributions.

7   Re-distribution of utility

In principle one could, after each removal and insertion, discard all utility and error values. Such a strategy is realized in the "growing grid" method [Fri95b] and enforces a number of adaptation steps per insertion which is proportional to the network size. The GNG algorithm [Fri95a], however, uses the following simple re-distribution scheme for the error which can also be applied to the utility value. New units are always inserted at the center of an existing edge. The error values of the units at the end points of the edge are diminished by a fraction $\alpha = 0.5$, and the new unit gets an initial error value interpolated from these two units. The advantage of this re-distribution is that the accumulated statistical information is not thrown away at each insertion and, therefore, less adaptation steps per insertion are needed. In particular a new unit can be inserted after a

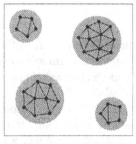 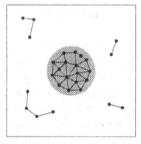 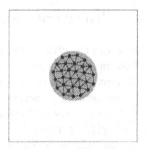

a) after 20000 signals   b) after 25000 signals   c) after 40000 signals

Fig. 3. A GNG-U network successfully tracks a non-stationary signal distribution $p(\xi)$.
a) Situation before the distribution changes. b) State of the network 5000 adaptation
steps after the distribution had changed. c) After 40000 adaptation steps all units have
been adapted or relocated to regions of high probability density. The value for k is 3.
The other parameters (including β) are as in figure 2.

constant number λ of adaptation steps. An identical re-distribution is done with
the utility values in the simulations shown below.

8 Simulation examples

The proposed utility criterion for removal of units was combined with the GNG
algorithm and the resulting method, which we call GNG-U (growing neural gas
with utility criterion) in the following, was successfully applied to various non-
stationary probability distributions. We report here on three experiments with
GNG-U. In figure 3 a simulation with the same non-stationary distribution as in
figures 1 and 2 is shown. The GNG-U algorithm successfully relocates all units
after the sudden change in the probability distribution.

In figure 4 a non-stationary one-dimensional distribution is used to illustrate
the behavior of GNG-U throughout a complete simulation (see the caption for
details).

Figure 5 illustrates that the utility measure U can also be used to reduce
the size of an existing network in a controlled way. To do this, one only has
to remove sequentially the unit with the minimum utility. For the distribution
used in figure 5, the remaining units distribute optimally to the existing clusters
of positive probability density. On should note that for pure size reduction the
utility *criterion* (8) is not needed. Instead, the utility measure U can be used
directly since in this case the only question is *which* unit to remove and not
whether to perform a removal at all. To detect good clusterings, one could observe
the mean quantization error during the process of eliminating units and look for
configurations where a removal of the next unit leads to a large increase of
quantization error.

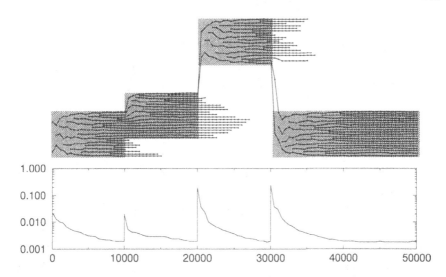

Fig. 4. A GNG-U network with maximum size set to 20 tracks a one-dimensional non-stationary probability distribution which is indicated by the shaded areas. After 10000, 20000, and 30000 signals the mean of the distribution was suddenly changed. It can be seen that moderate changes of the mean (like the change after 10000 signals) lead to a re-use of many units. More drastic changes, however, lead to deletions and subsequent insertions of units. Parameters are the same as in figure 2. The graph at the bottom displays the mean square quantization error.

9 Choice of parameter values

The proposed method requires the setting of two new parameters β and k. In the following we investigate the role of these parameters and give guidelines for suitable values.

9.1 Choice of β

The parameter β is used in equations 3 and 7 to let the error E_c or the utility U_c of a unit c decay exponentially fast with the number of presented input signals. Since for each input signal the values E_c and U_c of the winning unit are incremented (equations 2 and 6), both values represent an exponentially decaying average of past error or utility increments. In the following we restrict our discussion to E_c since the utility U_c can be analyzed in complete analogy.

Let us assume a certain mean error increment $\overline{\Delta E}$ per presented input signal. In this case the quantity E_c does only grow up to a value $\tilde{E}_c$ where the exponential decay is equally to the mean error increment. This can be characterized as a dynamic equilibrium for which the following holds:

$$\overline{\Delta E} = \beta \tilde{E}_c. \tag{9}$$

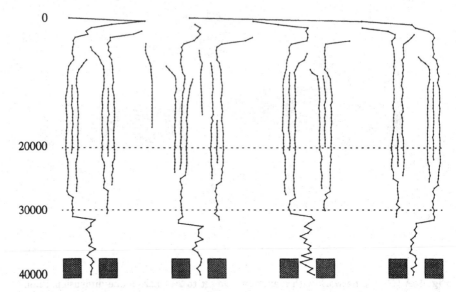

Fig. 5. Controlled reduction of network size using the utility measure. In this simulation, which involved 40000 signals a GNG-U network was allowed to grow to a size of 24 units. The resulting units were distributed evenly over the 8 segments (arranged in 4 groups of two) of uniform probability density. After 20000 adaptation steps the allowed number of units was reduced to 8 and the utility measure was used to successively remove the least useful unit. The result was a configuration that each unit handles one segment. After another 10000 adaptation steps the allowed number of units was reduced to 4. The evaluation of the utility measure lead to a configuration where each unit handles one of the 4 two-segment clusters which is the optimum in this case. Parameters are the same as in figure 2.

Thus, the equilibrium is reached at

$$\tilde{E}_c = \frac{\overline{\Delta E}}{\beta}. \tag{10}$$

and takes on larger values if β approaches zero from above.

Since we are in the context of non-stationary probability distributions, it is of interest, how the running average E_c is constituted. In particular we like to know how old the "mean input signal" is which is incorporated in E_c. This figure can be obtained by computing the following infinite sum:

$$\overline{age} = \sum_{i=0}^{\infty} i(1 - \beta)^i \tag{11}$$

which we approximate by a continuous integral:

$$\overline{age} \approx \int_0^{\infty} x(1 - \beta)^x \, dx = \int_0^{\infty} x e^{\ln(1-\beta)x} \, dx. \tag{12}$$

Substituting $\varepsilon := -\ln(1 - \beta)$ gives

$$\overline{\text{age}} \approx \int_0^\infty x e^{-\varepsilon x}\, dx = \left[\frac{e^{-\varepsilon x}}{\varepsilon^2}(-\varepsilon x - 1)\right]_0^\infty \tag{13}$$

$$= 0 - \frac{1}{\varepsilon^2}(-1) = \frac{1}{\varepsilon^2} \tag{14}$$

$$= \frac{1}{\ln(1 - \beta)^2}. \tag{15}$$

Apart from the question how many signals are incorporated in the exponential average, it is important to know, how fast the average is able to react to sudden changes in the probability distribution. In particular one could ask how long it takes for E_c (or equivalently U_c) to decay to 50% of its current value if the probability density in the Voronoi region of a unit c suddenly becomes zero. For the number n_{half} of input signals needed for this to happen holds

$$(1 - \beta)^{n_{\text{half}}} = 0.5. \tag{16}$$

or

$$n_{\text{half}} = \frac{\ln 0.5}{\ln(1 - \beta)} \tag{17}$$

The value of n_{half} gives an indication, how quickly a probability distribution may change in order to still be trackable by the described system.

In the following table we have computed (rounded) values of $\overline{\text{age}}$ and n_{half} for a number of possible settings for the parameter β.

β	$\overline{\text{age}}$	n_{half}
0.5	2	1
0.1	90	7
0.01	10,000	69
0.001	1,000,000	693
0.0005	4,000,000	1386
0.0001	10,000,000	6931

What is evident from the table is the fact that even though the number of input signals effectively used to compute the values E_c and U_c may be large ($\overline{\text{age}}$), the system is still able to detect sudden changes in the probability density rapidly (n_{half}). Of course this is due to the exponential decay which initially causes large changes of the accumulated values.

In our simulations we used a value $\beta = 0.0005$. Further experiments with the probability distribution used in figure 4, however, gave similar results for values in the whole range from 0.01 to 0.0005. Larger values (e.g. 0.1) lead to deletions even in completely stationary and well-quantized distributions. Lower values (e.g. 0.0001) lead to to few deletions with the result that the distribution was not well tracked.

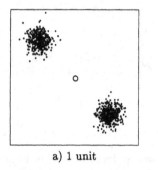 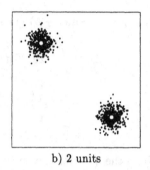

a) 1 unit   b) 2 units

Fig. 6. For a highly non-uniform probability density the insertion of a new unit can reduce the mean error considerably. a) shows a situation where one reference vector encodes two clusters with a large error. b) shows the same data set after one new unit has been inserted. The quantization error is much smaller and would become even zero if each cluster would collapse to its respective center.

9.2 Choice of k

The parameter k is used in the utility criterion (8) to decide whether a unit should be deleted and subsequently re-inserted at a different location in input space. Let us recall that the overall goal of the system is to reduce the expected quantization error (1). Thus we should only perform a deletion/insertion if it is likely that this will reduce the error.

Deleting a unit c in the first place increases the quantization error since all data vectors in the Voronoi region of this unit will be mapped to neighboring (but more distant) units. For the actual increase in error we have an excellent estimate through the utility variable U_c which is nothing else than an exponential average of hypothetical error values caused by the hypothetical removal of c.

Estimating how much the error will decrease after we (re-)insert the deleted unit somewhere in input space is much more difficult, however. In particular the amount of error reduction depends on the local structure of the probability density in the vicinity of the inserted unit. To illustrate the point, consider a reference vector which codes data belonging to two dense clusters (figure 6a). Inserting a new unit here may lead to a drastic decrease in error (figure 6b). In the case of locally uniform probability distributions, however, the effect of an insertion is much more limited (see figure 7).

From the above examples it is evident that a cautious estimate of the error reduction can be made by assuming locally uniform densities. Let us compute the corresponding change in error for the case of a 1-dimensional distribution. Let us assume that the unit q with maximum accumulated error encodes an interval in R^1 with length l and uniform probability density. The optimum position for the unit is in the centroid, i.e. in the center of the interval. The expected error in this case is

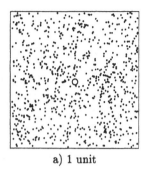 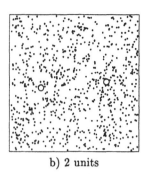

a) 1 unit   b) 2 units

Fig. 7. For a uniform probability density the insertion of a new unit has a relatively small effect on the mean error. a) shows a situation where one reference vector encodes an area of uniform probability density. b) shows the same data set after one new unit has been inserted. The mean distance between data points and reference vectors is still rather large. The situation would be similar for further insertions.

$$E_1 = 2 \int_0^{l/2} x^2 \, dx = 2 \left[\frac{1}{3} x^3 \right]_0^{l/2} = \frac{l^3}{12}. \tag{18}$$

If we now insert a second unit in the interval, after a while both units will be adapted symmetrically to positions $l/4$ and $3l/4$ in the interval. The resulting expected error is

$$E_2 = 4 \int_0^{l/4} x^2 \, dx = 4 \left[\frac{1}{3} x^3 \right]_0^{l/4} = \frac{l^3}{48} \tag{19}$$

which is a reduction of 75% w.r.t to E_1. One should note that the error reduction would be even stronger if the interval was part of a larger zone of uniform density since in this case also Voronoi regions of neighboring units would become smaller leading to additional error reduction.

From the above we can conclude that in the 1D-case we can expect the overall quantization error to decrease after a deletion/insertion, if for the unit i with minimum utility U_i and the unit q with maximum error E_q the following holds:

$$E_q * 0.75 > U_i \quad \text{or equivalently:} \quad \frac{E_q}{U_i} > \frac{4}{3}. \tag{20}$$

This means that for 1D data a value of $k \geq 2$ would be appropriate.

For higher-dimensional data the corresponding estimates become more elaborate since the shape of the Voronoi regions can be any convex polyhedron (depending on the positions of the neighboring reference vectors). However, it seems to be reasonable at this point to also assume a reduction of the mean distance from data points to reference vectors by a factor of 2 and thus a reduction of the local square error by a factor of 4. Again, the error reduction may be much larger in the case of non-uniform densities. We can thus generally use values for k larger or equal to 2 to acchieve error reduction.

10 Discussion and outlook

In this paper a novel on-line criterion for removal of units in self-organizing networks is proposed. The combination of this criterion and an existing incremental network model (growing neural gas) results in a new method, called GNG-U, which is able to closely track non-stationary distributions. This is a feature not found in known related models such as different variants of the self-organizing map or the neural gas algorithm and which could open up completely new application areas for these kinds of networks, e.g. for data mining.

For the new criterion a parameter k has to be chosen which indicates how large the ratio between accumulated error of a unit q and accumulated utility of a unit i must be, before unit i is removed and re-inserted at a place near q. Moreover, a parameter β must be chosen which controls the decay of local error estimates and thus the "reaction time" of the whole method. Both parameters have been theoretically analyzed and suitable values have been derived.

The proposed utility measure, which is the basis for the utility criterion, can also be used on its own to prune self-organizing networks to a desired size. When always the unit with minimum utility is removed, the increase in error is near-minimal. Observing the change in overall error as a function of the network size during pruning may give information on the number of natural clusters in the data.

References

[BV95]   H.-U. Bauer and Th. Villmann. Growing a hypercubical output space in a self-organizing feature map. TR-95-030, International Computer Science Institute, Berkeley, 1995.

[Fri94] B. Fritzke. Growing cell structures – a self-organizing network for unsupervised and supervised learning. *Neural Networks*, 7(9):1441–1460, 1994.

[Fri95a] B. Fritzke. A growing neural gas network learns topologies. In G. Tesauro, D. S. Touretzky, and T. K. Leen, editors, *Advances in Neural Information Processing Systems 7*, pages 625–632. MIT Press, Cambridge MA, 1995.

[Fri95b] B. Fritzke. Growing grid – a self-organizing network with constant neighborhood range and adaptation strength. *Neural Processing Letters*, 2(5):9–13, 1995.

[Fri97] B. Fritzke. The LBG-U method for vector quantization – an improvement over LBG inspired from neural networks. *Neural Processing Letters*, 5(1):35–45, 1997.

[Koh82] T. Kohonen. Self-organized formation of topologically correct feature maps. *Biological Cybernetics*, 43:59–69, 1982.

[Mar93] T. M. Martinetz. Competitive Hebbian learning rule forms perfectly topology preserving maps. In *ICANN'93: International Conference on Artificial Neural Networks*, pages 427–434, Amsterdam, 1993. Springer.

[MBS93] T. M. Martinetz, S. G. Berkovich, and K. J. Schulten. Neural-gas network for vector quantization and its application to time-series prediction. *IEEE Transactions on Neural Networks*, 4(4):558–569, 1993.

[RMS91] H. J. Ritter, T. M. Martinetz, and K. J. Schulten. *Neuronale Netze*. Addison-Wesley, München, 1991. 2. Auflage.

Development of Decision Support Algorithms for Intensive Care Medicine: A New Approach Combining Time Series Analysis and a Knowledge Base System with Learning and Revision Capabilities

Michael Imhoff[1], Ursula Gather[2] and Katharina Morik[3]

[1] Surgical Department, Community Hospital Dortmund,
Beurhausstraße 40, D-44137 Dortmund
[2] Department of Statistics and
[3] Department of Computer Science,
Dortmund University, D-44221 Dortmund

Abstract. The overwhelming flood of data in intensive care medicine precludes consistent judgement of medical interventions by humans. Therefore, computerized decision support is needed to assist the health care professional in making reproducible, high-quality decisions at the bedside. Traditional expert systems rely on a tedious, labor-intensive and time-consuming approach in their development which falls short of exploiting existing numerical and qualitative data in large medical databases. Therefore, we applied a new concept of combining time series analysis and a knowledge base system with learning and revision capabilities (MOBAL) for rapid development of decision support algorithms for hemodynamic management of the critically ill. This approach could be successfully implemented in an existing intensive care database handling time-oriented data to validate and refine the intervention rules. The generation of hypotheses for identified contradictions lead to conclusive medical explanations that helped to further refine the knowledge base. This approach will provide for a more efficient and timely development of decision support algorithms.

1 Introduction

In critical care an abundance of information is generated during the process of care. In the last several years a small number of clinical information systems (CIS) has become commercially available for use in intensive care. These systems provide for a complete medical documentation at the bedside. Their clinical usefulness and efficiency has been shown repeatedly in recent years [3, ?,?]. Databases with more than 2000 separate patient-related variables are now available for further analysis [5].

The multitude of variables presented at the bedside even without a CIS precludes medical judgement by humans. We can currently be confronted with more than

200 variables in the critically ill during a typical morning round [16]. We know, however, that an experienced physician may not be able to develop a systematic response to any problem involving more than seven variables [12]. Moreover, humans are limited in their ability to estimate the degree of relatedness between only two variables [10].

This problem is most pronounced in the evaluation of the measurable effect of a therapeutic intervention. Personal bias, experience, and a certain expectation toward the respective intervention may distort an objective judgement [2].

1.1 Data Situation

In modern intensive care, every minute numerous measurements are taken at the bedside. This generates a very high dimensional data space for each patient. However, the values of some vital signs are sometimes only recorded once every hour, while other vital signs are only recorded for a subset of patients. Hence, the overall high dimensional data space is sparsely populated. Moreover, the average time difference between intervention as charted and calculated hemodynamic effect can show a wide variation [7]. Even the automatic measurements can be noisy due to manipulation of measurement equipment, flushing of pressure transducers, or technical artifacts. To make it even worse, relevant demographic and diagnostic parameters may even not be recorded at all.

In summary, we have masses of noisy, high dimensional, sparse time series of numerical data. Medical experts explain the numerical data in qualitative terms of high abstraction. The background knowledge given by the expert covers functional models of the human body as well as expertise in the proper treatment of intensive care patients including effects of drugs and fluids. In the expert's reasoning, time becomes the relation between time intervals, abstracting from the exact duration of, e.g., an increasing heart rate and focusing on tendencies of other parameters (e.g., cardiac output) within overlapping time intervals. Thus we have complex qualitative background knowledge explaining both the patient's and the physician's behavior.

1.2 Decision Support Systems in Intensive Care Medicine

Using data from the most comprehensive singular clinical data repository at the LDS hospital, Salt Lake City, Utah, USA, the group of Morris [20] developed a rule-based decision support system (DSS) for respiratory care in acute respiratory distress syndrome. The development of this highly specialized system required more than 25 manyears.

It is a propositional rule base without a mechanism for consistency checking or matching rules and data. All currently known medical DSS are based on manually acquired expert knowledge which is refined in an iterative process of prospective trials. Numerical aspects of existing data from the processes that are to be controlled by these DSS are not integrated in the development of these expert systems.

This traditional approach to the development of knowledge bases and DSS has several serious shortcomings:

- Expert knowledge is not necessarily validated against clinical data.
- Many underlying pathophysiological mechanisms are not fully understood.
- Clinical experiments in intensive care are difficult or not feasible at all.
- Prospective trials for validation of the decision support algorithms, especially at an early stage of development, are costly, time-consuming and may not represent actual clinical practice.

With the advances of artificial intelligence and knowledge representation in recent years, consistency of rules and (patient) data can be checked automatically. Formalisms of first-order logic allow to represent relations between time intervals. Therefore, knowledge can be formalized and validated against highly multivariate data structures. However, this opportunity has not yet received appropriate attention. Until today only few applications of artificial intelligence to medical databases have been reported from the fields of pathology [11], infectious diseases [22], epidemiology [23], and cardiology [18]. None of these applications were done in the time domain analyzing temporal patterns of on-line monitoring data or other repeatedly measured variables in critical care.

Our new approach combines modeling of expert knowledge with data-driven methods. The knowledge base is validated against existing patients' data. This approach is meant to be significantly more effective than the tedious, time-consuming, and costly process of traditional DSS development.

2 Conceptual Framework

Looking again at the list of advantages of computer-assisted decision support in intensive care, we obtain a list of requirements for the system to be built. The system must base its decisions on explicit medical methods. We do not aim at modeling the hemodynamic system, the cardiac processes of patients. Neither do we aim at modeling the actual physician's behavior. Instead, the knowledge base must represent a therapy protocol which can be applied to measurements of the patient. While this may resemble the early days of knowledge acquisition for expert systems, the task at hand goes beyond classical medical knowledge acquisition, since the system has to cope with high dimensional data in real time. Its task is on-line monitoring, not heuristic classification or cover and differentiate. Moreover, the data consists of time series. Time stamped data do not necessarily require sequence analysis methods. For an application, we have to determine whether points in time, time intervals and their relations, or curves of measurements offer an adequate representation. These questions point at the problem of finding an adequate representation. Two sets of requirements on the capabilities of the representation can be distinguished:

- The representation must handle numerical data, valid in one point in time, and time series. For each point in time, it must classify whether and which therapy intervention is appropriate for the patient.

- The representation must handle relations of time intervals, interrelations of diverse drugs and relations between different parameters of the patient. It has to derive expected effects of medical interventions from medical knowledge and compare expected outcome with actual outcome.

The requirements are conflicting. While we know good formalisms for each of the sets, we are not aware of a representation that fulfills both sets of demands. Hence, we decided to break down the overall reasoning into several processes and find an appropriate representation for each of them, independently.

The patients' measurements at one point in time are used in order to recommend an intervention. This corresponds to clinical practice where for each point in time a recommendation for optimal treatment is needed. Of course, one of the recommendations may be not to change the current therapy. The recommendation of interventions constitutes a model of physicians' behavior. Its recommendations are input into the knowledge base. A recommended intervention is checked by calculating its expected effects on the basis of medical knowledge. Medical knowledge qualitatively describes a patient's state during a time interval and effects of drugs. It constitutes a model of the patient's hemodynamic system. The medical knowledge base uses relations between time intervals and their abstract characterizations. To this end, patient's measurements are abstracted with respect to their course over time. The abstraction mechanism handles curves of measurements. It is straightforward to determine appropriate representations from this conceptual framework: Data abstraction and state-action rules use numerical functions, the other modules use a restricted first-order logic. The integration of numerical and knowledge-based methods allows us to validate the processes carefully. In detail, the processes we have designed are:

Data abstraction: From a series of measurements of one vital sign of the patient, eliminate outliers and find level changes. This abstracts the measurements to qualitative propositions with respect to a time interval.

State-action rules: Given the numerical data describing vital signs of the patient and his or her current medication, find the appropriate intervention. An intervention is formalized as increasing, decreasing or not changing the dose of a drug. The decision is made every minute.

Action-effect rules: Given the state of a patient described in qualitative terms, medical knowledge about effects of substances, relations between different vital signs, interrelation between different substances, a sequence of interventions, and a current intervention, find the effects of the current intervention on the patient. The derivation of effects is made for each intervention.

Conflict detection: Given the expected effect of a medication for a patient and his or her actual state, find inconsistencies.

Conflict explanation: Given interventions with effects on the patient that follow the medical knowledge and those that are in conflict with medical knowledge, find characterizations which separate the two sets.

Following this conceptual framework we have developed a knowledge that, for medications affecting the cardiovascular system in the critically ill,

- provides explicit expert knowledge,
- validates and refines this knowledge against existing, i.e., historic, real-world data,
- generates explicit, exportable, and executable rules for measurements from the patient's cardiovascular system.

The knowledge base is coupled with an abstraction mechanism that detects outliers and level changes.

3 Methods

3.1 Data acquisition and data set

On a 16-bed surgical ICU all medication data was charted with a CIS, allowing the user one minute time resolution for all data. This data was transferred into a secondary SQL database and made available for further analysis.

The entire database comprises about 2000 independent variables. On-line monitoring data was acquired from 148 consecutive critically ill patients (53 female, 95 male, mean age 64.1 years), who had pulmonary artery catheters for extended hemodynamic monitoring, in one minute intervals from the CIS amounting to 679,817 sets of observations.

¿From the original database hemodynamic variables, vasoactive drugs, and some demographic data were selected for the analysis (table 1). This was done following medical reasoning and due to the limitation of the medical knowledge base to the cardiovascular system.

Table 1. Data set for knowledge base validation

Vital Signs (measured every minute)	Continously Given Drugs (changes charted at 1-min-resolution)	Demographic Attributes (charted once at admission)
Diastolic Arterial Pressure	Dobutamine	Broca-Index
Systolic Arterial Pressure	Adrenaline	Age
Mean Arterial Pressure	Glycerol trinitrate	Body Surface Area
Heart Rate	Noradrenaline	Emergency Surgery (y/n)
Central Venous Pressure	Dopamine	
Diastolic Pulmonary Pressure	Nifedipine	
Systlic Pulmonary Pressure		
Mean Pulmonary Pressure		

3.2 Development and Validation of the Knowledge-based System

The actual development and validation of the knowledge-based system was a process of four components:

- Development of a knowledge base with the help of a medical expert.
- Data abstraction of time oriented data with time series analysis.
- Knowledge discovery of state-action rules in the medical database using machine learning.
- Validation of the expert system against existing data from the CIS database using the MOBAL system and incorporating the data abstraction from the time series analysis. The knowledge discovery of state-action rules is described in [14]. Here, we present the integration of time series analysis with the knowledge base of action-effect rules.

Medical knowledge base. A medical expert defined the necessary knowledge. This knowledge is medical textbook knowledge for the cardiovascular system. It reflects direct pharmacological effects of a selected list of medical interventions on the basic hemodynamic variables. Any interaction of these interventions with other organ systems or of other organ systems with the cardiovascular system were ignored. An excerpt of intervention-effect relations is shown in table 2.

Table 2. Medical knowledge base for hemodynamic effects. +=increase of the respective variable or intervention; -=decrease of the respective variable or intervention; 0=no change

| Intervention | Heart Rate | Effect on hemodynamic variables | | | |
		Mean Arterial Pressure	Mean Pulmonary Artery Pressure	Central Venous Pressure	Cardiac Output
Dobutamine	+	+	+	+	0	+
	-	-	-	-	0	-
Adrenaline	+	+	+	+	0	+
	-	-	-	-	0	-
Noradrenaline	+	-	+	+	0	-
	-	+	-	-	0	+
Nitroglycerine	+	+	-	-	-	+
	-	-	+	+	+	-
Fluid intake/ output	+	-	+	+	+	+
	-	+	-	-	-	-

For modeling medical knowledge in terms of action-effect rules we chose a restricted first-order logic. Using the MOBAL system [15] it was extremely easy to write the according rules. It offers a compact representation of medical knowledge with a small number of rules, fulfilling the real-world demand for a knowledge base to be understandable by humans and accessible for expert validation. In addition, the knowledge base directly serves as background knowledge for learning refined rules and for doing knowledge revision. Discussions with experts in intensive care showed that the knowledge base is, in fact, understandable. The consistency checking of MOBAL allows to automatically detect cases where the actual patient state differs from the predicted effect of an intervention. Experts find it very useful to discuss a rule in the light of selected contradictory cases.Based on information from the medical expert we built a compact knowledge base modeling the effects of drugs. Not counting patients' records, the knowledge base consists of 39 rules and 88 facts. An example of facts and rules is shown in table 3.

Table 3. Excerpt from the knowledge base in MOBAL

% Facts:
contains(dobutrex,dobutamin).
med_ effect(dobutamin,1,10,hr,up).
opposite(up,down).
% Rules:
intervention(P,T1,T2,M,D1) &
intervention(P,T2,M,D2) &
contains(M,S) &
med_ effect(S,From1,To1,V,Dir) &
med_ effect(S,From2,To2,Dir) &
ne(From1,From2) & gt(D2,D1) &
lt(D1,To1) & ge(D1,From1) & lt(D2,To2) & ge(D2,From2)
→ interv_ effect(P,T2,T3,M,V,Dir).
% Patient Data:
level_ change(pat460, 160, 168, hr, up).
intervention(pat460, 159, 190, dobutrex, 8).

The example rule states that increasing the dose from D1 to D2 of a drug M leads to an increasing effect on the parameter V of a patient P. The time intervals in which a certain dose is given to the patient are immediate successors. The dose is changed significantly. Using unification, the effect of the substance, Dir, is propagated to the time interval following the intervention, i.e. T2 to T3.

Data abstraction. Time series analysis was employed for data abstraction of the time oriented variables. After previous studies [6] phase space models were used for this analysis.

Phase space models are based on a transformation of time series in an Euclidean space. This transformation is called Phase Space Embedding, a technique derived from the theory of nonlinear dynamic systems. Given a time series (x_t) with N observations, Packard [17] and Takens [19] constructed so-called phase space vectors x_t, which are defined by:

$$x_t = (x_{t+m-1},\ldots,x_{t+1},x_t), \quad x_t \in R, \quad t = 1,\ldots,N-m+1, \quad m \in N \backslash \{0\} \quad (1)$$

Here, m is called embedding dimension. Numerous rules exist for choosing m in nonlinear models. In most cases the components of the phase space vectors are not neighboring observations, but rather they are separated by a time delay [9, ?]. Focusing on stochastic processes it is necessary to take into account the dependencies of neighboring observations. Thus for the choice of only those preciding observations are considered, which have a direct influence on the present observation.

The components of x_t are chronological observations with a time delay (lag) always of one. This improves the identification of patterns as dependencies between consecutive observations are taken into consideration.

The identification procedure, that we developed, uses the differenced time series d_t, which is $d_t = y_t - y_{t-1}$, $t = 2, \ldots, N$. In a differenced series an abrupt level change will be represented by one outlier. The procedure focuses on the identification of these observations. On the basis of the movement of the phase space vectors, which contain such observations, a discrimination between different patterns is done. The vectors d_t, $t = 2, \ldots, N$ were analyzed in consecutive order whether they pointed into a distant region. If a vector lies in a distant region, i.e. the vector extrudes from the cloud, it can be discriminated between different patterns after observing further values (a detailed description of the methodology is given in [1]).

Validation. The inference engine of MOBAL derives from patient data expected effects.and compares them with actual effects. Overall, the patient data contain 8,200 interventions. 22,599 effects of the interventions were derived using forward chaining. In order to compare the predicted effects with the actual ones, we distinguish three types of conformity or contradiction. A predicted effect is

weakly conform with observed patient behavior, if no level change is observed, the patient's state remains stable;
strongly conform with observed patient behavior, if the observed level change has the same direction as is predicted by the rules;
contradictory with observed patient behavior, if a level change is observed into a direction opposite to the one predicted by the rules.

Note, that weak conformity is not in conflict with medical knowledge, but shows best therapeutic practice. Smooth medication keeps the patient's state stable and does not lead to oscillating reactions of the patient. The explanation of conflicts between prediction and actual outcome requires to investigate many hypotheses. For this task, we used the rule learning tool RDT of the system [13].

4 Results

When matching the derived effects with the actual ones, the system detected:

weak conformity: 13,364 effects (59.14%) took place in the restricted sense, that the patient's state remained stable.

strong conformity: 5,165 effects (22.85%) took place in the sense, that increasing or decreasing effects of drugs on vital signs matched corresponding level changes.

contradiction: 4,070 contradictions (18.01%) of the interventions, were detected. The observed level change of a vital sign went into the opposite direction of the knowledge-based prediction.

First, we started a knowledge revision process using concept formation using the methods of Stefan Wrobel [24]. A concept is learned that separates successful rule applications, i.e., those, where the rules are not in conflict with the observations, from rule applications that lead to a contradiction. However, no clear separation could be found. Hence, we weakened the task to filtering out influential aspects. For learning, we chose 5,466 interventions with their effects being classified as conform (including the weak conformity described above) and as not conform. Eleven predicates about the patient and the medications established the structured hypothesis space. Of all possible combinations, 121 hypotheses had to be tested. The findings were:

- The rule stating that lowering a dose of a drug that increases the observed vital sign should lower the respective vital sign is less reliable than the opposite rule.
- If combined with the age of the patient being around 55 years or the weight of the patient being small, the rule for effects of decreasing a medication is particularly unreliable.
- The weight of the patient alone has no impact on the reliability of action-effect rules.
- For elderly patients (65 year and older), the weight is an influential feature.
- To our surprise, the amount of reducing or increasing the dose is not a relevant aspect for explaining contradictions, neither alone nor in combination with other features.

Relational learning did a good job in generating and testing many hypotheses. However, the learning results clearly show that the decisive features that would distinguish successful rule applications from contradictory ones are not present in the data. This lead to careful inspection of contradictory cases.Medical explanation for the contradictions include the following aspects:

- Cardiac output measurements could explain many of the inconsistencies. Cardiac output measurements were not included in this data as they were acquired discontinuously at varying time intervals.
- Values for fluid intake and output (I/O) were also not included in initial data set for the same reason. Especially rapid fluid intake can significantly counteract or modify the effect of the vasoactive drugs in the current data sample. Both I/O and cardiac output values will be included in future validation sets.
- Depending on the actual state of disease and individual predisposition patients may show a great inter- and intraindividual variability in their response

to drug interventions. This is an effect that from a mechanistic perspective is very difficult to control.

- The increase of a vasoactive drug will also exert a more immediate effect than its decrease, because the halftime of a drug after decrease is typically longer. In theory, these pharmacodynamic and pharmacokinetic properties could also be modeled in the knowledge base.

5 Discussion

We present an approach to integrating statistical and knowledge-based methods for patient monitoring in intensive care. This application involves high dimensional time series data, demanding high quality decision support under real time constraints. It requires the integration of numerical data and qualitative knowledge. Validating the knowledge base is of particular importance.These properties make this case study a representative for a large number of applications in medicine and engineering.

The overall system is designed such that it can be applied at the hospital. The system exploits patients' data as given and outputs operational recommendations for interventions. The knowledge base could easily be validated against existing real-world data. The combination of knowledge-based and statistical approaches eases the development of medical decision support systems.

Our next steps are the addition of a limited number of variables, like cardiac output and I/O, to probably solve the majority of contradictions. A comparison with a hemodynamic knowledge base that is currently developed at the LDS hospital at Salt Lake City is planned. The LDS knowledge base does not take the stream of measurements as input, but reads vital signs on demand. It cannot be applied to past data and be evaluated with respect to them, because there is no component for checking consistency. We plan to transfer the knowledge base into our system so that it can be tested on patients' data. The impact of a stream of data (our approach) as opposed to some selected points in time when a vital sign is read (the LDS approach) will be investigated carefully.

Acknowledgments

This work has been supported, in part, by the Deutsche Forschungsgemeinschaft, SFB 475. We would like to thank Marcus Bauer for their continuing support with time series analysis. We would also like to thank Peter Brockhausen who linked RDT with an Oracle database and loaded diverse very large datasets.

References

1. Bauer M, Gather U, Imhoff M (in print): Analysis of high dimensional data from intensive care medicine. In: Payne R (ed) Proceeding in Computatinal Statistics, Springer-Verlag, Berlin

2. Guyatt G, Drummund M, Feeny D, Tugwell P, Stoddart G, Haynes R, Bennett K, LaBelle R (1986): Guidelines for the clinical and economic evaluation of health care technologies. Soc Sci Med **22**: 393-408.
3. Imhoff M (1995): A clinical information system on the intensive care unit: dream or night mare? In: Rubi J.A.G.: Medicina Intensiva 1995, XXX. Congreso SEMIUC. Murcia, 4.-6.10.1995
4. Imhoff M (1996): 3 years clinical use of the Siemens Emtek System 2000: Efforts and Benefits. Clinical Intensive Care **7 (Suppl.)**: 43-44
5. Imhoff M (1998): Clinical Data Acquisition: What and how? Journal für Anästhesie und Intensivmedizin **5**: 85-86
6. Imhoff M, Bauer M, Gather U, Löhlein D (1998): Statistical pattern detection in univariate time series of intensive care on-line monitoring data. Intensive Care Med **24**: 1305-1314
7. Imhoff M, Bauer M, Gather U (submitted): Time-effect relations of medical interventions in a clinical information system. KI-99, Bonn 13.-15.05.99
8. Imhoff M, Lehner JH, Löhlein D (1994): 2 years clinical experience with a clinical information system on a surgical ICU. In: Mutz N.J., Koller W., Benzer H.: 7th European Congress on Intensive Care Medicine. Monduzi Editore, Bologna. S. 163-166
9. Isham V (1993): Statistical aspects of chaos: a review. In: Barndorff-Nielsen OE, Jensen WS, Kendall WS (eds) Networks and Chaos - Statistical and Probabilistic Aspects. Chapman and Hall, London, pp 251-300
10. Jennings D, Amabile T, Ross L (1982): Informal covariation assessments: Data-based versus theory-based judgements. In: Kahnemann D, Slovic P, Tversky A, editors. Judgement under uncertainty: Heuristics and biases. Cambridge: Cambridge University Press: 211-230.
11. McDonald JM, Brossette S, Moser SA (1998): Pathology information systems: data mining leads to knowledge discovery. Arch Pathol Lab Med **122**: 409-411
12. Miller G (1956): The magical number seven, plus of minus two: Some limits to our capacity for processing information. Psychol Rev **63**: 81-97
13. Morik K, Brockhausen P (1997): A multistrategy approach to relational knowledge discovery in databases. Machine Learning Journal **27**: 287-312
14. Morik K, Brockhausen P, Joachims T (1999): Combining statistical learning with a knowledge-based approach - A case study in intensive care monitoring. In: Procs. Int. Conf. Machine Learning (in print).
15. Morik K, Wrobel S, Kietz JU, Emde W (1993): Knowledge acquisition and machine learning - Theory, methods, and applications. Academic Press, London
16. Morris A, Gardner R (1992): Computer applications. In: Hall J, Schmidt G, Wood L, editors. Principles of Critical Care. New York: McGraw-Hill Book Company: 500-14
17. Packard NH, Crutchfield JP, Farmer JD, Shaw RS (1980): Geometry from a time series. Phys Rev Lett **45**: 712-716
18. Soo VW, Wang JS, Wang SP (1994): Learning and discovery from a clinical database: an incremental concept formation approach. Artif Intell Med **6**: 249-261
19. Takens F (1980): Detecting strange attractors in turbulence. In: Rand DA, Young LS (eds) Dynamical systems and turbulence. Lecture Notes in Mathematics Vol. 898 Springer, Berlin, pp 366-381
20. Thomsen GE, Pope D, East TD, Morris AH, Kinder AT, Carlson DA, Smith GL, Wallace CJ, Orme JF Jr, Clemmer TP, et al. (1993): Clinical performance of a rule-based decision support system for mechanical ventilation in ARDS patients. Proc Annu Symp Comput Appl Med Care: 339-343

21. Tong H (1995): A personal overview of non-linear time series analysis from a chaos perspective, Scand J Statist **22**: 399-445
22. Tsumoto S, Ziarko W, Shan N, Tanaka H (1995): Knowledge discovery in clinical databases based on variable precision rough set model. Proc Annu Symp Comput Appl Med Care 1995: 270-274
23. da Veiga FA (1996): Structure discovery in medical databases: a conceptual clustering approach. Artif Intell Med **8**: 473-491
24. Wrobel S (1994): Concept formation and knowledge revision. Kluwer Academic Publishers, Dordrecht

Object Recognition with Shape Prototypes in a 3D Construction Scenario

Martin Hoffhenke and Ipke Wachsmuth

Universität Bielefeld
Technische Fakultät, Arbeitsgruppe Wissensbasierte Systeme
D-33594 Bielefeld
email: {martinh, ipke}@TechFak.Uni-Bielefeld.DE

Abstract. This paper is concerned with representations which enable a technical agent to recognize objects and aggregates from mechanical parts as they evolve in an ongoing construction task. The general goal is that the technical agent has a detailed understanding of the task situation such that it can execute instructions issued by a human user in a dynamically changing situation. A levelled approach for comprehensive shape representation is presented which is motivated by a cognitive model of pictorial shape representations (prototypes). In particular, our system is able to derive and represent spatial properties (such as orientation) and geometric features (e.g., axes or planes) that can be ascribed to the developing construct.

1 Introduction

The context of this work is a human-machine-integrated construction scenario. This paper is concerned with representations which enable a technical agent to recognize objects and aggregates from mechanical parts as they evolve in an ongoing construction task. The general goal is that the technical agent has a detailed understanding of the task situation such that it can serve as an intelligent assistant and execute instructions issued by a human user.

What are the general challenges that have to be met by a representation system employed by the agent? Firstly, the system needs to provide knowledge about the mechanical objects, their properties, and how they can be mounted, and also knowledge about the assembly groups that are being built up in a dynamically changing situation. As is typical in a construction scenario, aggregates of generic building parts form new meaningful objects in which the individual parts can change with respect to their function and role. Secondly, the system should be able to derive and represent spatial properties (such as orientation) and geometric features (e.g., axes or planes) that can be ascribed to the developing construct. A representation of shape could provide a more profound understanding of aggregate objects through enriching the representation by spatial attributes like intrinsic orientations, axis positions or directions (Fig. 1).

232

Fig. 1. General shape of an undercarriage. The line is the intrinsic axis which denotes the possible direction of movement restricted by the wheels. An intrinsic orientation (front and back) will be assigned to the undercarriage when mounted to a vehicle.

While the first aspect has been dealt with by way of propositional descriptions in some approaches (e.g., [10], [15]), the problem of representing generic shape has insufficiently been tackled in previous literature. For instance, in earlier work (e.g., [13], [4]), two- or three-dimensional shape models were employed in the recognition and classification of 2D camera percepts. But with respect to construction scenarios in which 3D camera percepts are provided, these approaches are still lacking solutions of the following problems:

- the representation of shape on different levels of abstraction to describe spatial structure at various levels of detail
- the ascription of additional spatial properties by which spatial references could be handled, as in the instruction 'Mount the bumper to the back side of the car'
- a greater complexity of recognition implied by the usage of 3D percepts

The present paper describes an approach to solve these problems, based on the idea of *levelled shape prototypes*. Our scenario is a complex construction task in which a human instructor supervises a robotic agent in the execution of various construction steps. The robot has stereo cameras and is thus able to utilize 3D percepts of the construction scene[1]. Our results have so far been evaluated in a 3D assembly simulation, the Virtual Constructor, that was also used by (Jung et al). The Virtual Constructor can perform all manipulations with virtual building blocks that are physically possible with their real counterparts.

The paper is organized as follows. In Section 2, we give a brief description of the task setting and how the Virtual Constructor is able to simulate construction steps based on a structural (propositional) representation. Section 3 describes the motivations and general goals and features for our novel approach of levelled shape prototypes. In Section 4, we present our approach in more detail and describe results how shape prototypes were used to build up spatial representations in the construction scenario. In Section 5 we conclude that the levelled approach has the advantage to enable both a very abstract shape recognition and an appropriate shape refinement by an integrated representation.

---

[1] To avoid misunderstandings: the percept is a threedimensional reconstruction of the current scene, not a bitplane.

2 A virtual construction scenario

The testbed we use for the evaluation of our shape prototypes is the Virtual Constructor, a tool able to simulate a construction scenario with a human being as instructor and a technical system as constructor. The instructor can manipulate the virtual scene containing a variety of different mechanical objects of a construction kit. Manipulations can be issued by simple natural language instructions (Fig. 2) or directly via the mouse input device. Scene manipulations can be: moving objects, connecting objects, disconnecting objects, and rotating objects within an aggregate.

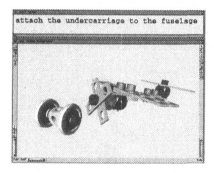

Fig. 2. Screenshot of the Virtual Constructor. The undercarriage will be attached to the airplane via a natural language instruction.

The Virtual Constructor uses a frame language to represent knowledge about the mechanical parts (how they can be connected with other parts) and knowledge about the aggregates which can be constructed. It allows to define a precise building-up of assembly groups by using containment, connection and simple spatial relations like parallelism of two bars. It also allows to define roles that specific building blocks can take on in the context of an assembly group. A bolt for example can become an axle in the context of an undercarriage. This way the bolt can be referred to in the following instructions *by role* like: 'Attach the axle to the green block'. Such a structural description of an assembly group is not very flexible against arbitrary but shape-invariant variations and it is bound to and restricted by a certain construction kit.

Shape prototypes as introduced in the following section are not for defining precise assembly groups out of a special construction kit. Rather, they enrich the construction system with knowledge about the rough appearance of objects belonging to a category. Shape prototypes describe the generic shape of complex objects with parametric geometry models. They are able to represent knowledge about spatial properties of objects like an intrinsic orientation, thus it is possible to handle instructions like: 'Attach the propeller to the front of the airplane'.

3 Object representation with shape prototypes

The idea we develop in this paper is based on a cognitive motivation. A variety of psychological investigations in the last 20 years have shown that human beings store and process information not only in an abstract-structural manner but also often make use of a pictorial concept of things and situations (e.g., [11], [2], [7]). Especially for the description of object categories, prototypical pictorial representations have been postulated. Pictorial representations seem advantageous in that they provide a more 'natural', i.e. intrinsic representation ([12]) of geometrical and spatial properties than propositional representations. It is assumed that by using intrinsic representations the classification of objects should be faster and more robust against variations and should give clues for a subsequent structural analysis, which is a motivation for our technical approach.

Fig. 3. Property inheritance from shape prototypes: the individual bolts inherit the fitting direction from the bolt shape prototype shown in the upper part.

3.1 Two aspects of shape prototypes

For an intuitive interaction with a human partner the technical agent needs to have a similar understanding of the common subject of their interaction. In the construction scenario aggregates often become meaningful for a human being by their overall shape. For example, the instructor calls two wheels an undercarriage when fitted together by an axle. This is the case when something *looks like* something known. To enable this *act of recognition* in a technical system it is necessary to model the prototypical shape properties of such objects. With structural representations (like frames) it is not possible to recognize objects which are similar by their rough appearance but which are built up in a varied way or with parts of some other construction kit. This is especially true for the recognition of the overall shape of multi-part objects. In contrast, shape prototypes just represent the rough shape and do not take the specific building parts into consideration.

Besides the aspect of recognition there is another important aspect to prototypes when used in the interaction with human beings, which is the *ascription of properties* even when they are not relevant for classification. But when for example speaking about the front side of an airplane this implies an intrinsic orientation (this means a spatial segmentation by the object's point of view). Another example is shown in Fig. 3. Each individual bolt has a fitting direction that can be inherited from the prototypical bolt shown in the top of the figure. Like classical property inheritance in propositional representations, our aim is a property inheritance of spatial attributes from a pictorial representation. Ascribing such spatial attributes to a part or a construct should enable a technical agent to handle instructions like:

- *Place the bolt upside down.*
- *Attach the propeller at the front side of the airplane.*
- *Show the bottom side of the airplane.*

Fig. 4. Examples of differently shaped entities of the category airplane.

3.2 Levels of abstraction

There are two conflicting requirements to be achieved by a pictorial shape representation. On the one hand it must be able to represent shape in the most abstract manner, in order to recognize a wide variety of objects that have a similar shape. On the other hand it must be able to distinguish between super- and subcategories by discriminating further shape properties. This means that shape prototypes must be able to describe categories at different levels of abstraction, like 'airplane' and 'jet'.

There are a variety of approaches to represent shape found in the literature. Some contributions just deal with 2D representations. For instance [8] uses a boundary representation; [3], [13], [14] are working with symmetry-axis representations. Some early contributions have suggested to use 3D representations ([9], [2]). Especially in image recognition (e.g., [4], [5]), 3D representations are often used to permit a viewer-independent description of objects. But these approaches are restricted to an isolated kind of representation and thus do not permit the modelling of category-adequate abstractions for shape prototypes.

Basically, the above-mentioned approaches restrict to a single type of object or even a single object. They do not provide a universal model of a shape prototype which realizes the conflicting requirements mentioned above. Thus we pursue an integrated approach of gradually abstracted levelled shape prototypes.

The *first requirement* for a shape prototype is to represent shape in the most abstract manner. This can be illustrated by the category airplane. The entities of this category are typically comprised by a wide variety of different shapes (Fig. 4). So, what is the characteristic shape of an airplane? Especially the model airplane built up with blocks of a toy construction kit (Fig. 4, bottom) illustrates that not constructional detail, but overall shape is typical for an airplane. Even with holes in the wings and propeller-blades and untypical features on the airplane's top the object is recognized as an airplane by human beings at first sight. The object is classified as an airplane by its typical spatial layout. Thus the three-dimensional skeleton-model shown in Fig. 5 provides an adequate shape representation for the most abstract shape description of an airplane.

Fig. 5. Shape representation of an abstract airplane (skeleton-model).

The *second requirement* for shape prototypes is the possibility of modelling further discriminating shape properties. A skeleton-model is not sufficient for this because differences often just occur in a two- or three-dimensional expression of a shape (for example, the characteristic delta-shape of the fighter plane in Fig. 4). Thus we need to represent also higher dimensional shape properties. This is why we developed a representation scheme that integrates shape descriptions at levels of different expressiveness.

4 A levelled approach for comprehensive shape representation

As was argued before, we cannot restrict ourselves to a single kind of shape representation, for the requirements on different levels of abstraction are too unlike by far. Thus the aim is to create a framework in which the various shape representations can be integrated.

There are two main aspects to be considered: First, the inclusion of the prototypes in a uniform conception to reach an inter-operability between them and, second, to save the individual properties of the different kinds of shape prototypes in their specific expressiveness. Note that the abstract expressiveness of a prototype is not only relevant for its (geometric) shape-model but also in the classification algorithm working on it.

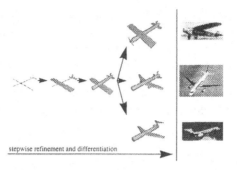

Fig. 6. Cascaded levels of shape prototypes: skeleton-model, plane-structure-model, volumetric model.

To keep the individual characteristics of each kind of prototype, they are embedded (with shape-model and classification algorithm) in a levelled architecture with a uniform input/output-structure and ordered like a cascade with respect to decreasing abstraction, as motivated by Fig. 6. This way the architecture takes care of an increasing differentiation of shape. In our approach we use three different levels of abstraction:

- The *skeleton model* represents objects on the most abstract level. Objects are described just by their major spatial extensions. All three objects in Fig. 7 can be recognized as an airplane on this level. Even the airplane with the swept back wings will be recognized, because on this level of abstraction the 'wing-line' does not stand for a straight line, but for a spatial extent at this position and direction.
- The second level (*plane-structure model*) takes also two-dimensional aspects (flat-shape elements) into account. This is why the left object in Fig. 7 would not be recognized as an airplane on this level (that is, it is not 'as good' an airplane as the other two objects).
- The third level (*volumetric model*) uses three-dimensional shape prototypes. Thus it is possible to discriminate shape elements of the objects by their volumetric shape. The middle and the right airplane in Fig. 7 can be distinguished between more special types of airplane like visualized in Fig. 6.

Any shape model on a more abstract level subsumes all less abstract models.

Fig. 7. Some examples of aggregates recognized as airplanes.

The classification process starts by comparing the percept with the shape prototype on the most abstract level (e.g., the skeleton airplane as represented in Fig. 9). If the shape properties of the prototype matches the percept (like visualized in Fig. 8), a hypothesis (e.g., Fig. 10) is generated. This hypothesis serves as additional information for the matching process with the prototype on the next lower level of abstraction. If it is possible to verify the hypothesis on this level, a new hypothesis with enriched information is generated and passed on to the next lower level.

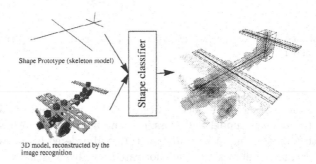

Fig. 8. Classification of the model airplane with the shape prototype of an airplane.

4.1 Object recognition with shape prototypes

We now describe in detail how the fully implemented shape recognizer for the skeleton model is used in object recognition. As an example Fig. 9 shows the skeleton prototype of the airplane that is visualized in Fig. 8 (upper left). The prototype contains several shape elements represented by line segments. Each line segment has a default position and some degrees of freedom (tolerance) for its relative position, rotation, and size (TRANSLATION, ROTATION and SCALE). Another important slot is the Significance-value. This value rates the expressiveness of a shape prototype as illustrated in Fig. 11. The level of abstraction of an expressive prototype is much lower for a low-structured object (like the wing) than for a high-structured object (like the airplanes in Fig. 6). So the skeleton model of the wing has a much smaller significance value than the airplane has on this level of abstraction.

The recognition process like visualized in Fig. 8 works as follows:

1. Transformation of the main axis of the prototype parallel to the main axis of the object by a principle-component-analysis. For a reduction of ambiguity we make use of the prototype's intrinsic top orientation as a clue for a preferred orientation.
2. For a first rough hypothesis the prototype is scaled to an object-similar size and centered.

```
Modelname: S_AIRPLANE
AbstractionLevel: Skeleton
Significance: 0.8
ShapeElements:
  (fuselage-1 FUSELAGE)
  (wing-1 WING)
  (elevatorunit-1 ELEVATORUNIT)
  (rudderassembly-1 RUDDERASSEMBLY)
LineSegments:
{
  fuselage-1 (CENTER (0, 0, 0) (1, 0, 0) 100) [
    FIX
  ]
  wing-1 (CENTER (10, 0, 0) (0, 0, 1) 90) [
    JOIN (fuselage-1
      STICK
      TRANSLATION ((+/-15unit) (0) (+/-15unit))
      ROTATION (() () ())
      SCALE (+/-15%)
    )
  ]
  rudderassembly-1 (START (-35, 0, 0) (0, 1, 0) 15) [
    JOIN (fuselage-1
      STICK
      TRANSLATION ((+/-10unit) (-5unit) (+/-1unit))
      ROTATION (() () ())
      SCALE (+/-50%)
    )
  ]
  elevatorunit-1 (CENTER (-35, 0, 0) (0, 0, 1) 35) [
    JOIN (rudderassembly-1
      STICK
      TRANSLATION ((+15unit) (0) (+/-10unit))
      ROTATION (() () ())
      SCALE (-50% +130%)
    )
  ]
}
Orientation:
{
  front (1, 0, 0)
  top (0, 1, 0)
}
```

Fig. 9. Coded description of the skeleton-model of an airplane visualized in Fig. 8, left-upper side (range values in TRANSLATION denote tolerances).

3. Search for shape elements of the object (building blocks) which are near and similar-oriented to the main shape element of the prototype (defined as FIX in Fig. 9). The prototype is corrected in position, orientation, and size accordingly.

4. Step by step now the other shape elements are searched for in the object with respect to their degrees of freedom in translation, rotation, and size.

```
Name: AIRPLANE-1
RelatedAggregate: AGGREGATE-9
IP-Type: AIRPLANE
AbstractionLevel: Skeleton
EstimateOfQuality: 0.8
ShapeElements:
{
  ( Name: fuselage-1
    IP-Type: FUSELAGE
    Line: (134.581, 92.3067, 186.292)
      (134.581, 92.3067, -27.7083)
    Parts: H5_BAR-na-2 H3_BAR-na-1 H3_BAR-na-4)
  ( Name: wing-1
    IP-Type: WING
    Line: (26.5817, 98.4755, 48.291)
      (242.586, 98.1379, 48.2924)
    Parts: H7_BAR-na-2)
  ( Name: elevatorunit-1
    IP-Type: ELEVATORUNIT
    Line: (58.0907, 124.427, 172.292)
      (211.095, 124.187, 172.292)
    Parts: H5_BAR-na-1)
  ( Name: rudderassembly-1
    IP-Type: RUDDERASSEMBLY
    Line: (134.621, 76.2946, 172.292)
      (134.621, 138.319, 172.292)
    Parts: BOLT-re-1 BOLT-re-2 BLOCK-re-4)
}
Orientation:
{
  front (0, 0, -1)
  top (0, 1, 0)
}
```

Fig. 10. Description of the instance of the airplane shape prototype in Fig. 8 (right).

If the shape prototype matches the perceived object, an instance of the prototype is created as exemplified in Fig. 10 (and visualized in Fig. 8, right). The instance contains information about the individual shape elements which are passed as clue to the recognizer on the next abstraction level, together with an estimate of quality. Such a clue consists of:

– the type of the object
– the position, orientation and size of the object and its individual shape elements

– the building blocks by which the individual shape elements are formed

This estimate of quality is composed by different factors like the activation of the related category and the significance value of the actual shape-prototype.

4.2 Property inheritance of spatial attributes

As described in Section 3.1, the technical agent needs knowledge about spatial properties that goes beyond the shape aspects important for recognition. So shape prototypes have to contain information that provides additional spatial knowledge about the prototype. As an example, Fig. 9 shows the coded skeleton shape prototype of an airplane. The prototype contains, beside the shape elements, spatial properties describing the intrinsic orientation of an airplane. In this example, the Orientation-slots define the front and the top direction. When an instance of a shape prototype is created, it inherits the spatial properties like the intrinsic orientation in Fig. 10 or the fitting direction in Fig. 3.

5 Conclusion and future work

In this paper we presented an integrated approach to model shape prototypes on different levels of abstraction. The object recognizer working on these prototypes is able to recognize and classify objects that are similar to a very general shape prototype on the one hand and, on the other hand, to differentiate among more special prototypes. The shape prototypes can be enriched with spatial attributes that are inherited to the instances of the prototypes. Thus it is possible to ascribe spatial properties like an intrinsic orientation to an object.

Fig. 11. An expressive prototype for a low-structured object like the wing is provided on a low level of abstraction.

First experiences show that our recognition system based on shape prototypes is up to two times faster than the frame-based recognizer provided with the Virtual Constructor (cf. Section 2). More importantly, it can recognize a diverse variety of different instances of a category like the different airplanes shown in Fig. 7. So far, the recognition time depends much on the complexity of the percept. We expect to get better results by preprocessing the percept with perceptual-grouping methods (e.g., [1]).

Another aspect of future work is the integration of a cognitive process model ([6]) which allows us to take category activation into account. This means that if, for example, some parts or aggregates are recognized as parts of an airplane, then other (generic) parts in this context will be classified more likely as airplane parts, too.

Acknowledgement

This work is partly supported by the the Collaborative Research Centre "Situated Artificial Communicators" (SFB 360) of the German National Science Foundation (DFG).

References

1. F. Ackermann, A. Maßmann, S. Posch, and D. Sagerer, G. aand Schlüter. Perceptual grouping of contour segments using markov random fields. *Int. Journal of Pattern Recognition and Image Analysis*, 7 (1):11–17, 1997.

2. I. Biederman. Recognition-by-components: a theory of human image understanding. *Psychological Review*, 94(2):115–147, 1987.

3. H. Blum and R. N. Nagel. Shape description using weighted symmetric axis features. *Pattern Recognition*, 10:167–180, 1978.

4. R. A. Brooks. *Model-based computer vision*. UMI Research Press, Ann Arbor, Michigan, 1984.

5. J. M. Ferryman, A. D. Worrall, G. D. Sullivan, and K. D. Baker. A generic deformable model for vehicle recognition. In *Proceedings of British Machine Vision Conference*, pages 127–136, University of Birmingham, 1995.

6. K. Kessler and G. Rickheit. Dynamische Konzeptgenerierung in konnektionistischen Netzen: Begriffsklärung, Modellvorstellungen zur Szenenrekonstruktion und experimentelle Ergebnisse. *Kognitionswissenschaft*, 8 (2), 1999.

7. S. M. Kosslyn. *Image and brain: the resolution of the imagery debate*. MIT Press, Cambridge (MA), 1994.

8. M. Leyton. A process-grammar for shape. *Artificial Intelligence*, 34:213–247, 1988.

9. D. Marr and H. K. Nishihara. Representation and recognition of the spatial organization of three-dimensional shapes. In *Proceedings of the Royal Society of London B*, volume 200, pages 269–294, 1978.

10. L. Padgham and P. Lambrix. A framework for part-of hierarchies in terminological logics. In *Principles of Knowledge Representation and Reasoning*, pages 485–496. Morgan Kaufmann, San Francisco (CA), 1994.

11. A. Paivio. *Imagery and verbal processes*. Lawrence Erlbaum Associates, Hillsdale (N.J.), 1979.

12. S. E. Palmer. Fundamental aspects of cognitive representations. In E. Rosch and B. B. Lloyd, editors, *Cognition and Categorization*, pages 259–303. Erlbaum, Hillsdale (NJ), 1978.

13. H. Rom and G. Medioni. Hierarchical decomposition and axial shape description. *IEEE Transactions on Pattern Analysis and Machine Intelligence*, 15 (10):973–981, 1993.

14. K. Siddiqi, A. Shokoufandeh, S. J. Dickinson, and S. W. Zucker. Shock graphs and shape matching. In *Proceedings of the Sixth International Conference on Computer Vision*, Bombay, India, 1998.

15. I. Wachsmuth and B. Jung. Dynamic conceptualization in a mechanical-object assembly environment. *Artificial Intelligence Review*, 10 (3-4):345–368, 1996.

Probabilistic, Prediction-Based Schedule Debugging for Autonomous Robot Office Couriers

Michael Beetz, Maren Bennewitz, and Henrik Grosskreutz

Dept. of Computer Science III,   Department of Computer Science V
University of Bonn,   Aachen Univ. of Technology,
D-53117 Bonn, Germany,   D-52056 Aachen, Germany
beetz,bennewit@cs.uni-bonn.de   grosskreutz@cs.rwth-aachen.de

Abstract. Acting efficiently and meeting deadlines requires autonomous robots to schedule their activities. It also requires them to act flexibly: to exploit opportunities and avoid problems as they occur. Scheduling activities to meet these requirements is an important research problem in its own right. In addition, it provides us with a problem domain where modern symbolic AI planning techniques could considerably improve the robots' behavior.

This paper describes PPSD, a novel planning technique that enables autonomous robots to impose order constraints on *concurrent percept-driven plans* to increase the plans' efficiency. The basic idea is to generate a schedule under simplified conditions and then to iteratively detect, diagnose, and eliminate behavior flaws caused by the schedule based on a small number of randomly sampled symbolic execution scenarios. The paper discusses the integration of PPSD into the controller of an autonomous robot office courier and gives an example of its use.

1 Introduction

Carrying out their jobs efficiently and meeting deadlines requires service robots to schedule their activities based on predictions of what will happen when the robot executes its intended course of action. Efficiency also requires robots to act flexibly: to exploit opportunities and avoid problems as they occur. Unfortunately, scheduling activities with foresight and acting flexibly at the same time is very hard. Even more so when jobs are underspecified or the robot is to carry out opportunistic plan steps which are triggered by enabling conditions. AI researchers have proposed several techniques to tackle these kinds of problems.

Constraint-based scheduling techniques have been successfully applied to very large scheduling problems. They gain efficiency by assuming that schedules can be checked without making detailed simulations of what will happen when a schedule gets executed. Sacrificing simulation-based plan checking will reduce the success rate of plans for robots that act in changing environments or carry out subplans that interact in subtle ways.

Decision-theoretic partial-order planning systems [KHW95,WH94] aim at synthesizing plans with maximal expected utility by weighing the utility of each possible outcome with the probability that the plan will produce it. Given realistic computational resources, the probability distribution over possible outcomes can often not be estimated accurately enough to determine the plans with the highest expected utility [Yam94].

Decision-theoretic planning based on solving (partially observable) Markov decision problems ((PO)MDPs) [BDH98] model the robot's operation and effects on the world as a finite state automaton in which actions cause stochastic state transitions. The robot is rewarded for achieving its goal quickly and reliably. A solution for such problems is a *policy*, a mapping from states to actions that maximizes the accummulated reward. The use of (PO)MDPs to generate concurrent plans that contain event-enabled subplans (if you notice that room A is open while you pass the door then pick up the red letter) requires to use fine-grained discretizations that result in huge state spaces. The grain size is thereby determined by the frequency of events that might trigger a subplan. Whether we can encode control problems of such complexity as (PO)MDP problems that are so compact that they can be solved effectively is still an open question.

The advantage of these approaches is that they compute optimal solutions with respect to the models they use because they reason exhaustively through all possible execution scenarios yielded by the models. They are therefore restricted to the use of simple models. We believe that one of the key impediments in applying AI planning techniques to improve robot behavior is the lack of realistic models of the operational principles of modern control systems and accurate models of the temporal structure of activities and their interactions.

Our proposed solution to scheduling flexible activity is PPSD *(Probabilistic Prediction-based Schedule Debugging)*, a method that generates candidate schedules fast ignoring the possibility of certain kinds of flaws and then iteratively debugs scheduled plans by revising them to eliminate flaws resulting from the simplifying assumptions (cf. [Sim92]). To forestall a wide range of behavior flaws typical for service robots we use models that represent *continuous processes, exogenous events, interferences between concurrent behavior, incomplete information*, and *passive sensors*. As a consequence, PPSD cannot afford to reason exhaustively about all possible consequences of the robot's plans. Instead it draws inferences based on a small number of scenarios randomly projected wrt the robot's probabilistic belief state in order to guess whether and if so which flaws will probably occur. This weaker kind of inference can be drawn fast and with reliability varying with the available resources [BM97].

Compared with the scheduling approaches discussed above we can categorize PPSD as follows. PPSD uses constraint-based scheduling techniques for generating initial schedules under idealizing assumptions. The resulting candidate plans are then tested using symbolic prediction techniques that generate qualitatively accurate predictions of the robot's behavior. The main advantage of PPSD over the (PO)MDP planning approaches is its parsimony in the consumption of computational resources. PPSD uses a very compact representation of actions and states by reasoning about action models that describe continuous behavior, yet predicting only those state transitions that might affect the course of plan execution [BG98]. In addition, PPSD is less likely to waste computation time on reasoning about very unlikely plan outcomes. As a consequence, farsighted scheduling of the operation of autonomous service robots in realistic environments lie well within the applicability scope of PPSD, even of its prototypical implementation discussed in this paper. Of course, those advantages over the other approaches can only be obtained by contending ourselves with scheduling methods that produce with high probability "pretty good" schedules.

Using PPSD we have implemented a special purpose planner that revises concurrent reactive plans for robots performing sets of complex, interacting, and user-specified tasks. This planner that is employed by the controller of an autonomous robot office courier is a specific example where modern AI planning technology can contribute to autonomous robot control by improving the robot's behavior.

This paper makes three important contributions: It (1) describes PPSD, a novel scheduling technique; (2) shows how the technique can be instantiated for an autonomous robot office courier; and (3) describes an experiment where the technique improves the robot's performance. In the remainder of the paper we proceed as follows. After an illustrative example, we characterize the control problem abstractly, and describe the three contributions listed above in detail.

2 An Illustrative Example

To illustrate the kinds of control problems we address, consider the following situation in the environment pictured in Fig. 1. A robot is to deliver a letter in a yellow enve-lope from room A-111 to A-117 (*cmd-1*) and another letter for which the envelope's color is not known from A-113 to A-120 (*cmd-2*). The robot has already tried to ac-complish *cmd-2* but because it recognized A-113 as closed (using its range sensors) it revised its intended course of action into achieving *cmd-2* opportunistically. That is, if it later detects that A-113 is open it will interrupt its current activity and reconsider its intended course of action under the premise that the steps for accomplishing *cmd-2* are executable.

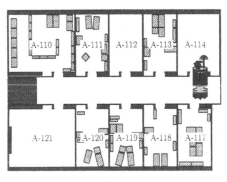

Fig. 1. Environment.

To perform its tasks fast the robot should schedule the pick-up and deliver actions to minimize execution time and assure that letters are picked up before deliv-ered. To ensure that the schedules will work, the robot has to take into account how the state of the robot and the world changes as the robot carries out its sched-uled activities. Aspects of states (state variables) that the robot has to consider when scheduling its activities are loca-tions of the letters. Constraints on the state variables that schedules have to sat-isfy are that they only ask the robot to pick up letters that are at that moment at the robot's location and that the robot does not carry two letters in envelopes with the same color.

Suppose our robot standing in front of room A-117 has received evidence that A-113 has been opened in the meantime. This requires the robot to reevaluate its options for accomplishing its jobs with respect to its changed belief state. Executing its current plan without modifications might yield mixing up letters because the robot might carry two letters in envelopes with the same color. The different options are: (1) to pick up the letter if it notices during plan execution that room A-113 is open, (2) to immediately ask (via email) for the letter from A-113 to be put into an envelope that is not yellow (to exclude mixing ups when taking the opportunity later); (3) to constrain later parts of the schedule such that no two yellow letters will be carried even when the letter in A-113 turns out to be yellow; and (4) to skip the opportunity. Which option the robot should take depends on its belief state with respect to the states of doors and locations of letters. To find out which schedules will probably work and which ones might yield mixing up letters, the robot must apply a model of the world dynamics to the state variables.

3 The Prediction-Based Schedule Debugger

PPSD is to solve the following computational task: Given a probabilistic belief state, a set of jobs, and a concurrent reactive partially ordered plan. Find a totally ordered plan, that accomplishes all jobs, and has a high performance (i.e., it is flawless and fast). To accomplish the computational task fast we only consider plans that result from the given plan through a sequence of revisions (addition of steps and ordering constraints).

In our particular example the probabilistic belief state comprises probability distributions about which envelopes are on which desks and the current state of doors in the form of probability distributions over the values of predefined random variables (e.g., $P(\text{nr-of-yellow-envelopes-on-desk-3} = 2) = 0.7$). We use a *performance measure* that is roughly proportional to the number of flaws in the schedule. Resource consumption (in our case, the travel time) has a smaller impact.

A concurrent reactive partially ordered plan consists of a set of plan steps STEPS, a partial order O on these steps and concurrent threats POLICIES with higher priority than STEPS. POLICIES specify constraints on the execution of the plan which might overwrite the order constraints within the plan. For example, the robot courier uses a policy to reschedule the plan whenever it detects that a door previously assumed to be closed is open and vice versa. To specify such flexible and efficient behavior without policies, in particular those that reschedule plans dynamically, the robot would have to use a "universal" plan in which decisions for all possible situations the robot might run into are precomputed.

The computational structure of prediction-based schedule debugging is a cycle that consists of the following steps: First, call a *schedule generator* to produce a candidate schedule O'. The schedule generator uses heuristics and domain-specific methods for quickly computing schedules that minimize travel time. It does not take into account the state changes of the robot and the world that occur as the activities get executed. Second, randomly project the plan ⟨STEPS,O'⟩ constrained by a set of policies with respect to the belief state BS N times to produce the execution scenarios ES. Apply the flaw detection module to the execution scenarios ES and collect all flaws FLAWS that occur at least K times in the N scenarios. Next, RULE is set to a revision rule that is applicable to one of the flaws in FLAWS. The new plan ⟨STEPS,O⟩ results from the application of RULE to the current plan. The loop is iterated until no probable flaw in the schedule is left. The following pseudo code sketches the main computational steps of the PPSD algorithm.

algorithm **PPSD**(PLAN, BS, N)

```
1   ⟨STEPS,O⟩ ← PLAN
2   loop
3        O' ← GENERATE-SCHEDULE(⟨STEPS,O⟩)
4        ES ← RANDOMLY-PROJECT(⟨STEPS,O'⟩, BS, N)
5        FLAWS ← DETECT-SCHEDULE-FLAWS(ES,K)
6        RULE ← CHOOSE(RULES(CHOOSE(FLAWS)))
7        ⟨STEPS,O⟩ ← APPLY(RULE)
8   until FLAWS = {}
```

There is, in general, no guarantee that eliminating the cause for one flaw won't introduce new flaws or that debugging will eventually reduce the number of flaws [Sim92].

PPSD is implemented within the XFRM planning framework [McD92]. In some ways, XFRM is like a tool for building expert systems: it is an empty shell for building transformational planners for robot controllers written in RPL. XFRM provides powerful and general tools for the prediction-based revision of concurrent reactive plans:

- The **Reactive Plan Language (RPL)** [McD91] allows programmers to specify concurrent reactive plans. RPL comes with an execution system that runs on autonomous robots [Bee99]. It also represents controllers as syntactic objects that can be inspected and manipulated by plan revision methods [McD92].
- The **projection module** PTOPA (see [McD94]) that generates symbolic representations of possible execution scenarios. PTOPA takes as its input an RPL plan, rules for generating exogenous events, and a set of probabilistic rules describing the effects of exogenous events and concurrent reactive control processes. PTOPA randomly samples execution scenarios from the probability distribution that is implied by the rules. An execution scenario describes how the execution of a robot controller might go, that is, how the environment changes as the plan gets executed. It is represented as a timeline, a linear sequence of dated events, which cause new world states. World states are characterized by a set of propositions.
- **XFRMML**, a declarative notation for formulating queries about projected execution scenarios, as well as plan transformation rules. Using XFRMML, PPSDs can reason about plans and apply plan transformation rules to eliminate schedule flaws.

To realize a PPSD debugger within the XFRM framework, a programmer has to do several things:

1. Specify a domain-specific fast heuristic schedule generator and implement it as an XFRMML plan revision rule.
2. Specify a causal model of the world and the robot controller using the PTOPA rule language, which is used to predict the state transitions caused by executing the robot controller. The causal model includes PTOPA rules that use the robot's belief state to sample the initial state of the timeline.
3. Provide XFRMML rules that represent what behavior flaws are and how plans can be revised to eliminate them.
4. Choose an appropriate parameterization of the flaw detection module (see below).
5. Provide a mechanism for updating the belief state, that is the probability distributions over the values of the random variables that are needed by the causal model.

The Flaw Detection Module A main factor that determines the performance of PPSD for a particular application is the *flaw detector*. In general, different kinds of flaw detectors differ with respect to (1) the time resources they require; (2) the reliability with which they detect flaws that should be eliminated; and (3) the probability that they hallucinate flaws (that is, that they signal a flaw that is so unlikely that eliminating the flaw would decrease the expected utility).

PPSD should classify a flaw as to be eliminated if the probability of the flaw wrt the agent's belief state is greater than θ. PPSD should classify a flaw as hallucinated if the probability of the flaw wrt the agent's belief state is smaller than τ. We assume that flaws with probability between τ and θ have no large impact on the robot's performance.

To be more specific, consider a schedule flaw f that occurs in the distribution of execution scenarios of a given scheduled plan with respect to the agent's belief state with probability p. Further, let $X_i(f) = 1$ represent the event that behavior flaw f occurs in the *i*th execution scenario ($X_i(f) = 0$ else).

The random variable $Y(f,n) = \sum_{i=1}^{n} X_i(f)$ represents the number of occurrences of the flaw f in n execution scenarios. Define a probable schedule flaw detector DET such that DET$(f,n,k) = true$ iff $Y(f,n) \geq k$, which means that the detector classifies a flaw f as to be eliminated if and only if f occurs in at least k of n randomly sampled scenarios.

Now we can build a model for the schedule flaw detector. Since the occurrence of schedule flaws in randomly sampled execution scenarios are independent from each other, the value of $Y(f)$ can be described by the binomial distribution $b(n,p)$. Using $b(n,p)$ we can compute the likelihood of overlooking a schedule flaw f with probability p in n scenarios: $P(Y(f) < j) = \sum_{k=0}^{j-1} \binom{n}{k} * p^k * (1-p)^{n-k}$.

In our prototypical implementation, we choose θ starting at 50% and τ smaller than 5% and typically use DET$(f,3,2)$, DET$(f,4,2)$, or DET$(f,5,2)$. Projecting an execution scenario takes about three to eight seconds depending on the complexity of the plan and the number of probabilistically occurring exogenous events. This enables the robot to perform a schedule debugging step in less than thirty seconds (on average). Fig. 2(left) shows the probability that the flaw detector DET$(f,n,2)$ for $n = 3,...,5$ will detect a schedule flaw with probability θ. The probability of a flaw less likely than τ to be eliminated is smaller than 2.3% (for all $n \leq 5$).

	Prob. of Flaw θ				
	50%	60%	70%	80%	90%
DET$(f,3,2)$	50.0	64.8	78.4	89.6	97.2
DET$(f,4,2)$	68.8	81.2	91.6	97.3	99.6
DET$(f,5,2)$	81.2	91.3	96.9	99.3	99.9

	θ					
	1%	10%	20%	40%	60%	80%
$\tau =.1\%$	1331	100	44	17	8	3
$\tau =1\%$	$\perp$	121	49	17	8	3
$\tau =5\%$	$\perp$	392	78	22	9	3

Fig. 2. Reliability of the probable schedule flaw detector (left). Number of scenarios to get 95% accurateness (right).

As projection gets faster, another question arises: how many execution scenarios (n) must be projected to recognize flaws of probability $\geq \theta$ with probability at least β? Or put another way, which flaw detector $DET(n, k)$ should we use? Instead of determining the optimal detector $DET(n, k)$ for given parameters τ, θ and β, we will consider the detectors $DET(n(\theta + \tau)/2, n)$ that will signal a flaw if $Y > n(\theta + \tau)/2$. Although $n(\theta + \tau)/2$ is not the optimal k, it is a reasonable guess. Note, if $1 - \theta \geq \tau$, which is often the case, the probability of overlooking a flaw with probability θ is greater than the probability of hallucinating a flaw of probability τ. The less probable a flaw becomes, the more probable it will be overlooked. Thus, in the remainder we will only consider the probability of overlooking a flaw with probability θ, using $DET(n(\theta + \tau)/2, n)$.

When n is large (particularly, $np(1 - p) > 9$) the binomial distribution $b(n, p)$ can be approximated a by a normal distribution $N(\sigma^2 = np(1 - p), \mu = np)$. In addition, the normal distribution can be restated as a standard normal distribution: $W(\xi \leq x) = \Phi((x - \mu)/\sigma)$; Φ beeing the standard normal distribution function.

So, how probable is it that we overlook a flaw using $DET(n(\theta + \tau)/2, n)$? A flaw with probability θ is overlooked if $Y \leq n(\theta + \tau)/2$ with Y distributed according to $N(n\theta, n\theta(1 - \theta))$. The probability, written as a standard normal distribution, is thus $\Phi(\frac{n(\theta+\tau)/2-\mu}{\sigma})$. To use β as an upper bound, $\frac{n(\theta+\tau)/2-\mu}{\sigma}$ must be $\leq \lambda_\beta$, with λ_β such that $\Phi(\lambda_\beta) \leq \beta$. Replacing σ with $\sqrt{n\theta(1 - \theta)}$ and μ with $n\theta$, we need to determine the n that implies $\frac{n((\theta+\tau)/2-\theta)}{\sqrt{n\theta(1-\theta)}} \geq \lambda_{beta}$. Solving this equation for n yields $n \geq 4\lambda_{beta}^2\theta(1 - \theta)/(\tau - \theta)^2$. Fig. 2(right) shows the number of necessary projections to achieve $\beta = 95\%$ ($\lambda_\beta = 1.65$) (the difference compared to Fig. 2 result from the approximation).

4 Online Scheduling of Office Delivery Jobs

This section applies PPSD to the control of an autonomous robot office courier, called RHINO, operating in the environment shown in Fig. 1. The robot courier uses a library of routine plans that specify among other things that objects are delivered by specifying that RHINO is to navigate to the pickup place, wait until the letter is loaded, navigate to the destination of the delivery, and wait for the letter to be unloaded. The plans are specified as concurrent reactive plans written in RPL.

The remainder of the section describes the domain and task specific instantiations of the PPSD components.

RHINO's Schedule Generator The algorithm for generating schedules is simple. Essentially, it sorts the navigation tasks (going to a target location) (counter)clockwise to get a candidate schedule. After this initial sort the scheduler iteratively eliminates and collects all steps s such that s has to occur after s' with respect to the required ordering constraints but occurs before s' in the candidate schedule. Then the steps s are iteratively inserted into the candidate schedule such that the ordering constraints are satisfied and the cost of insertion is minimal. While this greedy algorithm is very simple and fast it tends to produce fast schedules because the benign structure of the environment.

The Causal Model As the causal models of the concurrent reactive control routines, in particular the navigation behavior, we use a variation of the one proposed by Beetz and Grosskreutz [BG98] which makes probabilistically approximately accurate predictions. The predictions are represented compactly and generated within a few seconds. The probability distribution over which scenarios are generated is implied by the robot's belief state.

The Flaw Detector is realized through plan steps that generate "fail" events. Thus to detect schedule flaws an XFRMML query has to scan projected scenarios for the occurrence of failure events. For example, to detect deadline violations we run a monitoring process that sleeps until the deadline passes concurrently with the scheduled activity. When it wakes up it checks whether the corresponding command has been completed. If not a deadline violation failure event is generated.

Revision Rules. For schedule revision we use revision rules that are very similar to the ones originally developed by Beetz and Bennewitz [BB98]. The main difference is that the applicability of their rules is checked on real situation that occur during the execution of scheduled activity. The ones used in this paper include in addition rules that are triggered by predicted situations and applied to predicted execution scenarios. These revision rules have the advantage that they can revise the schedule *while* the scheduled plan is executed. Revision rules applied by the PPSD of the robot courier include one that adds ordering constraints to avoid holding two letters of the same color and one that inserts an email action asking the sender of a letter to use a particular envelope into the plan.

Belief State Manager. The belief state of the robot is updated through the interpretation of email messages, rules for updating the beliefs about dynamic states as time passes, and sensor data. For example, the rule "if office A-111 is closed it typically stays closed for about fifteen minutes" specifies that if the robot has not received any evidence about the door of room A-111 for fifteen minutes, the probability distribution for the door state is reset to the a priori probility. Email messages may contain information about probility distributions or facts that change conditional probabilities.

5 The Example Revisited

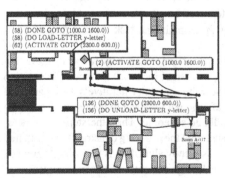

Fig. 3. A Possible projected execution scenarios for the initial plan.

Recall the example from the second section in which the robot standing at its initial position has perceived evidence that door A-113 has been opened. Therefore its belief state assigns probability p for the value true of random variable open-A113. The belief state also contains probabilities for the colors of letters on the desk in A-113.

Wrt. this belief state, different scenarios are possible. The first one, in which A-113 is closed, is pictured in Fig. 3. Points on the trajectories represent predicted events. The events without labels are actions in which the robot changes its heading (on an approximated trajectory) or events representing sensor updates generated by passive sensing processes. For example, a passive sensor update event is generated when the robot passes a door. In this scenario no intervention by prediction-based debugging is necessary and no flaw is projected.

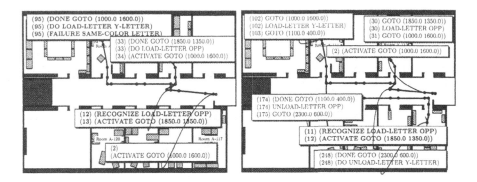

Fig. 4. The other possible execution scenarios for the initial plan.

In the scenarios in which office A-113 is open the controller is projected to recognize the opportunity and to reschedule its enabled plan steps as described above [1]. The resulting schedule asks the robot to first enter A-113, and pickup the letter for cmd-2, then enter A-111 and pick up the letter for cmd-1, then deliver the letter for cmd-2 in A-120, and the last one in A-117. This category of scenarios can be further divided into two categories. In the first subcategory shown in Fig. 4(left) the letter to be picked up is yellow. Performing the pickup thus would result in the robot carrying two yellow letters and therefore an execution failure is signalled. In the second subcategory shown in Fig. 4(right) the letter has another color and therefore the robot is projected to succeed by taking for all these scenarios the same course of action. Note, that the possible flaw is introduced by the reactive rescheduling because the rescheduler doesn't consider how the state of the robot will change in the course of action, in particular that a state may be caused in which the robot is to carry two letters with the same color.

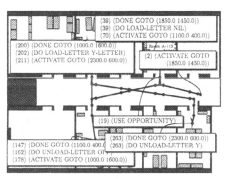

Fig. 5. Projected scenario for the revised plan.

In this case, PPSD will probably detect the flaw if it is probable with respect to the robot's belief state. This enables the debugger to forestall the flaw, for instance, by introducing an additional ordering constraint, or by sending an email that increases the probability that the letter will be put in a particular envelope. These are the revision rules introduced in the last section. Fig. 5 shows a projection of a plan that has been revised by adding the ordering constraint that the letter for A-120 is delivered before entering A-111.

Fig. 6(left) shows the event trace generated by the initial plan and *executed* with the RHINO control system [TBB+98] for the critical scenario without prediction based schedule debugging; Fig. 6(right) the one with the debugger adding the additional order-

---

[1] Another category of scenarios is characterized by A-113 becoming open after the robot has left A-111. This may also result in an execution failure if the letter loaded in A-113 is yellow, but is not discussed here any further.

252

ing constraint. This scenario shows that reasoning about the future execution of plans in PPSD is capable of improving the robot's behavior.

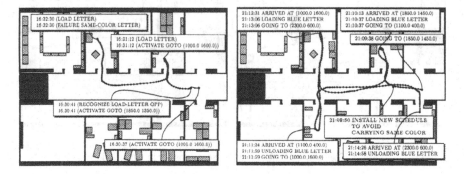

Fig. 6. Trajectory without PPSD (left). Trajectory when the flaw is forestalled by PPSD (right).

6 Experimental Results

We have carried out three kinds of experiments to verify that PPSD-based scheduling can improve the behavior of autonomous robots. In the first one we have validated that PPSD-like plan revisions can be carried out reliably while the robot is operating. We have implemented a high-level controller for an interactive museums tourguide robot. The tourguide robot, called MINERVA, has operated for a period of thirteen days in the Smithsonian's National Museum of American History.[2] In this period, it has been in service for more than ninetyfour hours, completed 620 tours, showed 2668 exhibits, and travelled over a distance of more than fortyfour kilometers. MINERVA used plan revisions for the installment of new commands, the deletion of completed plans, and tour scheduling. The MINERVA experiment demonstrates that SRCs can (1) reliably control an autonomous robot over extended periods of time and (2) reliably revise plans during their execution.

The second experiment evaluated PPSD in a series of 12 randomly generated scenarios with about six delivery jobs most of them added asynchronously. In these scenarios no plan revisions were necessary. Each scenario was tested in five runs (taking about twenty to twentyfive minutes for PPSD- and situation-based scheduling) As expected PPSD-based scheduling did on average not worse than situation-based scheduling methods.

In the last experiment we made up five scenarios in which the computation of good schedules required foresight. In those scenarios PPSD outperformed situation-based scheduling by about eight percent (in this experiment, if the robot recognizes that it is about to load two letters of the same color, it puts the current job aside; thus confusing two letters never occurs. In the example pictured in Fig. 6(left) the robot would leave the room *without* loading the second letter. PPSD forestalls such detours.) We compared the expected durations for the paths taken in the scenario in order to eliminate the large

---

[2] See http://www.cs.cmu.edu/~minerva for details.

variations in navigation time caused by different load averages for computers, interference by the reactive collision avoidance, etc.

7 Related Work

Probabilistic prediction-based schedule debugging is a planning approach rooted in the tradition of transformational planners, like HACKER [Sus77] and GTD [Sim92] that diagnoses "bugs" or, in our case, plan failures, in order to revise plans appropriately. In spirit, the control strategy of PPSD is very similar to the Generate/Test/Debug strategy proposed by Simmons [Sim92]. Our approach, like the XFRM system [McD92], differs from other transformational planners in that it tries to debug a simultaneously executed plan instead of constructing a correct plan. Also, we reason about full-fledged robot plans and are able to diagnose a larger variety of failures.

A number of approaches have been applied to activity scheduling in autonomous agent control. McDermott [McD92] has developed a prediction-based scheduler for location specific plan steps, which can probabilistically guess locations at which plan steps are executed if the locations are not specified explicitly. The main contribution of our approach is that McDermott's approach has been applied to a simulated agent in a grid-based world whereas ours controls a physical autonomous robot. Pell *et al.* [PBC+97] (re)schedule the activities of an autonomous spacecraft. While their system must generate working schedules we are interested in good schedules wrt. a given utility/cost function. In addition, scheduling activities of autonomous service robots typically requires faster and resource adaptive scheduling methods. Our scheduling approach differs from the one proposed by McVey *et al.* [MADS97] in that theirs generates real-time guaranteed control plans while our scheduler optimizes wrt. a user defined objective function.

8 Conclusions

In this paper we have developed a scheduling technique that enables autonomous robot controllers to schedule flexible plans, that is plans that allow autonomous robots to exploit unexpected opportunities[3] and to reschedule dynamically. The technique makes informed scheduling decisions by reasoning through concurrent sensor-driven plans that even reschedule themselves during execution. PPSD uses a fast heuristic schedule generator that might propose flawed schedules and then iteratively detects and eliminates schedule flaws based on a small number of randomly sampled execution scenarios.

Besides the technique itself, the paper gives a specific example in which modern AI planning technology can contribute to autonomous robot control by improving the robot's behavior. The planning techniques, in particular the temporal projection and the plan revision techniques can do so because they (1) extend standard scheduling techniques and reactive plan execution techniques with means for predicting that the execution of a scheduled activity *will result* in a behavior flaw; (2) predict states relevant for making scheduling decisions that the robot *won't be able* to observe; (3) uses information about predicted states *before* the robot can observe them.

---

[3] By unexpected opportunities we mean states like an open door. The robot knows that doors can be open and closed but it does not know which door is open when. This is the unexpected part.

References

[BB98]   M. Beetz and M. Bennewitz. Planning, scheduling, and plan execution for autonomous robot office couriers. In R. Bergmann and A. Kott, editors, *Integrating Planning, Scheduling and Execution in Dynamic and Uncertain Environments*, volume Workshop Notes 98-02. AAAI Press, 1998.

[BDH98]   C. Boutilier, T. Dean, and S. Hanks. Decision theoretic planning: Structural assumptions and computational leverage. *Journal of AI research*, 1998.

[Bee99]   M. Beetz. Structured reactive controllers — a computational model of everyday activity. In *Proceedings of the Third International Conference on Autonomous Agents*, 1999. to appear.

[BG98]   M. Beetz and H. Grosskreutz. Causal models of mobile service robot behavior. In R. Simmons, M. Veloso, and S. Smith, editors, *Fourth International Conference on AI Planning Systems*, pages 163–170, Morgan Kaufmann, 1998.

[BM97]   M. Beetz and D. McDermott. Fast probabilistic plan debugging. In *Recent Advances in AI Planning. Proceedings of the 1997 European Conference on Planning*, pages 77–90. Springer Publishers, 1997.

[KHW95]   N. Kushmerick, S. Hanks, and D. Weld. An algorithm for probabilistic planning. *Artificial Intelligence*, 76:239–286, 1995.

[MADS97] C. McVey, E. Atkins, E. Durfee, and K. Shin. Development of iterative real-time scheduler to planner feedback. In *"Proceedings of the 15th Int. Joint Conf. on Artificial Intelligence (IJCAI-87)"*, pages 1267–1272, 1997.

[McD91]   D. McDermott. A reactive plan language. Research Report YALEU/DCS/RR-864, Yale University, 1991.

[McD92]   D. McDermott. Transformational planning of reactive behavior. Research Report YALEU/DCS/RR-941, Yale University, 1992.

[McD94]   D. McDermott. An algorithm for probabilistic, totally-ordered temporal projection. Research Report YALEU/DCS/RR-1014, Yale University, 1994.

[PBC+97] B. Pell, D. Bernard, S. Chien, E. Gat, N. Muscettola, P. Nayak, M. Wagner, and B. Williams. An autonomous spacecraft agent prototype. In *Proceedings of the First International Conference on Autonomous Agents*, 1997.

[Sim92]   Reid Simmons. The role of associational and causal reasoning in problem solving. *AI Journal*, 53, 1992.

[Sus77]   G. Sussman. *A Computer Model of Skill Acquisition*, volume 1 of *Aritficial Intelligence Series*. American Elsevier, New York, NY, 1977.

[TBB+98] S. Thrun, A. Bücken, W. Burgard, D. Fox, T. Fröhlinghaus, D. Hennig, T. Hofmann, M. Krell, and T. Schmidt. Map learning and high-speed navigation in RHINO. In D. Kortenkamp, R.P. Bonasso, and R. Murphy, editors, *AI-based Mobile Robots: Case studies of successful robot systems*. MIT Press, Cambridge, MA, 1998.

[WH94]   M. Williamson and S. Hanks. Utility-directed planning. In *Proc. of AAAI-94*, page 1498, 1994.

[Yam94]   E. Yampratoom. Using simulation-based projection to plan in an uncertain and temporally complex world. Technical Report 531, University of Rochester, CS Deptartment, 1994.

Collaborative Multi-robot Localization

Dieter Fox[†], Wolfram Burgard[‡], Hannes Kruppa[††], Sebastian Thrun[†]

[†] School of Computer Science   [‡] Computer Science Department III   [††] Department of Computer Science
Carnegie Mellon University   University of Bonn   ETH Zurich
Pittsburgh, PA 15213   D-53117 Bonn, Germany   CH-8092 Zurich, Switzerland

Abstract. This paper presents a probabilistic algorithm for collaborative mobile robot localization. Our approach uses a sample-based version of Markov localization, capable of localizing mobile robots in an any-time fashion. When teams of robots localize themselves in the same environment, probabilistic methods are employed to synchronize each robot's belief whenever one robot detects another. As a result, the robots localize themselves faster, maintain higher accuracy, and high-cost sensors are amortized across multiple robot platforms. The paper also describes experimental results obtained using two mobile robots. The robots detect each other and estimate their relative locations based on computer vision and laser range-finding. The results, obtained in an indoor office environment, illustrate drastic improvements in localization speed and accuracy when compared to conventional single-robot localization.

1 Introduction

Sensor-based robot localization has been recognized as one of the fundamental problems in mobile robotics. The localization problem is frequently divided into two sub-problems: *Position tracking*, which seeks to compensate small dead reckoning errors under the assumption that the initial position of the robot is known, and *global self-localization*, which addresses the problem of localization with no a priori information about the robot position. The latter problem is generally regarded as the more difficult one, and recently several approaches have provided sound solutions to this problem. In recent years, a flurry of publications on localization—which includes a book solely dedicated to this problem [2]—document the importance of the problem. According to Cox [8], "Using sensory information to locate the robot in its environment is the most fundamental problem to providing a mobile robot with autonomous capabilities."

However, virtually all existing work addresses localization of a *single* robot only. At first glance, one could solve the problem of localizing N robots by localizing each robot *independently*, which is a valid approach that might yield reasonable results in many environments. However, if robots can detect each other, there is the opportunity to do better. When a robot determines the location of another robot relative to its own, both robots can refine their internal believes based on the other robot's estimate, hence improve their localization accuracy. The ability to exchange information during localization is particularly attractive in the context of global localization, where each sight of another robot can reduce the uncertainty in the estimated location dramatically.

The importance of exchanging information during localization is particularly striking for heterogeneous robot teams. Consider, for example, a robot team where some robots are equipped with expensive, high accuracy sensors (such as laser range-finders), whereas others are only equipped with low-cost sensors such as ultrasonic range finders. By transferring information across multiple robots, high-accuracy sensor information can be leveraged. Thus, collaborative multi-robot localization facilitates the amortization of high-end, high-accuracy sensors across teams of robots. Thus, phrasing the problem of localization as a collaborative one offers the opportunity of improved performance from less data.

This paper proposes an efficient probabilistic approach for collaborative multi-robot localization. Our approach is based on *Markov localization* [23, 27, 16, 6], a family of probabilistic approaches that have recently been applied with great practical success to single-robot localization [4, 3, 30]. In contrast to previous research, which relied on grid-based or coarse-grained topological representations, our approach adopts a sampling-based representation [10, 12], which is capable of approximating a wide range of belief functions in real-time. To transfer information across different robotic platforms, probabilistic "detection models" are employed to model the robots' abilities to recognize each other. When one robot detects another the individual believes of the robots are synchronized, thereby reducing the uncertainty of both robots during localization. While our approach is applicable to any sensor capable of (occasionally) detecting other robots, we present an implementation that integrates color images and proximity data for robot detection.

In what follows, we will first introduce the necessary statistical mechanisms for multi-robot localization, followed by a description of our sampling-based Monte Carlo localization technique in Section 3. In Section 4 we present our vision-based method to detect other robots. Experimental results are reported in Section 5. Finally, related work is discussed in Section 6, followed by a discussion of the advantages and limitations of the current approach.

2 Multi-robot Localization

Throughout this paper, we adopt a probabilistic approach to localization. Probabilistic methods have been applied with remarkable success to single-robot localization [23, 27, 16, 6], where they have been demonstrated to solve problems like global localization and localization in dense crowds.

Let us begin with a mathematical derivation of our approach to multi-robot localization. Let N be the number of robots, and let d_n denote the data gathered by the n-th robot, with $1 \leq n \leq N$. Each d_n is a sequence of three different types of information:

1. **Odometry measurements**, denoted by a, specify the relative change of the position according to the robot's wheel encoders.

2. **Environment measurements**, denoted by o, establish the reference between the robot's local coordinate frame and the environment's frame of reference. This information typically consists of range measurements or camera images.

3. **Detections**, denoted by r, indicate the presence or absence of other robots. Below, in our experiments, we will use a combination of visual sensors (color camera) and range finders for robot detection.

2.1 Markov Localization

Before turning to the topic of this paper—collaborative multi-robot localization—let us first review a common approach to single-robot localization, which our approach is built upon: Markov localization (see [11] for a detailed discussion). Markov localization uses only dead reckoning measurements a and environment measurements o; it ignores detections r. In the absence of detections (or similar information that ties the position of one robot to another), information gathered at different platforms cannot be integrated. Hence, the best one can do is to localize each robot individually, i.e. independently of all others.

The key idea of Markov localization is that each robot maintains a belief over its position. Let $Bel_n^{(t)}(L)$ denote the belief of the n-th robot at time t. Here L denotes the random variable representing the *robot position* (we will use the terms *position* and *location* interchangeably), which is typically a three-dimensional value composed of a robot's x-y position and its orientation θ. Initially, at time $t = 0$, $Bel_n^{(0)}(L)$ reflects the initial knowledge of the robot. In the most general case, which is being considered in the experiments below, the initial position of all robots is unknown, hence $Bel_n^{(0)}(L)$ is initialized by a uniform distribution.

At time t, the belief $Bel_n^{(t)}(L)$ is the posterior with respect to all data collected up to time t:

$$Bel_n^{(t)}(L) = P(L_n^{(t)} \mid d_n^{(t)}) \tag{1}$$

where $L_n^{(t)}$ denotes the position of the n-th robot at time t, and $d_n^{(t)}$ denotes the data collected by the n-th robot *up to* time t. By assumption, the most recent sensor measurement in $d_n^{(t)}$ is either an odometry or an environment measurement. Both cases are treated differently, so let's consider the former first:

1. **Sensing the environment:** Suppose the last item in $d_n^{(t)}$ is an environment measurement, denoted $o_n^{(t)}$. Using the Markov assumption (and exploiting that the robot position does not change when the environment is sensed), the belief is updated using the following *incremental* update equation:

$$Bel_n^{(t)}(L = l) \longleftarrow \alpha\, P(o_n^{(t)} \mid L_n^{(t)} = l)\, Bel_n^{(t-1)}(L = l) \tag{2}$$

Here α is a normalizer which ensures that $Bel_n^{(t)}(L)$ sums up to one. Notice that the posterior belief of being at location l after incorporating $o_n^{(t)}$ is obtained by multiplying the observation likelihood $P(o_n^{(t)} \mid L_n^{(t)} = l)$ with the prior belief. This likelihood is also called the *environment perception model* of robot n. Typical models for different types of sensors are described in [11, 9, 18].

2. **Odometry:** Now suppose the last item in $d_n^{(t)}$ is an odometry measurement, denoted $a_n^{(t)}$. Using the Theorem of Total Probability and exploiting the Markov property, we obtain the following incremental update scheme:

$$Bel_n^{(t)}(L = l) \longleftarrow \int P(L_n^{(t)} = l \mid a_n^{(t-1)}, L_n^{(t-1)} = l')\, Bel_n^{(t-1)}(L = l')\, dl' \tag{3}$$

Here $P(L_n^{(t)} = l \mid a_n^{(t-1)}, L_n^{(t-1)} = l')$ is called the *motion model* of robot n. In the remainder, this motion model will be denoted as $P(l \mid a_n, l')$ since it is assumed to be independent of the time t. It is basically a model of robot kinematics annotated with uncertainty and it generally has two effects: first, it shifts the probabilities according to the measured motion and second it convolves the probabilities in order to deal with possible errors in odometry coming from slippage etc. (see e.g. [12]).

These equations together form the basis of Markov localization, an incremental probabilistic algorithm for estimating robot positions. As noticed above, Markov localization has been applied with great practical success to mobile robot localization. However, it is only designed for single-robot localization, and cannot take advantage of robot detection measurements.

2.2 Multi-robot Markov Localization

The key idea of multi-robot localization is to integrate measurements taken at different platforms, so that each robot can benefit from data gathered by robots other than itself. At first glance, one might be tempted to maintain a single belief over all robots' locations, i.e.,

$$L = \{L_1, \ldots, L_N\} \tag{4}$$

Unfortunately, the dimensionality of this vector grows with the number of robots: Since each robot position is three-dimensional, L is of dimension $3N$. Distributions over L are, hence, exponential in the number of robots. Thus, modeling the joint distribution of the positions of all robots is infeasible for larger values of N.

Our approach maintains *factorial* representations; i.e., each robot maintains its own belief function that models only its own uncertainty, and occasionally, e.g., when a robot sees another one, information from one belief function is transfered from one robot to another. The factorial representation assumes that the distribution of L is the product of its N marginal distributions:

$$P(L_1^{(t)}, \ldots, L_N^{(t)} \mid d^{(t)}) = P(L_1^{(t)} \mid d^{(t)}) \cdot \ldots \cdot P(L_N^{(t)} \mid d^{(t)}) \tag{5}$$

Strictly speaking, the factorial representation is only approximate, as one can easily construct situations where the independence assumption does not hold true. However, the factorial representation has the advantage that the estimation of the posteriors is conveniently carried out locally on each robot. In the absence of detections, this amounts to performing Markov localization independently for each robot. Detections are used to provide additional constraints between the estimated pairs of robots, which will lead to refined local estimates.

To derive how to integrate detections into the robots' beliefs, let us assume the last item in $d_n^{(t)}$ is a detection variable, denoted $r_n^{(t)}$. For the moment, let us assume this is the only such detection variable in $d^{(t)}$, and that it provides information about the location of the m-th robot relative to robot n (with $m \neq n$). Then

$$
\begin{aligned}
Bel_m^{(t)}(L = l) &= P(L_m^{(t)} = l \mid d^{(t)}) \\
&= P(L_m^{(t)} = l \mid d_m^{(t)}) \, P(L_m^{(t)} = l \mid d_n^{(t)}) \\
&= P(L_m^{(t)} = l \mid d_m^{(t)}) \int P(L_m^{(t)} = l \mid L_n^{(t)} = l', r_n^{(t)}) P(L_n^{(t)} = l' \mid d_n^{(t-1)}) \, dl' \quad (6)
\end{aligned}
$$

which suggests the incremental update equation:

$$Bel_m^{(t)}(L = l) \longleftarrow Bel_m^{(t)}(L = l) \int P(L_m^{(t)} = l \mid L_n^{(t)} = l', r_n^{(t)}) \, Bel_n^{(t)}(L = l') \, dl' \quad (7)$$

In this equation the term $P(L_m^{(t)} = l \mid L_n^{(t)} = l', r_n^{(t)})$ is the robot perception model. A typical example of such a model for visual robot detection is described in Section 4. Of course, Eq. (7) is only an approximation, since it makes certain independence assumptions (it excludes that a sensor reports "I saw a robot, but I cannot say which one"), and strictly speaking it is only correct if there is only a single r in the entire run. However, this gets us around modeling the joint distribution $P(L_1, \ldots, L_N \mid d)$, which is computationally infeasible as argued above. Instead, each robot basically performs single-robot Markov localization with these additional probabilistic constrains, hence estimates the marginal distributions $P(L_n|d)$ separately.

The reader may notice that, by symmetry, the same detection can be used to constrain the n-th robot's position based on the belief of the m-the robot. The derivation is omitted since it is fully symmetrical.

3 Monte Carlo Localization

The previous section left open how the belief is represented. In general, the space of all robot positions is continuous-valued and no parametric model is known that would accurately model arbitrary beliefs in such robotic domains. However, practical considerations make it impossible to model arbitrary beliefs using digital computers.

3.1 Single Robot MCL

The key idea here is to approximate belief functions using a Monte Carlo method. More specifically, our approach is an extension of Monte Carlo Localization (MCL), which was shown to be an extremely efficient and robust technique for single robot position estimation (see [10, 12] for more details). MCL is a version of Markov localization that relies on a sample-based representation and the sampling/importance re-sampling algorithm for belief propagation [25]. MCL represents the posterior beliefs $Bel_n(L)$ by a set $S = \{s_i \mid i = 1..K\}$ of K weighted random samples or *particles*[1]. Samples in MCL are of the type

$$s_i = \langle \langle x_i, y_i, \theta_i \rangle, p_i \rangle \quad (8)$$

where $\langle x_i, y_i, \theta_i \rangle$ denote a robot position, and $p_i \geq 0$ is a numerical weighting factor, analogous to a discrete probability. For consistency, we assume $\sum_{i=1}^{K} p_i = 1$. In analogy with the general Markov localization approach outlined in Section 2, MCL proceeds in two phases:

1. **Robot motion.** When a robot moves, MCL generates K new samples that approximate the robot's position after the motion command. Each sample is generated by

---

[1] A sample set constitutes a discrete distribution. However, under appropriate assumptions (which happen to be fulfilled in MCL), such distributions smoothly approximate the "correct" one at a rate of $1/\sqrt{K}$ as K goes to infinity [29].

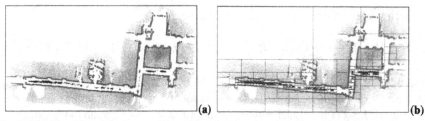

Fig. 1. (a) Map of the environment along with a sample set representing the robot's belief during global localization, and (b) its approximation using a density tree.

randomly drawing a sample from the previously computed sample set, with likelihood determined by their p-values. Let l' denote the position of such a sample. The new sample's position l is then generated by producing a single, random sample from $P(l \mid a, l')$, using the action a as observed. The p-value of the new sample is K^{-1}. An algorithm to perform this re-sampling process efficiently in $O(K)$ time is given in [7].

2. **Environment measurements** are incorporated by re-weighting the sample set, which is analogous to Bayes rule in Markov localization. More specifically, let $\langle l, p \rangle$ be a sample. Then, in analogy to Eq. (2) the updated sample is $\langle l, \alpha P(o \mid l)p \rangle$ where o is a sensor measurement, and α is a normalization constant that enforces $\sum_{i=1}^{K} p_i = 1$. The incorporation of sensor readings is typically performed in two phases, one in which p is multiplied by $P(o \mid l)$, and one in which the various p-values are normalized.

3.2 Multi-robot MCL

The extension of MCL to collaborative multi-robot localization is *not* straightforward. This is because under our factorial representation, each robot maintains its own, local sample set. When one robot detects another, both sample sets are synchronized according to Eq. (7). Notice that this equation requires the multiplication of two *densities* which means that we have to establish a correspondence between the individual samples in $Bel(L_m)$ and the density representing the robot detection.

To remedy this problem, our approach transforms sample sets into density functions using *density trees* [17, 22]. These methods approximate sample sets using piecewise constant density functions represented by a tree. The resolution of the tree is a function of the densities of the samples: the more samples exist in a region of space, the more fine-grained the tree representation. Figure 1 shows an example sample set along with the tree generated from this set. Our specific algorithm grows trees by recursively splitting in the center of each coordinate axis, terminating the recursion when the number of samples is smaller than a pre-defined constant. After the tree is grown, each leaf's density is given by the quotient of the sum of the weights p of all samples that fall into this leaf, divided by the volume of the region covered by the leaf. The latter amounts to maximum likelihood estimation of (piecewise) constant density functions.

To implement the update equation above, our approach approximates the density

$$\int P(L_m^{(t)} = l \mid L_n^{(t)} = l', r_n^{(t)}) \, Bel_n^{(t)}(L = l') \, dl' \tag{9}$$

using samples, just as described above. The resulting sample set is then transformed into a density tree. These density values are then multiplied into the weights (importance

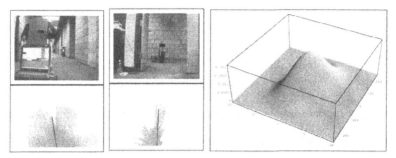

Fig. 2: Examples of successful robot detections and Gaussian density representing the robot perception model. The x-axis represents the deviation of relative angle and the y-axis the uncertainty in the distance between the two robots.

factors) of the samples in $Bel(L_m)$, effectively multiplying both density functions. The result is a refined density for the m-th robot, reflecting the detection and the belief of the n-th robot.

4 Visual Robot Detection

To implement collaborative multi-robot localization, robots must possess the ability to sense each other. The crucial component is the detection model $P(L_m = l \mid L_n = l', r_n)$ which describes the conditional probability that robot m is at location l, given that robot n is at location l' and perceives robot m with measurement r_n. In this section, we briefly describe one possible detection method which integrates camera and range information to estimate the relative position of robots.

Our implementation uses camera images to detect other robots and extracts from these images the relative direction of the other robot. After detecting another robot and its relative angle, it uses laser ranger finder scans to determine its distance. Figure 2 shows two examples of camera images taken by one of the robots. Each image shows another robot, marked by a unique, colored marker to facilitate the recognition. Even though the robot is only shown with a fixed orientation in this figure, the markers can be detected regardless of a robot's orientation. The small black rectangles, superimposed at the center of each marker in the images in Figure 2, illustrate the center of the marker as identified by this visual routine. The bottom row in Figure 2 shows laser scans for the example situations depicted in the top row of the same figure. Each scan consists of 180 distance measurements with approx. 5 cm accuracy, spaced at 1 degree angular distance. The dark line in each diagram depicts the extracted location of the robot in polar coordinates, relative to the position of the detecting robot. The scans are scaled for illustration purposes.

The Gaussian distribution shown in Figure 2 models the error in the estimation of a robot's location. Here the x-axis represents the angular error, and the y-axis the distance error. This Gaussian has been obtained through maximum likelihood estimation based on training data (see [13] for more details). As is easy to be seen, the Gaussian is zero-centered along both dimensions, and it assigns low likelihood to large errors. Please note that our detection model additionally considers a 6.9% chance to erroneously detecting a robot when there is none.

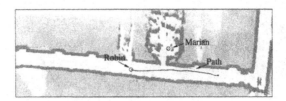

Fig. 3: Map of the environment along with a typical path taken by Robin during an experiment.

5 Experimental Results

Our approach was evaluated using two Pioneer robots (Robin and Marian) marked optically by a colored marker, as shown in Figure 2. The central question driving our experiments was: *Can cooperative multi-robot localization significantly improve the localization quality, when compared to conventional single-robot localization?*

Figure 3 shows the setup of our experiments along with a part of the occupancy grid map [31] used for position estimation. Marian operates in our lab, which is the cluttered room adjacent to the corridor. Because of the non-symmetric nature of the lab, the robot knows fairly well where it is (the samples representing Marian's belief are plotted in Figure 4 (a)). Figure 3 also shows the path taken by Robin, which was in the process of global localization. Figure 5 (a) represents the typical belief of Robin when it passes the lab in which Marian is operating. Since Robin already moved several meters in the corridor, it developed a belief which is centered along the main axis of the corridor. However, the robot is still highly uncertain about its exact location within the corridor and even does not know its global heading direction. Please note that due to the lack of features in the corridor the robots generally have to travel a long distance until they can resolve ambiguities in the belief about their position.

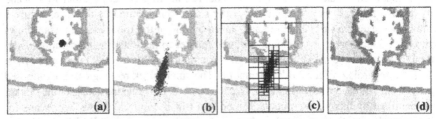

Fig. 4. Detection event: (a) Sample set of Marian as it detects Robin in the corridor. (b) Sample set reflecting Marian's belief about Robin's position (see *robot detection model* in Eq. (7)). (c) Tree-representation of this sample set and (d) corresponding density.

The key event, illustrating the utility of cooperation in localization, is a detection event. More specifically, Marian, the robot in the lab, detects Robin, as it moves through the corridor (see right camera image and laser range scan of Figure 2 for a characteristic measurement of this type). Using the detection model described in Section 4, Marian generates a new sample set as shown in Figure 4 (b). This sample set is converted into a density using density trees (see Figure 4 (c) and (d)). Marian then transmits this density to Robin which integrates it into its current belief. The effect of this integration on

(a) **(b)**

Fig. 5. Sample set representing Robin's belief (a) as it passes Marian and (b) after incorporating Marian's measurement.

Robin's belief is shown in Figure 5 (b). It shows Robin's belief after integrating the density representing Marian's detection. As this figure illustrates, this single incident almost completely resolves the uncertainty in Robin's belief.

We conducted ten experiments of this kind and compared the performance to conventional MCL for single robots which ignores robot detections. To measure the performance of localization we determined the true locations of the robot by measuring the starting position of each run and performing position tracking off-line using MCL. For each run, we then compared the estimated positions (please note that here the robot was not told it's starting location) with the positions on the reference path. The results are summarized in Figure 6.

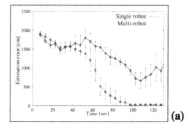**(a)** 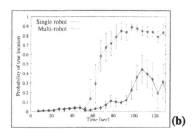**(b)**

Fig. 6. Comparison between single-robot localization and localization making use of robot detections. The x-axis represents the time and the y-axis represents (a) the estimation error and (b) the probability assigned to the true location.

Figure 6 (a) shows the estimation error as a function of time, averaged over the ten experiments, along with their 95% confidence intervals (bars). Figure 6 (b) shows the probability assigned to the true locations of the robot, obtained by summing over the weighting factors of the samples in an area ±50 cm and ±10 degrees around the true location. As can be seen in both figures, the quality of position estimation increases much faster when using multi-robot localization. Please note that the detection event typically took place 60-100 seconds after the start of an experiment.

Obviously, this experiment is specifically well-suited to demonstrate the advantage of detections in multi-robot localization, since the robots' uncertainties are somewhat orthogonal, making the detection highly effective. A more thoroughly evaluation of the benefits of MCL will be one topic of future research.

6   Related Work

Mobile robot localization has frequently been recognized as a key problem in robotics with significant practical importance. A recent book by Borenstein, Everett, and Feng [2] provides an overview of the state-of-the-art in localization.

Almost all existing approach address single-robot localization only. Moreover, the vast majority of approaches is incapable of localizing a robot globally; instead, they are designed to track the robot's position by compensating small odometric errors. Thus, they differ from the approach described here in that they require knowledge of the robot's initial position. Furthermore, they are not able to recover from global localizing failures. Probably the most popular method for tracking a robot's position is Kalman filtering [15, 20, 21, 26, 28], which represents the belief by a uni-modal Gaussian distribution. These approaches are unable to localize robots under global uncertainty. Recently, several researchers proposed *Markov localization*, which enables robots to localize themselves under global uncertainty [6, 16, 23, 27]. Global approaches have two important advantages over local ones: First, the initial location of the robot does not have to be specified and, second, they provide an additional level of robustness, due to their ability to recover from localization failures. Among the global approaches those using metric representations of the space such as MCL and [6, 5] can deal with a wider variety of environments than the methods relying on topological maps. For example, they are not restricted to orthogonal environments containing pre-defined features such as corridors, intersections and doors.

The issue of cooperation between multiple mobile robots has gained increased interest in the past. In this context most work on localization has focused on the question of how to reduce the odometry error using a cooperative team of robots [19, 24, 1]. While these approaches are very successful in reducing the odometry error, none of them incorporates environmental feedback into the estimation. Even if the initial locations of all robots are known, they ultimately will get lost although at a slower pace than a comparable single robot. The problem addressed here differs in that we are interested in collaborative localization in a global frame of reference, not just reducing odometry error.

7 Conclusions

In this paper, we presented a probabilistic method for collaborative mobile robot localization. At its core, our approach uses probability density functions to represent the robots' estimates as to where they are. To avoid exponential complexity in the number of robots, a factorial representation is advocated where each robot maintains its own, local belief function. A fast, universal sampling-based scheme is employed to approximate beliefs. The probabilistic nature of our approach makes it possible that teams of robots perform *global localization*, i.e., they can localize themselves from scratch without initial knowledge as to where they are.

During localization, detections are used to introduce additional probabilistic constraints between the individual belief states of the robots. As a result, our approach makes it possible to amortize data collected at multiple platforms. This is particularly attractive for heterogeneous robot teams, where only a small number of robots may be equipped with high-precision sensors.

Experimental results, carried out in a typical office environment, demonstrate that our approach can reduce the uncertainty in localization significantly, when compared to conventional single robot localization. Thus, when teams of robots are placed in a known environment with unknown starting locations, our approach can yield much faster localization at approximate equal computation costs and relatively small communication overhead.

The approach described here possesses several limitations that warrant future research. First, in our current system, only "positive" detections are processed. *Not seeing another robot* is also informative, and the incorporation of such negative detections is generally possible in the context of our statistical framework. Another limitation of the current approach arises from the fact that our detection approach must be able to identify individual robots. The ability to integrate over the beliefs of all other robots is a natural extension of our approach although it increases the amount of information communicated between the robots. Furthermore, the collaboration described here is purely passive, in that robots combine information collected locally, but they do not change their course of action so as to aid localization as, for example, described in [14]. Finally, the robots update their belief instantly whenever they perceive another robot. In situations in which both robots are highly uncertain at the time of the detection it might be more appropriate to delay the update and synchronize the beliefs when one robot has become more certain about its position.

Despite these open research areas, our approach provides a sound statistical basis for information exchange during collaborative localization, and empirical results illustrate its appropriateness in practice. While we were forced to carry out this research on two platforms only, we conjecture that the benefits of collaborative multi-robot localization increase with the number of available robots.

References

1. J. Borenstein. Control and kinematic design of multi-degree-of-freedom robots with compliant linkage. *IEEE Transactions on Robotics and Automation*, 1995.
2. J. Borenstein, B. Everett, and L. Feng. *Navigating Mobile Robots: Systems and Techniques.* A. K. Peters, Ltd., Wellesley, MA, 1996.
3. W. Burgard, A. B. Cremers, D. Fox, D. Hähnel, G. Lakemeyer, D. Schulz, W. Steiner, and S. Thrun. Experiences with an interactive museum tour-guide robot. *Artificial Intelligence*, 2000. accepted for publication.
4. W. Burgard, A.B. Cremers, D. Fox, D. Hähnel, G. Lakemeyer, D. Schulz, W. Steiner, and S. Thrun. The interactive museum tour-guide robot. In *Proc. of the National Conference on Artificial Intelligence (AAAI)*, 1998.
5. W. Burgard, A. Derr, D. Fox, and A.B. Cremers. Integrating global position estimation and position tracking for mobile robots: the Dynamic Markov Localization approach. In *Proc. of the IEEE/RSJ International Conference on Intelligent Robots and Systems (IROS)*, 1998.
6. W. Burgard, D. Fox, D. Hennig, and T. Schmidt. Estimating the absolute position of a mobile robot using position probability grids. In *Proc. of the National Conference on Artificial Intelligence (AAAI)*, 1996.
7. J. Carpenter, P. Clifford, and P. Fernhead. An improved particle filter for non-linear problems. Technical report, Department of Statistics, University of Oxford, 1997.
8. I.J. Cox and G.T. Wilfong, editors. *Autonomous Robot Vehicles.* Springer Verlag, 1990.
9. F. Dellaert, W. Burgard, D. Fox, and S. Thrun. Using the condensation algorithm for robust, vision-based mobile robot localization. In *Proc. of the IEEE Computer Society Conference on Computer Vision and Pattern Recognition (CVPR)*, 1999.
10. F. Dellaert, D. Fox, W. Burgard, and S. Thrun. Monte carlo localization for mobile robots. In *Proc. of the IEEE International Conference on Robotics & Automation (ICRA)*, 1999.

11. D. Fox. *Markov Localization: A Probabilistic Framework for Mobile Robot Localization and Naviagation.* PhD thesis, Dept. of Computer Science, University of Bonn, Germany, December 1998.

12. D. Fox, W. Burgard, F. Dellaert, and S. Thrun. Monte carlo localization: Efficient position estimation for mobile robots. In *Proc. of the National Conference on Artificial Intelligence (AAAI)*, 1999.

13. D. Fox, W. Burgard, H. Kruppa, and S. Thrun. A monte carlo algorithm for multi-robot localization. Technical Report CMS-CS-99-120, Carnegie Mellon University, 1999.

14. D. Fox, W. Burgard, and S. Thrun. Active markov localization for mobile robots. *Robotics and Autonomous Systems*, 25:195–207, 1998.

15. J.-S. Gutmann and C. Schlegel. AMOS: Comparison of scan matching approaches for self-localization in indoor environments. In *Proc. of the 1st Euromicro Workshop on Advanced Mobile Robots*. IEEE Computer Society Press, 1996.

16. L.P. Kaelbling, A.R. Cassandra, and J.A. Kurien. Acting under uncertainty: Discrete bayesian models for mobile-robot navigation. In *Proc. of the IEEE/RSJ International Conference on Intelligent Robots and Systems (IROS)*, 1996.

17. D. Koller and R. Fratkina. Using learning for approximation in stochastic processes. In *Proc. of the International Conference on Machine Learning (ICML)*, 1998.

18. K. Konolige. Markov localization using correlation. In *Proc. of the International Joint Conference on Artificial Intelligence (IJCAI)*, 1999.

19. R. Kurazume and N. Shigemi. Cooperative positioning with multiple robots. In *Proc. of the IEEE/RSJ International Conference on Intelligent Robots and Systems (IROS)*, 1994.

20. F. Lu and E. Milios. Globally consistent range scan alignment for environment mapping. *Autonomous Robots*, 4:333–349, 1997.

21. P.S. Maybeck. The Kalman filter: An introduction to concepts. In Cox and Wilfong [8].

22. A.W. Moore, J. Schneider, and K. Deng. Efficient locally weighted polynomial regression predictions. In *Proc. of the International Conference on Machine Learning (ICML)*, 1997.

23. I. Nourbakhsh, R. Powers, and S. Birchfield. DERVISH an office-navigating robot. *AI Magazine*, 16(2), Summer 1995.

24. I.M. Rekleitis, G. Dudek, and E. Milios. Multi-robot exploration of an unknown environment, efficiently reducing the odometry error. In *Proc. of the International Joint Conference on Artificial Intelligence (IJCAI)*, 1997.

25. D.B. Rubin. Using the SIR algorithm to simulate posterior distributions. In M.H. Bernardo, K.M. an DeGroot, D.V. Lindley, and A.F.M. Smith, editors, *Bayesian Statistics 3*. Oxford University Press, Oxford, UK, 1988.

26. B. Schiele and J.L. Crowley. A comparison of position estimation techniques using occupancy grids. In *Proc. of the IEEE International Conference on Robotics & Automation (ICRA)*, 1994.

27. R. Simmons and S. Koenig. Probabilistic robot navigation in partially observable environments. In *Proc. of the International Joint Conference on Artificial Intelligence*, 1995.

28. R. Smith, M. Self, and P. Cheeseman. Estimating uncertain spatial relationships in robotics. In I. Cox and G. Wilfong, editors, *Autonomous Robot Vehicles*. Springer Verlag, 1990.

29. M.A. Tanner. *Tools for Statistical Inference*. Springer Verlag, New York, 1993. 2nd edition.

30. S. Thrun, M. Bennewitz, W. Burgard, A.B. Cremers, F. Dellaert, D. Fox, D. Hähnel, C. Rosenberg, N. Roy, J. Schulte, and D. Schulz. MINERVA: A second generation mobile tour-guide robot. In *Proc. of the IEEE International Conference on Robotics & Automation (ICRA)*, 1999.

31. S. Thrun. Learning metric-topological maps for indoor mobile robot navigation. *Artificial Intelligence*, 99(1):27–71, 1998.

Object Classification Using Simple, Colour Based Visual Attention and a Hierarchical Neural Network for Neuro-Symbolic Integration

Hans A. Kestler, Steffen Simon, Axel Baune, Friedhelm Schwenker, and Günther Palm

Department of Neural Information Processing, University of Ulm, 89069 Ulm, Germany
{kestler, simon, abaune, schwenker,
palm}@neuro.informatik.uni-ulm.de

Abstract. An object classification system built of a simple colour based visual attention method, and a prototype based hierarchical classifier is established as a link between subsymbolic and symbolic data processing. During learning the classifier generates a hierarchy of prototypes. These prototypes constitute a taxonomy of objects. By assigning confidence values to the prototypes a classification request may also return symbols with confidence values.

For performance evaluation the classifier was applied to the task of visual object categorization of three data sets, two real–world and one artificial. Orientation histograms on subimages were utilized as features. With the currently very simple feature extraction method, classification accuracies in the range of 69% to 90% were attained.

1 Introduction

The object classifier is part of the autonomous vehicle developed by the Collaborative Research Center 527 at the University of Ulm, Germany. The main topic of this project is to study the organisation of useful interactions between methods of subsymbolic and symbolic information processing, in particular neural networks and knowledge based systems, in an autonomous vehicle [6]. It is intended to construct an artificial sensori-motor system that – within the next years – should be able to react quickly to unexpected changes in an office environment, to make strategic plans, and to gather experience from its environment, i.e. to exhibit adaptive or learning behaviour. For many of the planned tasks an object classificatation and recognition system, which provides a link to higher symbolic information processing layers, is indispensable.

Even more so, such an object classifier system has to find one or more regions within the camera image that contain objects of interest, as objects typically do not appear centered on the image, furthermore multiple objects within a single image need to be handled. As the system imposes strong computational requirements, the colour based mechanism of visual attention is simple and the classification is based on prototypes. This motivated the development of the presented hierarchical neural object classifier system.

Prototypes, which represent both symbols and points in feature space, are the basis of the link to the symbolic information processing layers. By utilizing a nearest neighbour rule, dichotomic regions of the feature space may be assigned to these prototypes. By further refining these regions, a hierarchical classifier structure is generated, which in turn allows the assignment of confidence values for sub- and super-classes.

A brief overview of the preprocessing via orientation histograms will be given in the next section. After that, a simple colour based method for visual attention will be introduced. The subsequent sections will center on the hierarchical classification of object images, both real and artificially generated.

2 Preprocessing

The image is subdivided into $n \times n$ non–overlapping subimages and for each subimage the orientation histogram of eight orientations (range: $0 - 2\pi$, dark/light edges) is calculated [8] from the gray valued image. The orientation histograms of all subimages are concatenated into the characterizing feature vector (Fig. 1).

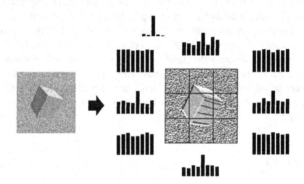

Fig. 1. Elements of the feature extraction method. The grey valued image (left) is convolved with the masks S_x and S_y resulting in the gradient image (right, center; absolute value of the gradient). Orientation histograms (encircling the gradient image) of non–overlapping subimages constitute the feature vector.

The gradient of an image $f(x, y)$ at location (x, y) is the two dimensional vector

$$\begin{pmatrix} G_x \\ G_y \end{pmatrix} = \begin{pmatrix} \frac{\partial f}{\partial x} \\ \frac{\partial f}{\partial y} \end{pmatrix} \approx \begin{pmatrix} f * S_x \\ f * S_y \end{pmatrix}$$

($*$ denotes the convolution operation). Gradient directions for the data described in Sect. 5 were calculated with modified 5×5 Sobel operators

$$S_x = \begin{pmatrix} -1 & -1 & -2 & -1 & -1 \\ -1 & -2 & -3 & -2 & -1 \\ 0 & 0 & 0 & 0 & 0 \\ 1 & 2 & 3 & 2 & 1 \\ 1 & 1 & 2 & 1 & 1 \end{pmatrix} \qquad S_y = \begin{pmatrix} -1 & -1 & 0 & 1 & 1 \\ -1 & -2 & 0 & 2 & 1 \\ -2 & -3 & 0 & 3 & 2 \\ -1 & -2 & 0 & 2 & 1 \\ -1 & -1 & 0 & 1 & 1 \end{pmatrix}$$

and 3×3 Sobel operators. The gradient directions are calculated with respect to the x-axis:

$$\alpha(x, y) = \text{atan2}\left(f * S_y, f * S_x\right)$$

The *atan2* function corresponds to the *atan* but additionally uses the sign of the arguments to determine the quadrant of the result. The eight bins of the histogram all have equal size $(2\pi/8)$. The histogram values are calculated by counting the number of angles falling into the respective bin. Histograms are normalized to the size of their subimages.

3 Regions of Interest

Visual classification of objects implies their reasonable delimitation from background. A possible way to achieve this is to locate regions of interest (ROI), that is regions of single objects in the image, and to preprocess subimages containing these regions.

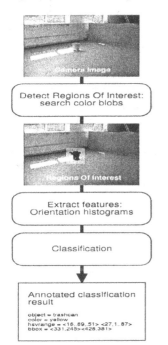

Fig. 2. Succession of processing steps: The pixels in the image center mark a colour blob found as a possible region of interest (ROI). The light rectangular area shows the resulting bounding box of the ROI.

In contrast to high–level methods for handling visual attention, like saliency-based methods [3, 4], we currently use a simple colour-based approach for the detection of possible ROIs in the actual camera image as computational requirements are tight. After downsampling the original camera image (by a factor depending on the image size, 4 in the case of a 768×576 image), the resulting image (192×144 pixel) is transformed from RGB (red, green, blue) to HSV-colour space (hue, saturation, value) [10]. We use the HSV colour space, because this colour model makes it very easy to describe colour ranges independent of saturation and lightness. Downsampling has the effect of a low pass filtering and also reduces the computational complexity. After downsampling and colour space transformation, we mark each pixel inside a valid colour range (Table 1) and then search for regions of connected pixels of the same colour range (colour blobs, see Fig. 2). For every found blob the colour range and number of pixels is determined. If the number of pixel of a blob is larger then a threshold value (200 pixel in the case of a 768×576 image), a bounding box is calculated.

In addition to these, the following heuristics for merging regions are applied: (a) In an initial step, bounding boxes which are contained in larger ones are merged. This is typically applied to objects, whose colour range does not match one of the valid ranges, but matches two. In this case one part of the object belongs to another blob than the rest. (b) After that, the distances between centers of regions within the same colour range are calculated.

If the distance between two regions is less then 48 pixel, they will be merged. This is useful for merging small colour blobs, for example bottles, whose colour regions are usually separated by the label of the bottle.

In a last step, bounding boxes are rescaled to match the original image size and are enlarged (32 pixel in the case of a 768×576 image). From the so extracted region the feature vector is calculated by the method of Sect. 2.

4 Hierarchical Classifier

Neural networks are used in different applications for classification tasks [1]. Typically, in this kind of application the neural network performs a mapping from the feature space $\mathbb{R}^n$ into a finite set of classes $C = \{1, 2, \ldots, l\}$. During the training phase of a supervised learning algorithm the parameters of the network are adapted by presenting a set of training examples to the network. The training set $S := \{(x^\mu, t^\mu) \mid \mu = 1, \ldots, M\}$ in a supervised learning scenario consists of feature vectors $x^\mu \in \mathbb{R}^n$ each labeled with a class membership $t^\mu \in C$. After the training procedure – in the recall phase – unlabeled observations $x \in \mathbb{R}^n$ are presented to the trained network. A class $z \in C$ is returned as the estimated output corresponding to the input x.

A hierarchical network architecture is used, as misclassifications of patterns typically take place only between a subset of all classes and not between the whole set of classes [9]. Thus, it should be possible to construct specialized classifiers that handle only a part of the patterns. This is in contrast to tree–based classifiers such as CART [2] which primarily divide the feature space into axis–parallel subregions. This motivation lead us to the construction of a hierarchical neural network based on a first level network for the coarse classification of the patterns and a set of networks each recursively refining the discrimination of patterns from their confusion class (see Sect. 4.2).

4.1 Supervised Learning Vector Quantization (LVQ)

An LVQ network is a competitive neural network consisting of a single layer of k neurons. The neurons have binary valued outputs, where $y_j = 1$ stands for an active neuron and $y_j = 0$ for a silent one. The synaptic weight vectors $c_1, \ldots, c_k \in \mathbb{R}^n$ of an LVQ network divide the input space into k disjoint regions $R_1, \ldots, R_k \subset \mathbb{R}^n$, where each set R_j is defined by

$$R_j = \{x \in \mathbb{R}^n \mid \|x - c_j\| = \min_{i=1\ldots,k} \|x - c_i\|\}. \tag{1}$$

Here $\| \cdot \|$ denotes the Euclidean norm. Such a partition is called a *Voronoi tesselation* of the input space. The weight vector c_j is also called a *prototype vector* of region R_j. When presenting an input vector $x^\mu \in \mathbb{R}^n$ to the net, all neurons $j = 1, \ldots, k$ determine their Euclidean distance $d_j = \|x - c_j\|$ to the input x. The competition between the neurons is realized by searching for the prototype with minimum distance: $d_{j^*} = \min_{j=1\ldots,k} d_j$. In LVQ-networks each prototype vector c_j is labeled with its class membership $y_j \in C$. In LVQ1 training – the simplest version of LVQ-training –

only the neuron j^* – the *winning neuron* – is adapted according to the learning rule:

$$\Delta c_{j^*} = \begin{cases} \eta(t)(x - c_{j^*}) & : \quad \text{class}(x) = \text{class}(c_{j^*}) \\ -\eta(t)(x - c_{j^*}) & : \quad \text{class}(x) \neq \text{class}(c_{j^*}) \end{cases} \tag{2}$$

where class(.) denotes the class membership of a prototype or training sample. $\eta > 0$ is a decreasing sequence of learning rates and t the number of presented training samples. This competitive learning algorithm has been suggested ([5]) for vector quantization and classification tasks. From this algorithm, Kohonen derived the OLVQ1, LVQ2 and LVQ3 network training procedures, which are useful algorithms for the fine-tuning of a pre-trained LVQ1 network. Here, OLVQ1 is used which realizes prototype depending learning rates which are adapted during training.

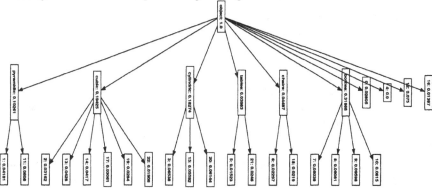

Fig. 3. Confidence tree resulting from the classification of a bottle. The floating point numbers denote confidence values. Labels are given as integer numbers: 0 cone, 1 pyramid, 2 cube, 3 cylinder, 4 bowl, 5 table, 6 chair, 7-10 different bottles, 11 tetrahedron, 12 octahedron, 13 dodec-ahedron, 14 gem, 15 coffee mug, 16 column, 17 clipped column, 18 office chair, 19 monitor, 20 bucket, 21 office table and 22 drawer

4.2 Calculation of Confusion Classes

Based on the confusion matrix V, resulting from the nearest neighbour classification of a prototype, and a threshold $\Theta > 0$ the Boolean $l \times l$-matrix V' is calculated:

$$V'_{ij} = \begin{cases} 1 & : \quad V_{ij} > \Theta \\ 0 & : \quad \text{otherwise} \end{cases}$$

An entry $V'_{ij} = 1$ indicates a confusion from target class j to class i. A set of confusion classes $K_1, \ldots, K_p$ is defined by V'. Each confusion class K_ν is a nonempty subset of the set of classes C. The number p of confusion classes depends on V'. The set of confusion classes is defined by the following relations:

1. $V'_{ij} = 1, i \neq j$ implies $\exists p : \{i, j\} \subset K_p$ (symmetry)
2. $V'_{ij} = V'_{jk} = 1, i \neq j, j \neq k$ implies $\exists p : \{i, j, k\} \subset K_p$ (transitivity)

At classification the nearest prototype either maps the input pattern x^μ to a confusion class K_ν or to its class–label $z \in C$, i.e. a leaf of the classification hierarchy.

In the case where confusions appear between many different classes, the transitivity condition may lead to the trivial solution of a single confusion class containing all classes. If the training data is distorted by a lot of noise, this effect can be controlled by setting Θ to a larger positive value. This was used as a stopping criterion in the classifier construction process because this means, that the classifier hierarchy cannot further be refined; the resulting subclassifier would be identical to the current one.

4.3 Training and Recall

Training. In this section the training algorithm of the hierarchical neural network classifier is described. The recursive training phase of this network with a training set S and a set of classes C contains the following steps:

```
train HC(S,C):
        train LVQ(S,C)
        V:=ConfusionMat(S,C)
        Calculate confusion classes and sub-training sets:
            (S_i,K_i); i=1,...,p
        FOREACH i:=1,...,p: train HC(S_i,K_i)
```

In the first step of this training procedure a Learning Vector Quantization (LVQ) network [5] is trained with the training set S. After this, the classification performance of the LVQ network is tested on the training set. The result of this classification is represented by a $l \times l$ confusion matrix V. In every recursion step for each confusion class K_i, a finer tesselation of the feature space is generated by LVQ–learning. The training set S_i is given by the subset of training samples $x^\mu \in S$ with label $t^\mu \in K_i$.

Recall. In order to achieve a simple subsymbolic–symbolic coupling together with the traversing of a taxonomy, confidence- or belief–values b_j are assigned to every prototype c_j (or confusion class K_j) during classification of an unknown input pattern $x \in \mathbb{R}^n$, see Fig. 3. Via these belief–values it is possible, depending on the classification task, to traverse the taxonomy which has been built in the training phase. The confidence values are assigned as follows:

- Determine the Euclidean distance of the unknown input pattern x to every prototype.
- Select the nearest class specific prototypes j (one for every class).
- Assign a confidence value b_j to each of these prototypes:

$$b_j = \frac{\Delta_{max} - \Delta_j}{\sum_{i=1}^{l}(\Delta_{max} - \Delta_i)}$$

Δ_{max} denotes the maximal distance of the class specific prototypes to the unknown input pattern x, Δ_j is the distance of the nearest class specific prototype j to the feature vector x. Obviously $\sum_{j=1}^{l} b_j = 1$ and $b_j \in [0,1]$.

- Assign the sum of the confidence of the individual prototypes to their super–prototype (in Fig. 3. "bottles")

In the case of a perfect match of the unknown input pattern x with the prototype k, i.e. $\Delta_k = 0$, b_k is assigned the maximal belief value.

Here, the classification process was realized as stepwise sequential decision making according to the confidence values of the respective prototypes until a leaf of the hierarchy was reached. This is illustrated by the following example on the artificial data set (Sect. 5: Depending on the requested confidence, the classification result could either be a class label or a confusion class. If a threshold depending on the class probability (in this case $\frac{1}{23}$ for uniform distributed data) is used, a decision criterion would be available.

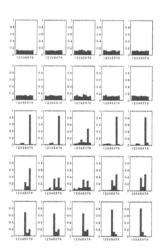

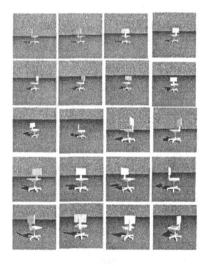

Fig. 4. Example of an orientation histogram: office chair

Fig. 5. Example of the variation inside the class "office chair"

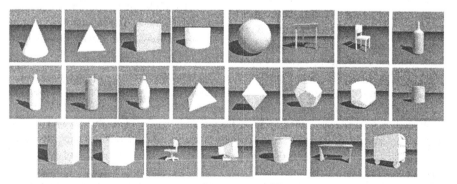

Fig. 6. Examples of all 23 classes. Labels are from top left to right bottom: cone, pyramid, cube, cylinder, bowl, table, chair, bottle 1 to 4, tetrahedron, octahedron, dodecahedron, gem, coffee mug, column, clipped column, office chair, monitor, bucket, office table and drawer

If $\Theta = \frac{2}{23}$ is selected as a threshold, in Fig. 3 the result would be *class 9 = bottle 3*, because $b_9 = 0.09598 > \Theta$. In case of a threshold of $\Theta = \frac{3}{23}$ the result would be super–class *bottle*.

5 Data

Three different data sets were used in performance evaluation of the classifier, artificially generated and real–world.

Artificial data: 5920 Images of 23 different objects were generated with the raytracing software PoV-Ray [7] (cone, pyramid, cube, cylinder, bowl, table, chair, bottle 1 to 4, tetrahedron, octahedron, dodecahedron, gem, coffee mug, column, clipped column, office chair, monitor, bucket, office table and drawer, class 0 to 5: 250 images; class 6 to 22: 260 images, see Fig. 6). Images (256 × 256 pixel) were created through a virtual helical movement of the observer around the object (constant distance). The azimuth angle, was varied in steps of 36°(0°to 360°) with reference to the object.

The declination angle was varied in steps of 1.5°(0°to 15°). A point light source made the same circular movement 45°ahead of the observer but at a constant declination angle of 45°. During this "movement" the objects were rotated at random (uniform distribution) around their vertical axis, the scaling of the objects (x,y,z -axis) was varied uniformly distributed in the range of 0.7 to 1.3 and the ambient light was varied at random (uniform) in the range of 0.2 to 0.6.An example of the variation within one class is given in Fig. 5. From all images, 5 × 5 orientation histograms (modified 5 × 5 Sobel operator, see Sect. 2) were concatenated into a feature vector. **Real–world data A: Manual determination of regions** Camera images were recorded for four different 3-D objects (0 cone, 1 pyramid, 2 cube, 3 cylinder), see Figs. 7 and 8.

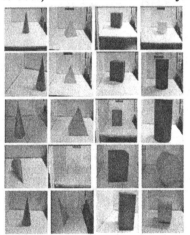

Fig. 7. Examples of all four classes of the real–world data A. The images give an impression of the variation within one class (columns) and across classes (rows).

The test scenes were acquired under two different illumination conditions: (1) natural illumination by sunlight passing through a nearby window, and (2) mixed natural and artificial lighting. In the latter condition an overhead neon lamp was added as an additional light source. The objects were registered in an upright or lying position. The 1315 images were labeled by hand (class 0: 229, class 1: 257, class 2: 385, class 3: 444), and had an initial resolution of 240×256 pixels, which was then reduced to 200×200 pixels by manually cutting out the region of interest. Features were calculated from empirically determined, concatenated 8×8 orientation histograms (modified 5 × 5 Sobel operator, see Sect. 2).

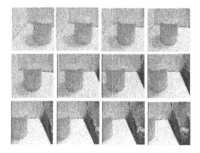

Fig. 8. Example of a recording sequence of the real–world data A which was used for labeling.

Table 1. Color ranges used for detecting regions of interest in real–world data set B. Hue (H) values are in the range 0...359, saturation (S) and value (V) in the interval [0, 1].

Colour	lower bound			upper bound		
	H	S	V	H	S	V
yellow	16	0,59	0,26	68	1	0,91
red	345	0,2	0,14	15	1	0,91
blue	200	0,47	0,26	240	1	0,91

Real–world data B: Automatic determination of regions Camera images were recorded for five different 3-D objects (coke bottle, orange juice bottle, cylinder, ball and bucket) with an initial resolution of 768×576 pixels. To these five objects twelve classes were assigned (bottle full/empty, bottle lying/upright, cylinder lying/upright). The test scenes were acquired under mixed natural and artificial lighting. Regions of interest where calculated from 653 images using the colour blob detection method described in Sect. 3. These regions where checked and labeled by hand, 399 images remained for training (coke bottle empty upright: 7, coke bottle empty lying: 15, coke bottle full upright: 6, coke bottle full lying: 14, juice bottle empty upright: 4, juice bottle empty lying: 9, juice bottle full upright: 14, juice bottle full lying: 62, cylinder upright: 64, cylinder lying: 141, bucket: 42, ball: 21). Features were calculated from concatenated 3×3 histograms (3×3 Sobel operator, see Sect. 2). Figure 9 gives an impression of the variability within one class of the original camera images. Regions of interest are detected using three colour ranges, one for red (bucket, cylinder, ball, label of coke bottle), blue (cylinder) and yellow (cylinder, bucket, orange juice). See Table 1 for details about the used colour ranges. The images in Fig. 10 show the automatically extracted regions of interest for some of the *bucket*-images, they correspond to those of Fig. 9.

6 Results

The classifier was tested on the set of artificial data, containing 5920 images of twenty-three objects. The top-level network started with seven prototypes/class. This number was increased by 50% in each recursion step, so the resulting second level networks

Fig. 9. Examples of class bucket of the real–world data set **B**. The images give an impression of the variation within one class.

contained 10 prototypes/class. Based on the obtained confusions, multiple confusion classes were generated (Fig. 3.). Table 2 shows the cross–validation errors of the runs. The results given in Table 2 show a very stable performance across different cross–validation runs.

In the case of the real–world data set A, a classification accuracy in the range of 93.5% to 94.6% (97.1% to 97.3% on the training set) was attained, see Table 3. These are the results of three five–fold cross–validation runs with eight prototypes/class.

Classification results of the real–world data set B are based on three ten–fold cross–validation runs, see Table 4. Due to the small number of samples, only five proto-types/class are used. In this case a classification accuracy in the range of 68.7% to 72.4% (94.8% to 95.2% on the training set) was attained.

Table 2. Cumulated results of three five-fold cross–validation runs on the artificial data. The absolute number of samples classified, the number of misclassified samples, the mean, the standard deviation and the minimum and maximum misclassification rates of the individual cross–validation runs are given.

	cross validation 1		cross validation 2		cross validation 3	
	training set	test set	training set	test set	training set	test set
number of samples	23680	5920	23680	5920	23680	5920
misclassified	2379	1099	2311	1117	2298	1073
mean error	0.10047	0.18564	0.09759	0.18868	0.09704	0.18125
std. dev. of error	0.00263	0.01510	0.00426	0.01094	0.00297	0.01564
minimum error	0.09671	0.17145	0.09438	0.17230	0.09396	0.15710
maximum error	0.10367	0.20608	0.10367	0.20101	0.10030	0.19595

Table 3. Cumulated results of three five-fold cross–validation runs on the four-class real–world data A (Fig. 7, 1315 images). The absolute number of samples classified, the number of misclassified samples, the mean, the standard deviation and the minimum and maximum misclassification rates of the individual cross–validation runs are given.

	cross validation 1		cross validation 2		cross validation 3	
	training set	test set	training set	test set	training set	test set
number of samples	5260	1315	5260	1315	5260	1315
misclassified	144	85	143	70	154	79
mean error	0.02738	0.06464	0.02719	0.053231	0.02928	0.06008
std. dev. of error	0.00677	0.00892	0.00883	0.02048	0.00246	0.00867
minimum error	0.01616	0.05703	0.01711	0.02661	0.02662	0.04943
maximum error	0.03423	0.07605	0.03992	0.07985	0.03327	0.07224

Table 4. Cumulated results of three ten-fold cross–validation runs on the real world data set B. The absolute number of samples classified, the number of misclassified samples, the mean, the standard deviation and the minimum and maximum misclassification rates of the individual cross–validation runs are given.

	cross validation 1		cross validation 2		cross validation 3	
	training set	test set	training set	test set	training set	test set
number of samples	3591	399	3591	399	3591	399
misclassified	184	110	174	111	177	125
mean error	0.05124	0.27571	0.04845	0.27827	0.04929	0.31301
std. dev.	0.01781	0.08082	0.01429	0.07338	0.01385	0.08279
minimum error	0.02786	0.125	0.03621	0.2	0.02786	0.20512
maximum error	0.08357	0.35	0.07520	0.45	0.07520	0.45

Fig. 10. Calculated regions of interest of the camera images of Fig. 9.

7 Conclusions

In this study we presented a system for object classification consisting of a simple colour based visual attention method and a very simple hierarchical classifier. This hierarchical classifier leads to a subsymbolic–symbolic coupling via a hierarchy of prototypes as a first step to more detailed investigations. The results obtained are both remarkable and astonishing, that with such a simple preprocessing (orientation histograms) procedure and a conceptually easy hierarchical classifier these classification accuracies were possible on data sets with a high intra–class variability (see Figs. 5, 7 and 9). How symbols are assigned to intermediate confusion classes still needs to be defined. The classification accuracies on artificial data and real–world data (set A) proved to be very stable across the different cross–validation runs. The high accuracy on the training set B, but low accuracy on the cross–validation run suggest that there are too few samples to describe the problem domain sufficiently. Another problem are flat hierarchies observed with data sets with a very high intra– and extra–class variability. This is caused by the transitivity relation described in Sect. 4.2. This might be overcome by additional features as colour, corners etc., which may be fused to the existing ones.

Even though the simplicity of the methods used, calculation times are still a problem for the deployment in a real–time environment. The calculation of ROIs within an high resolution camera image sized 768 × 576 pixel currently takes about 1.1 seconds on a PowerMacintosh 7500/200 (64 MByte RAM, 200 MHz PPC604e processor). On the same computer a classification throughput of about 300 classifications/second is obtained. Although these results are promising, further research in speeding up the image processing is neccessary.

Acknowledgement

We thank Florian Vogt for help with data aquisition and labeling. This project is part of the Collaborative Research Center (SFB 527) at the University of Ulm and is supported by the German Science Foundation (DFG).

References

1. C.M. Bishop. *Neural Networks for Pattern Recognition.* Clarendon Press, Oxford, 1995.
2. Leo Breiman, J. H. Friedman, R. A. Olshen, and C. J. Stone. *Classification and Regression Trees.* Wadsworth Publishing Company, Belmont, California, U.S.A., 1984.
3. L. Itti, C. Koch, and E. Niebur. A model of saliency-based visual attention for rapid scene analysis. *IEEE Transactions on Pattern Analysis*, 20(11):1254–1259, 1998.
4. C. Koch and S. Ullman. Shifts in selective visual attention: towards the underlying neural circuitry. *Human Neurobiology*, 4:219–227, 1985.
5. T. Kohonen. *Self Organizing Maps.* Springer Verlag, 1995.
6. G. Palm and G. Kraetzschmar. SFB 527: Integration symbolischer und subsymbolischer Informationsverarbeitung in adaptiven sensorimotorischen Systemen. In M. Jarke, K. Pasedach, and K. Pohl, editors, *Informatik '97 – Informatik als Innovationsmotor*, pages 111–120. Springer Verlag, 1997.
7. Persistence of vision raytracer (pov-ray). Webpage. http://www.povray.org/.
8. M. Roth and W.T. Freeman. Orientation histograms for hand gesture recognition. Technical Report 94-03, Mitsubishi Electric Research Laboratorys, Cambridge Research Center, 1995.
9. J. Schürmann. *Pattern Classification.* Wiley, New York, 1996.
10. A. R. Smith. Color gamut transform pairs. In R. L. Phillips, editor, *5th annual conf. on Computer graphics and Interactive Techniques*, pages 12–19, New York, 1978. ACM.

A Flexible Architecture for Driver Assistance Systems*

Uwe Handmann, Iris Leefken und Christos Tzomakas

Institut für Neuroinformatik, Ruhr Universität Bochum, 44780 Bochum

Introduction. The problems encountered in building a driver assistance system are numerous. The collection of information about real environment by sensors is erroneous and incomplete. When the sensors are mounted on a moving observer it is difficult to find out whether a detected motion was caused by ego-motion or by an independent object moving. The collected data can be analyzed by several algorithms with different features designed for different tasks. To gain the demanded information their results have to be integrated and interpreted. In order to achieve an increase in reliability of information a stabilization over time and knowledge about important features have to be applied. Different solutions for driver assistance systems have been published. An approach proposed by Rossi et. al. [8] showed an application for a security system. An application being tested on highways has been presented by Bertozzi and Broggi [1]. Dickmanns et al. presented a driving assistance system based on a 4D-approach [2]. Those systems were mainly designed for highway scenarios, while the architecture presented by Franke and Görzig [3] has been tested in urban environment.

Architecture. In contrast, the content of this paper concentrates on a flexible, modular architecture of a driver assistance system working on evaluation and integration of the actual information gained from different sensors. The modules of the architecture are represented by the object-related analysis, the scene interpretation and the behavior planning (fig. 1). The accumulated knowledge is organized in the knowledge base. The presented architecture is able to handle different tasks. New requirements to the system can be integrated easily. The proposed architecture is intended to produce different kinds of behavior according to given tasks. Information about the actual state of the environment is perceived by the system's sensors. The data collected by each sensor have to be processed and interpreted to gain the desired information for the actual task.

Object-related Analysis. The object-related analysis can be subdivided into a sensor information processing and a representational part. In the sensor information processing part the collected sensor data are preprocessed (e.g. segmentation, classification of regions of interest (ROI) or lane detection) and interpreted according to their capabilities. The processing can be performed for each sensor as well as information from different sensors can be fused [7]. Objects are extracted by segmentation, classification and tracking (fig. 2). The results of the sensor information processing stage are stabilized in movement-sensitive representations by introducing the time dimension. In this sense, a ROI is accepted

---

* Extended abstract of the paper *Eine flexible Architektur für Fahrerassistenzsysteme* [6] accepted by the DAGM-99 for the common part of the DAGM-99 and KI-99.

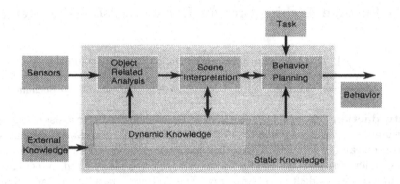

Fig. 1. Architecture for a driver assistance system

Fig. 2. Vision-based object detection, object classification and object tracking

as a valid hypothesis only if it has a consistent history. This is implemented by spatio-temporal accumulation using different representations with predefined sensitivities. The sensitivities are functions of the objects' supposed relative velocity and of the distance to the observer (fig. 3). In order to apply a time

Fig. 3. Image and representation: prediction and object detection of oncoming vehicles

stabilization to these regions and to decide whether they are valid or not, a prediction of their position in the knowledge integration part is realized. A competition between the different representations and a winner-takes-all mechanism ensures reliable object detection. An implementation of an object-related analysis on vision data has been presented in [4]. The results are passed to the scene interpretation.

Scene Interpretation. The scene interpretation interprets and integrates the different sensor results to extract consistent, behavior-based information. The scene interpretation is subdivided into a behavior-based representational and a scene analysis part. Objects and lane information are transformed to world coordinates with respect to the moving observer. The positions of the detected objects are determined in a bird's eye view of the driving plane. This dynamically organized representation is shown in fig. 4. The transformation rules follow the given position of the CCD camera in the car and the position of the car on the lane. The physical laws are given by the transformation equations for the camera and physical considerations of the movement and position of potential objects. The transformation also depends on constant data (e.g. length of a vehicle according to its classification). The scene analysis sustains the driver assistance

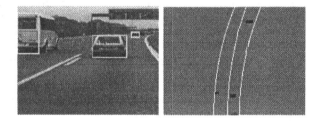

Fig. 4. Image with objects and bird's eye view

by evaluating the actual traffic condition as well as the scenery. According to the actual traffic condition and the planned behavior a risk-factor for actions is estimated. The determination of the traffic condition is performed by evaluating the information from scene interpretation. This is done by counting the objects, evaluating their relative speed and the movement according to their class. The scenario can be determined using GPS and street maps for investigating the kind of street, e.g. highway, country road or urban traffic. According to these scenarios different objectives have to be taken into consideration. The determined traffic condition as well as the classified scenario are proposed to the behavior planning.

Behavior Planning. The behavior planning depends on the given task and on the scene interpretation. Different solutions for the planning task are possible. A rule-based fuzzy-logic approach is described in [10]. An expert system is shown in [9]. In the present system an intelligent cruise control system was integrated. Behavior planning for the observer results in advices to the driver which are not only based on the intention to follow the leader but on regards concerning the safety of the own vehicle. This means that the object cannot be followed or might be lost in case of other objects or obstacles endangering the observer. The signal behavior for the main tasks is determined by a flow diagram shown in [5].

Knowledge Base. The knowledge needed for the evaluation of the data and for information management is given by the efforts of the task of driver assistance, by physical laws and traffic rules. An improvement of the results can be achieved by the information of the knowledge base. In the knowledge base static and dynamic knowledge is represented. Static knowledge is known in advance independently of the scenery of movement (e.g. physical laws, traffic rules). Dynamic knowledge (e.g. actual traffic situation, scenery) is knowledge changing with the actual information or with the task to be performed (e.g. objects in front of the car). Dynamic knowledge can also be influenced by external knowledge like GPS-information.

Conclusion. The proposed architecture has been tested on a simulation surface.

References

1. M. Bertozzi and A. Broggi. GOLD: a Parallel Real-Time Stereo Vision System for Generic Obstacle and Lane Detection. In IEEE, editor, *IEEE Transactions on Image Processing*, volume 4(2), pages 114–136, 1997.
2. E.D. Dickmanns et al. Vehicles capable of dynamic vision. In *15th International Joint Conference on Artificial Intelligence (IJCAI)*, pages 1–16, Nagoya, Japan, 1997.
3. S. Goerzig and U. Franke. ANTS - Intelligent Vision in Urban Traffic. In *IV'98, IEEE International Conference on Intelligent Vehicles 1998*, pages 545–549, Stuttgart, Germany, 1998.
4. U. Handmann, T. Kalinke, C. Tzomakas, M. Werner, and W. von Seelen. An Image Processing System for Driver Assistance. In *IV'98, IEEE International Conference on Intelligent Vehicles 1998*, pages 481 – 486, Stuttgart, Germany, 1998.
5. U. Handmann, I. Leefken, and C. Tzomakas. A Flexible Architecture for Intelligent Cruise Control. In *ITSC'99, IEEE Conference on Intelligent Transportation Systems 1999*, Tokyo, Japan, 1999.
6. U. Handmann, I. Leefken, and C. Tzomakas. Eine flexible Architektur für Fahrerassistenzsysteme. In *Mustererkennung 1999*, Heidelberg, 1999. Springer-Verlag. DAGM'99.
7. U. Handmann, G. Lorenz, T. Schnitger, and W. von Seelen. Fusion of Different Sensors and Algorithms for Segmentation. In *IV'98, IEEE International Conference on Intelligent Vehicles 1998*, pages 499 – 504, Stuttgart, Germany, 1998.
8. M. Rossi, M. Aste, R. Cattoni, and B. Caprile. The IRST Driver's Assistance System. Technical Report 9611-01, Instituto per la Ricerca Scientificia e Technologica, Povo, Trento, Italy, 1996.
9. R. Sukthankar. *Situation Awareness for Tactical Driving*. Phd thesis, Carnegie Mellon University, Pittsburgh, PA, United States of America, 1997.
10. Qiang Zhuang, Jens Gayko, and Martin Kreutz. Optimization of a Fuzzy Controller for a Driver Assistant System. In *Proceedings of the Fuzzy-Neuro Systems 98*, pages 376 – 382, München, Germany, 1998.

A Theory for Causal Reasoning

Camilla Schwind

LIM, Faculté des Sciences de Luminy, Université de la Méditerranée
163, Avenue de Luminy, Case 901, 13288 Marseille, Cedex 9, France

The objective of this paper is to study causal inference as an issue of common sense reasoning. It is well known that causal inference cannot be represented by classical implication which has a number of properties not correct for causal implication, such as monotonicity, contraposition or transitivity. In several fields of artificial intelligence, causality has been represented in different ways, by conditional-type operators [5, 9, 10], by modal operators [2, 14, 6], by meta-rules [13, 3] and by means of non-logical predicates [7].

We define causal rules as pairs of propositional formulas, written as $\alpha \approx\!\!\!> \beta$ and meaning "If α is true then this causes β to be true. Inference rules for causal rules allow to deduce new causal rules. The following inference rules are defined using assertions about causal rules ($\alpha \approx\!\!\!> \beta$) and about classical propositional validity ($\models \gamma$):

$$
\begin{array}{ll}
LLE & \text{if } \alpha \approx\!\!\!> \beta \text{ and } \models \alpha \leftrightarrow \gamma \text{ then } \gamma \approx\!\!\!> \beta \\
RW & \text{if } \alpha \approx\!\!\!> \beta \text{ and } \models \beta \rightarrow \gamma \text{ then } \alpha \approx\!\!\!> \gamma \\
AND & \text{if } \alpha \approx\!\!\!> \beta \text{ and } \alpha \approx\!\!\!> \gamma \text{ then } \alpha \approx\!\!\!> \beta \wedge \gamma \\
OR & \text{if } \alpha \approx\!\!\!> \beta \text{ and } \gamma \approx\!\!\!> \beta \text{ then } \alpha \vee \gamma \approx\!\!\!> \beta
\end{array}
$$

Definition 1 *The* causal closure $\overrightarrow{C}$ *of a set of causal rules C is the smallest set of causal rules closed under the application of the three inference rules LLE, AND and OR. A set of causal rules is called* closed *iff* $\overrightarrow{C} = C$

Given a set of causal rules C and an inference rule c, we denote by C^c the c-clusure of C .

Causal rules apply to a propositional base ($\mathcal{B}$), and allow to infer new propositional formulas from $\mathcal{B}$, which are caused in $\mathcal{B}$. We define an ordering $\prec$ on the set of causal rules, where $c \prec c'$ means that c is preferred to c' whenenver both can be applied. A causal rule $\alpha \approx\!\!\!> \beta$ *applies* to a propositional base $\mathcal{B}$ whenever $\alpha \in \mathcal{B}$ and allows to infer β, provided β is not already in $\mathcal{B}$ and $\neg\beta$ cannot be inferred. If two causal rules c and c', can both be applied to $\mathcal{B}$, and $c \prec c'$, then only c, which is preferred according to the order $\prec$, will be applied.

Definition 2 *A causal base CB is a triplet $(\mathcal{B}, C, \prec)$, where $\mathcal{B}$ is a set of propositional formulas, C is a set of causal rules and $\prec$ is a strict partial order on C . A causal base $CB = (\mathcal{B}, C, \prec)$ is closed whenever C is closed.*

In order to express what can be inferred from a knowledge base by means of causal rules, we introduce the notion of extension. There are known variants of default logic, where a notion of priority of defaults based on an ordering relation exists. In this present work, we adopt Brewka's formalism for priority in defaut logic [1].

Definition 3 *Let B be a set of propositional formulas and $c = (\alpha \approx\!\!> \beta)$ a causal rule. We say that c can be applied to B iff $\alpha \in B$ and $\beta \notin B$ and $\neg\beta \notin B$*

Definition 4 *Let $CB = (B, C, \prec)$ be a (prioritized) causal base, $\ll$ a strict total order containing $\prec$. E is an extension of CB iff $E = \bigcup E_i$, where*

1. $E_0 = Th(B)$
2. $E_{i+1} = Th(E_i)$ *if no causal rule can be applied to E_i*
3. $E_{i+1} = Th(E_i) \cup \{\beta\}$ *if $c = \alpha \approx\!\!> \beta$ can be applied to E_i and there is no $c' \in C$ such that c' can be applied to E_i and $c' \ll c$*

The set of extensions of a causal theory CB is denoted $\mathcal{E}(CB)$. The set of causal rules applied for the generation of an extension E is denoted $GR(E)$.

The following example shows how priority can block unintuitive extensions.

Example 1 $CB = (B, C, \prec)$
$C = \{rain \approx\!\!> wet, rain \wedge umbrella \approx\!\!> \neg wet, rain \wedge umbrella \wedge broken \approx\!\!> wet\}$
$B = \{rain, umbrella, broken\}$
$(rain \wedge umbrella \wedge broken \approx\!\!> wet) \prec (rain \wedge umbrella \approx\!\!> \neg wet) \prec (rain \approx\!\!> wet)$
Here, priority is based on specificity: the causal rule with a more specific precondition is preferred to one with a less specific precondition. CB has one extension $E = Th(B \cup \{wet\})$. Only the most specific causal rule is applied, namely "rain $\wedge$ umbrella $\wedge$ broken $\approx\!\!> wet$" and the set of generating rules is $GR(E) = \{rain \wedge umbrella \wedge broken \approx\!\!> wet\}$.

The theory $CB_1 = (B_1, C, \prec)$ where $B_1 = \{rain, umbrella\}$ with the same set of causal rules C has the extension $E_1 = Th(B \cup \{\neg wet\})$. Here, the most preferred rule is "rain $\wedge$ umbrella $\approx\!\!> \neg wet$" and the set of generating rules is $GR(E_1) = \{rain \wedge umbrella \approx\!\!> \neg wet\}$.

The theory $CB_2 = (B_2, C, \prec)$ where $B_2 = \{rain\}$ with the same set of causal rules C has the extension $E_2 = Th(B \cup \{wet\})$. Here, the most preferred rule is "rain $\approx\!\!> wet$" and the set of generating rules is $GR(E_1) = \{rain \approx\!\!> wet\}$.

Without priorities, we would get two extensions for CB and also for CB_1, namely $Th(B \cup \{\neg wet\})$ and $Th(B \cup \{wet\})$. In CB, *wet* holds *because* it rains and the umbrella is broken, whereas in CB_2, *wet* holds *because* it rains.

The notion of extension of a causal base defines an equivalence relation on causal bases, according to extensions: two causal bases are equivalent if they have the same extensions.

Definition 5 *Two causal bases CB and CB' are equivalent, $CB \sim CB'$, if $\mathcal{E}(CB) = \mathcal{E}(CB')$.*

It is easy to see that $\sim$ is an equivalence relation. As shows the following theorem, LLE and AND are implicitly applied with respect to extensions.

Theorem 1 *Let* $(\mathcal{B}, \mathcal{C}, \prec)$ *be a causal base. Then*

1. *If* $\alpha \approx\!\!\!> \beta \in \mathcal{C}$ *and* $\models \alpha \leftrightarrow \gamma$ *then* $(\mathcal{B}, \mathcal{C} \cup \{\gamma \approx\!\!\!> \beta\}, \prec) \sim (\mathcal{B}, \mathcal{C}, \prec)$
2. *If* $\alpha \approx\!\!\!> \beta \in \mathcal{C}$ *and* $\alpha \approx\!\!\!> \gamma \in \mathcal{C}$, *then* $(\mathcal{B}, \mathcal{C} \cup \{\alpha \approx\!\!\!> (\beta \wedge \gamma)\}, \prec) \sim (\mathcal{B}, \mathcal{C}, \prec)$

The following two examples show that neither OR nor RW are implicitly applied as are LLE and AND.

Example 2 *We consider a suitcase with two latches, which is closed whenever at least one of the latches is down.*

$$CB = (\mathcal{B}, \mathcal{C}, \prec)$$
$$\mathcal{B} = \{down_1 \vee down_2\}$$
$$\mathcal{C} = \{down_1 \approx\!\!\!> closed, down_2 \approx\!\!\!> closed\}$$

Since neither $down_1$ nor $down_2$ follow from $\mathcal{B}$, none of the causal rules can be applied. CB has the only extension $Th(\mathcal{B})$. The causal theory without the inference rule OR behaves like default theory and it is well known that in default logics reasoning by case cannot be taken into account. If we apply the inference rule OR to the causal rules $down_1 \approx\!\!\!> closed$ and $down_2 \approx\!\!\!> closed$, we obtain the rule $down_1 \vee down_2 \approx\!\!\!> closed$. If we add this derived rule to $\mathcal{C}$, we get one extension containing $closed$, since $\mathcal{B}$ contains the precondition of the new rule, namely $down_1 \vee down_2$.

Example 3

$$CB = (\mathcal{B}, \mathcal{C})$$
$$\mathcal{C} = \{D \approx\!\!\!> B, C \approx\!\!\!> \neg B\}$$
$$\mathcal{B} = \{D, A \vee B \to C\}$$

CB has the only extension $Th(\mathcal{B} \cup \{B, C\})$. But $CB' = (\mathcal{B}, \mathcal{C} \cup \{D \approx\!\!\!> A \vee B\})$, obtained by applying RW to $D \approx\!\!\!> B$ has the extensions $Th(\mathcal{B} \cup \{B, C\})$ and $Th(\mathcal{B} \cup \{\neg B, C\})$. As an immediate consequence of theorem 1 and the previous two examples we get

Corollary 1 *A causal base* $CB = (\mathcal{B}, \mathcal{C})$ *is closed iff* $\mathcal{C}$ *is closed with respect to RW and OR.*

The priority ordering $\prec$ on the set of causal rules extends as follows to RW-derived rules.

Definition 6 *Let* $\mathcal{C}$ *be a set of causal rules and* $c, c' \in \mathcal{C}$ *such that* $c \prec c'$. *Then* $\forall c'' \in RW(c), c'' \prec c'$. *The extended relation is denoted by* $\prec^{RW}$

The derived rule $OR(c, c')$ is not inserted into the $\prec$ −relation, because it is possible that for two rules $c_1 = \alpha \approx\!\!\!> \beta$ and $c_2 = \gamma \approx\!\!\!> \beta$, there is a causal rule c, such that $c_1 \prec c$ and $c \prec c_2$ (see example 1). Therefore it seems impossible to relate $OR(c_1, c_2)$ to c.

The definitions of applicability and of extension can be extended taking directly into account the inference rules OR and RW (see [12]).

In [11], an exhaustive comparison of approaches to actions and causality can be found [10, 13, 8].

The causal consequence notion defined by us in this paper is nonmonotonic, not transitive and not reflexive. It admits conjunction of consequences and reasoning by case without admitting contraposition.

The next step of this research is to study causality on the object language level, which would permit to consider imbricated causal implications like e.g. (*A CAUSES B*) *CAUSES C*.

References

1. Gerd Brewka. Reasoning about priorities in default logic. In *Proceedings of the 11th National Conference on Artificial Intelligence*, pages 940–945. AAAI Press / The MIT Press, 1993.
2. Hector Geffner. Causal theories for nonmonotonic reasoning. In *Proceedings of the 12th International Joint Conference on Artificial Intelligence*, pages 524–530, Australia, 1991. Morgan Kaufmann.
3. E. Giunchiglia and V. Lifschitz. An action language based on causal explanation: Preliminary report. In R. Miller and M. Shanahan, editors, *Fourth Symposium on Logical Formalizations of Commonsense Reasoning*, January 1998.
4. *Proceedings of the 14th International Joint Conference on Artificial Intelligence*, Montreal, Canada, 1995. Morgan Kaufmann.
5. Frank Jackson. A causal theory of counterfactuals. *Australian Journal of Philosophy*, 55(1):321–137, 1977.
6. V. Lifschitz. On the logic of causal explanation. *Artificial Intelligence*, 96:451–465, 1997.
7. Fangzhen Lin. Embracing causality in specifying the indirect effects of actions. In IJCAI'95 [4], pages 1985–1991.
8. Fangzhen Lin. Embracing causality in specifying the indeterminate effects of actions. In *Proceedings of the 133th National Conference on Artificial Intelligence*, pages 1985–1991. AAAI Press / The MIT Press, August 1996.
9. N. McCain and H. Turner. A causal theory of ramifications and qualifications. In IJCAI'95 [4], pages 1978–1984.
10. Norman McCain and Hudson Turner. Causal theories of action and change. In *Proceedings of the 14th National Conference on Artificial Intelligence*, pages 460–465. AAAI Press / The MIT Press, July 1997.
11. C. Schwind. Causality in action theories. *Linkping electronic articles in computer and information science*, 4(4), 1999.
12. C. Schwind. Non-monotonic causal logic. Technical report, LIM, april 1999.
13. Michael Thielscher. Ramification and causality. *Artificial Intelligence*, 89(1-2):317–364, 1997.
14. Hudson Turner. A logic of universal causation. *Artificial Intelligence*, to appear, 1999.

Systematic vs. Local Search for SAT

Holger H. Hoos[1] and Thomas Stützle[2]

[1] University of British Columbia, Computer Science Department,
2366 Main Mall, Vancouver, BC, V6T 1Z4 Canada
[2] Université Libre de Bruxelles, IRIDIA,
Avenue Franklin Roosevelt 50, CP 194/6, 1050 Brussels, Belgium
On leave from FG Intellektik, FB Informatik, TU Darmstadt, Germany

Abstract. Traditionally, the propositional satisfiability problem (SAT) was attacked with systematic search algorithms, but more recently, local search methods were shown to be very effective for solving large and hard SAT instances. Generally, it is not well understood which type of algorithm performs best on a specific type of SAT instances. Here, we present results of a comprehensive empirical study, comparing the performance of some of the best performing stochastic local search and systematic search algorithms for SAT on a wide range of problem instances. Our experimental results suggest that, considering the specific strengths and weaknesses of both approaches, hybrid algorithms or portfolio combinations might be most effective for solving SAT problems in practice.

1 Introduction

(SAT) is a central problem in logic, artificial intelligence, theoretical computer science, and many applications. Therefore, much effort has been invested on improving known solution methods and designing new techniques for its solution which led to a continuously increasing ability to solve large SAT instances.

Traditionally, SAT instances are solved by systematic search algorithms of which the most efficient ones are recent variants of the Davis-Putnam (DP) procedure. These algorithms systematically examine the entire solution space defined by the given problem instance to prove that either a given formula is unsatisfiable or that it has a model. Only recently it was found that stochastic local search (SLS) algorithms can be efficiently applied to find models for hard, satisfiable SAT instances. SLS algorithms start typically with some randomly generated truth assignment and try to reduce the number of violated clauses by iteratively flipping some variable's truth value. While SLS algorithms cannot prove unsatisfiability, in the past they have been found to outperform systematic SAT algorithms on hard subclasses of satisfiable SAT instances.

Both techniques are largely different and comprehensive comparisons involving both types of algorithms are rather rare. In particular, it is not well understood which type of algorithm should be applied to specific types of formulae. To reduce this gap in our current knowledge we compare systematic and local search algorithms on a wide range of benchmark instances. To limit the computational burden of the study we focussed our investigation to the best performing systematic and local search algorithms currently available. In particular, from the competitive systematic search algorithms available we tested SATZ [9] and REL_SAT [1], from the SLS algorithms we used those based on the WalkSAT architecture [10].

2   Empirical Methodology

Because SLS algorithms cannot be used to prove the unsatisfiability of a given SAT formula, we restrict our comparisons to satisfiable instances. The benchmark instances are taken from a benchmark suite containing a broad variety of problem types and mainly focussed on benchmark instances which are relatively hard for both types of search techniques. This benchmark suite comprises Random-3-SAT instances sampled from the phase transition region, test-sets obtained by SAT-encodings of randomly generated hard Graph Colouring problems (GCPs), SAT-encoded instances from AI planning domains, and satisfiable instances from the Second DIMACS Challenge benchmark collection. All problem instances and algorithms used are publically available through SATLIB (http://www.informatik.tu-darmstadt.de/AI/SATLIB).

For our empirical study we measured and compared algorithmic performance in terms of CPU-time as well as by using machine and implementation independent operation counts. Additionally, where possible, we compare the scaling behaviour of both types of algorithms based on implementation-independent cost measures. For all WalkSAT algorithms, the operations counted are local search steps. In preliminary runs we optimised the parameter settings for the SLS algorithms and, due to the inherent randomness of the algorithm, we determine the expected search cost for each instance by running each algorithm at least 100 times. An analogous technique is used for REL_SAT, which also involves randomised decisions. For SATZ, we measure the number of search steps; since the algorithm is completely deterministic, the search cost per instance can be determined from a single run.

3   Experimental Results

In the following, we outline the most significant results of our empirical study. A complete description can be found in [6].

Hard Random-3-SAT problems have been frequently used for empirically investigating the behaviour of SAT algorithms. Comparing the performance of R-Novelty[+], the best performing SLS algorithm for this subclass of SAT [5,7], with that of SATZ, one of the best-performing systematic algorithms for this problem class (which is superior to REL_SAT on hard Random-3-SAT instances [9]), we find the following situation (see Figure 1). For a major part of the test-set, R-Novelty[+] is significantly more efficient with respect to absolute CPU-time than SATZ; in particular, for about 95% of the test-set, R-Novelty[+] is

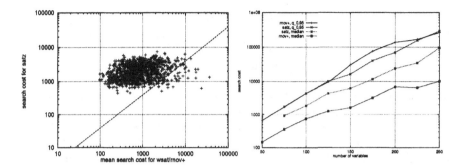

Fig. 1. *Left:* Correlation between mean local search cost and systematic search cost across Random-3-SAT test-set of 1,000 instances with 100 variables, 430 clauses each; each data point corresponds to one instance; search cost measured in expected number of flips per solution for R-Novelty[+], and number of search steps for SATZ. The line indicates points of equal CPU-time for the two algorithms. *Right:* Scaling of search cost with problem size for Random-3-SAT test-sets from the solubility phase transition region ($\geq$ 100 instances each).

instance	#vars	#clauses	WalkSAT strategy	avg.flips	secs	SATZ #steps	secs	REL_SAT #lab. vars	secs
bw_large.b.cnf	1,087	13,772	Tabu	152,104	1.62	544	< 0.1	8,818.5	0.48
bw_large.c.cnf	3,016	50,457	Tabu	$2.5 \cdot 10^6$	72.9	3,563	2.74	252,646	23.77
logistics.c.cnf	1,141	10,719	Rnov+	66,013	0.66	179,679	1.76	$6.5 \cdot 10^6$	344.94
logistics.d.cnf	4,713	21,991	Nov+	121,391	1.96	$4.4 \cdot 10^7$	87.58	31,550.4	1.28

Table 1. Comparison of solution times of WalkSAT, SATZ, and REL_SAT on SAT-encoded instances of the Blocks World and Logistics Planning problems. The computation times are measured on a 300MHz Pentium II PC with 320M RAM under Linux.

up to one order of magnitude faster than SATZ. Notably, while for different SLS algorithms for SAT, the same instances tend to be hard for all algorithms [5, 7], no such correlation is observed between R-Novelty[+] and SATZ. Next we investigated the scaling behaviour of search cost with problem size. Figure 1 shows the scaling of the median and the 0.95 percentile of the mean search cost per instance measured across test-sets of hard Random-3-SAT instances. We observed that for the median search cost the advantage of R-Novelty[+] over SATZ increases significantly with problem size. However, when comparing the 0.95 percentiles, such differences cannot be observed. Further analysis suggests that median search cost for R-Novelty[+] and SATZ might grow polynomially with problem size while the 0.95 percentile clearly shows exponential growth.

In analogous experiments with SAT-encoded, randomly generated 3-colourable hard GCPs we found a very weak negative correlation between the search cost for Novelty[+], the best performing SLS algorithm for these instances, and SATZ. When comparing CPU-times, Novelty[+] is more efficient than SATZ on a significant part of the test-set of graphs with 100 vertices. Yet, a scaling analysis revealed that SATZ appears to scale better with problem size. Further analysis showed that, different from the Random-3-SAT instances, for the GCP, all percentiles seem to show exponential scaling with problem size. When applied to SAT-encoded AI Planning instances (see Table 1 for some computational results) or instances from the DIMACS Benchmark set, the relative performance of SLS and systematic algorithms depends very strongly on the instance class. For example, for the planning instance from the Blocks World planning domain, both SATZ and REL_SAT show a significantly better performance than the best WalkSAT variant, while SLS algorithms appear to be preferable on the instances of the Logistics domain.

4 Conclusions

In this paper, we summarised our extensive comparative study of some of the best currently known systematic and local search algorithms for SAT. For the full version of this study, the interested reader is referred to [6]. Our results clearly indicate that currently, none of the two approaches dominates the other w.r.t. performance over the full benchmark suite. Instead, we observed a tendency that SLS algorithms show superior performance for problems containing random structure (as in Random-3-SAT) and rather local constraints (as in the DIMACS Graph Colouring instances or the logistics planning domain), while systematic search might have advantages for structured instances with more global constraints. This view is supported by recent findings in studying a spectrum of graph colouring instances with varying degree of structural regularity [4].

The results we obtained suggest that combining the advantages of systematic and local search methods is a promising approach. Such combinations could either be in the form of algorithm portfolios [3], or as truly hybrid algorithms. In this context, it would be interesting to extend our investigation to recently introduced randomised variants of systematic search methods, which show improved performance over pure systematic search under certain conditions [2]. Another issue which should be included in future research on SAT algorithms is the investigation of polynomial simplification strategies (like unit propagation, subsumption, restricted forms of resolution), which, when used as preprocessing steps, have been shown to be very effective in increasing the efficiency of SLS and systematic search methods in solving structured SAT instances [8].

References

1. R.J. Bayardo Jr. and R.C. Schrag. Using CSP Look-back Techniques to Solve Real World SAT Instances. In *Proceedings of AAAI'97*, pages 203–208, 1997.
2. C.P. Gomes, B. Selman, and H. Kautz. Boosting Combinatorial Search Trough Randomization. In *Proceedings of AAAI'98*, pages 431–437. MIT press, 1998.
3. C.P. Gomes and B. Selman. Algorithm Portfolio Design: Theory vs. Practice. In *UAI'97*, Morgan Kaufmann, 1997.

4. I.P. Gent, H.H. Hoos, P. Prosser, and T. Walsh. Morphing: Combining Structure and Randomness. To appear in *Proceedings of AAAI-99*, 1999.
5. H.H. Hoos. *Stochastic Local Search — Methods, Models, Applications*. PhD thesis, Department of Computer Science, Darmstadt University of Technology, 1998. Available at http://www.cs.ubc.ca/spider/hoos/publ-ai.html
6. H.H. Hoos and T. Stützle. Local Search Algorithms for SAT: An Empirical Evaluation. Technical Report TR-99-06. Department of Computer Science, University of BC, Canada, 1999. Available at http://www.cs.ubc.ca/spider/hoos/publ-ai.html
7. H.H. Hoos and T. Stützle. Local Search Algorithms for SAT: An Empirical Evaluation. Submitted to *Journal of Automated Reasoning*, 1999.
8. H. Kautz and B. Selman. Pushing the Envelope: Planning, Propositional Logic, and Stochastic Search. In *Proceedings of AAAI'96*, pages 1194–1201. MIT Press, 1996.
9. C.M. Li and Anbulagan Look-Ahead Versus Lock-Back for Satisfiability Problems. In *Proceedings of CP'97*, pages 341–355. Springer Verlag, 1997.
10. D. McAllester, B. Selman, and H. Kautz. Evidence for Invariants in Local Search. In *Proceedings of AAAI'97*, pages 321–326, 1997.

Information Environments for Software Agents

Stefan Haustein

University of Dortmund, Artificial Intelligence Unit
`haustein@ls8.cs.uni-dortmund.de`

Abstract. For biological agents, the perceived world does not consist of other agents only. Also software agents can have a shared world model they operate on. In this paper we discuss information systems as an important part of software agents' environment and try to identify the corresponding demands, leading us to build an information system for software agents on top of a conventional database. Thus we were able to add "agentified" access to the advantages of current information systems like reliability and speed.

1 Introduction

The COMRIS project aims to develop, demonstrate and experimentally evaluate a scalable approach to integrating the inhabited information spaces schema with a concept of software agents. The COMRIS vision of co-habited mixed-reality information spaces emphasizes the co-habitation of software and human agents in a pair of closely coupled spaces, a virtual and a real one. The COMRIS project uses a conference center as the thematic space and concrete context of work. Each participant wears his personal assistant, an electronic badge and ear-phone device, wirelessly hooked into an Intranet. This personal assistant – the COMRIS parrot – realizes a bidirectional link between the real and the virtual space. It observes what is going on around its host (whereabouts, activities, other people around), and it informs its host about potentially useful encounters, ongoing demonstrations that may be worthwhile attending, and so on. This information is gathered by several personal representatives, the software agents that participate in the virtual conference on behalf of a real person. Each of these has the purpose to represent and further a particular interest or objective of the real participant, including those interests that this participant is not explicitly attending to.

In the COMRIS and similar agent scenarios, information systems can take over two important roles: Firstly, information systems can serve as the world abstraction for software agents. Thus a virtual environment that does not consist of other agents only can be created: Passive objects like the COMRIS conference schedule and other general conference information are not modeled as agents but stored in an agentified information system. Using an web interface, also human users can access this information. Secondly, information systems can serve as a "blackboard" [HR85] [KHM99] system for software agents. COMRIS information gathering agents are helping the personal representative agents (PRA) [PANS98] improving the matchmaking process between participants. These two types of agents do not communicate directly: In order to become more independent from the current access characteristics of the world wide web the gatherers are already collecting information about the known participants before the conference starts. The results are stored persistently in the information system and are available immediately when demanded by PRAs. The persistent storage capabilities can also be used to save the state of an agents easily.

2 Existing Solutions

The first solution we considered was a set of tables in a conventional relational database. Normalization [Ull88] provides a more or less clear and unambiguous way to come from the conference ontology to a set of tables. The first problem we encountered was that several concepts of our domain ontology are inherited from other concepts. Inheritance of concepts suggests that an Object–Oriented Database Management System (OODBMS) could be a more suitable solution. Also with introduction of sets as table fields [Wag98], it is possible to encode relations to sets containing references to related objects instead of table joins, which overcomes a general performance problem in traditional databases [LP83] The disadvantage of having more possibilities to model relations is that all persons involved in modeling would not only have to agree on a domain ontology, but also on the concrete implementation in an object–oriented database. Already [The90] explicitly addresses a similar point: "Performance indicators have almost nothing to do with data models and must not appear in them".

The Object Database Management Group (ODMG) standard [Cat97] seems clearer in this point by removing SQL concepts like tables and rows and replacing them by classes and instances, but ODMG carries around its own baggage from its persistent object storage roots. The conceptual overload raises not only negotiation problems but also makes it more difficult to provide agents with domain independent capabilities like general reasoning or learning. Additionally, also knowledge about the "implementation" of the ontology into a database schema would have to be maintained somehow[1].

Frame systems [Min85] [BS85] are the initial root of object oriented programming and are theoretically also able to serve as an information system for our agents. A very relevant example for a frame system is Ontobroker [DEFS99]. Ontobroker is an ontology–based tool for accessing distributed and semi-structured information developed at the AIFB institute of the University of Karlsruhe. Ontobroker stores knowledge in a central frame system and provides a query interface to the stored data. Data collection from the web is delegated to wrappers, accessing external databases, and the "Ontocrawler". The "Ontocrawler" is a program that is able to extract information from specially annotated web pages. Our problem – not only concerning Ontobroker but frame systems in general – is that we are performing all reasoning in the agents, and therefore we do not need advanced reasoning and classification capabilities in the information

---

[1] A good example that it makes sense to reduce conceptional overload may be the JAVA programming language compared to C++

system, mixing up the storage and application layer. In frame systems we have a better logical foundation than in OODBMS, but we are not willing to trade off performance.

3 The COMRIS Information Layer

Because of the raised problems, we did not want to use an existing information system without any adaption. So we decided to build our information system on top of a conventional database. The additions we made are an agent communication interface and the introduction of relations as first class members of the data model: Thus the characteristics of relations can be specified directly like in class diagrams of the Unified Modeling Language (UML) [FS97] instead of explicitly using container classes like the ODMG standard. The corresponding subset of UML class diagrams may be used to specify the ontology and thus also the data model. So the modeling process is able to concentrate on the "real work" instead of problems like choosing the "right" data structures etc.

For the agent communication, a direct mapping from the ontology to an XML encoding is used, embedded in an XML encoded FIPA ACL frame [Fou97]. XML can be parsed easily in several programming languages. Using style sheets, XML can be displayed by any web browser in the near future, enabling the software agents to send messages to "human agents" directly. An XML structure can be extended easily by adding XML-attributes without changing the general XML element structure, thus enabling agents using probabilistic data models to communicate with non-probabilistic services without the need of two completely different content languages.

Utilizing the simplicity of the data model, we were able to develop an instance browser and a web interface for the COMRIS information layer without great effort. The instance browser can be used to inspect and modify the whole content of the information system avoiding possibilities of syntactical errors completely. It also provides the advantages of a graphical user interface like selection lists for relations etc. The web interface of the information system can be used to make information like the conference schedule etc. accessible for human users directly.

4 Conclusion

With an agentified interface, information systems can play an important role in the environment of software agents. Beneath being a part of the perceived world of the agents, information systems can provide services like persistent storage and blackboard functionality that helps agent designers to focus on their main objectives and design agents as lean and powerful entities. In multi–agent systems, where agents are designed by different vendors, a simple data model directly derived from the domain ontology can reduce negotiation demands.

A more detailed technical documentation of the COMRIS information layer is available at:

http://www-ai.cs.uni-dortmund.de/FORSCHUNG/VERFAHREN/IL

298

Acknowledgements

The research reported in this paper is supported by the ESPRIT LTR 25500 COMRIS project.

References

[BS85]   R. J. Brachman and J.G. Schmolze. An overview of the KL-ONE knowledge representation system. *Cognitive Science*, 9(2):171–216, 1985.

[Cat97]   R. G. G. Cattell, editor. *The Object Database Standard: ODMG 2.0.* Morgan Kaufmann, 1997.

[DEFS99]   Stefan Decker, Michael Erdmann, Dieter Fensel, and Rudi Studer. *Semantic Issues in Multimedia Systems, Kluwer Academic Publisher, Boston, 1999*, chapter Ontobroker: Ontology Based Access to Distributed and Semi-Structured Information. Kluwer Academic Publisher, Boston, 1999.

[Fou97]   Foundation for Intelligent Physical Agents. Agent communication language, 1997. http://www.fipa.org/spec/f8a22.zip.

[FS97]   M. Fowler and K. Scott. *UML Distilled. Applying the Standard Object Modelling Language.* Adison-Wesley-Longman, 1997.

[HR85]   B. Hayes-Roth. A blackboard architecture for control. *Artificial Intelligence*, 26(3), 1985.

[KHM99]   Martin Kollingbaum, Tapio Heikkilae, and Duncan McFarlane. Persistent agents for manufacturing systems. In *AOIS 1999 Workshop at the Third International Conference on Autonomous Agents*, 1999.

[LP83]   R. Lorie and W. Plouffe. Complex objects and their use in design transactions. In *Proc. ACM SIGMOD Database Week – Enginieering Design Applications*, May 1983.

[Min85]   Marvin Minsky. A framwork for representing knowledge. In R. J. Brachman and H. J. Levesque, editors, *Readings in Knowledge Representation*, chapter III, pages 245–263. Morgan Kaufman, Los Altos, CA, 1985.

[PANS98]   Enric Plaza, Josep Llus Arcos, Pablo Noriega, and Carles Sierra. Competing agents in agent-mediated institutions. *Personal Technologies Journal*, 2(3):1–9, 1998.

[The90]   The Committee for Advanced DBMS Function. Third-generation data base system manifesto. Technical report, University of California, Berkeley, April 1990.

[Ull88]   Jeffrey D. Ullman. *Principles of database and knowledge–base systems*, volume 1. Computer Science Press, Rockville, MD, 1988.

[Wag98]   Gerd Wagner. *Foundations of Knowledge Systems with Applications to Databases and Agents.* Kluwer Academic Publisher, 1998.

Improving Reasoning Efficiency for Subclasses of Allen's Algebra with Instantiation Intervals[1]

Jörg Kahl, Lothar Hotz, Heiko Milde, Stephanie E. Wessel

Laboratory for Artificial Intelligence, University of Hamburg
Vogt-Koelln-Str. 30, D-22527 Hamburg
kahl@kogs.informatik.uni-hamburg.de

1 Introduction

A common way to express incomplete qualitative knowledge about an order of occurring events is Allen's interval algebra $\mathcal{A}$ [1]. Unfortunately, the main reasoning tasks of $\mathcal{A}$, consistency checking ISAT and strongest implied relation computation ISI, are intractable [10]. To overcome this, some tractable subclasses of $\mathcal{A}$ have been identified, e.g., the continuous algebra $\mathcal{C}$ [9], the pointisable algebra $\mathcal{P}$ [7], and the ORD-Horn algebra $\mathcal{H}$ [8]. These subalgebras form a strict hierarchy with $\mathcal{H}$ as the largest tractable subalgebra that contains all thirteen basic relations [8]. To solve ISAT($\mathcal{H}$) and ISI($\mathcal{H}$), several approaches like constraint propagation [8], linear programming [3], or graph oriented approaches [2] have been presented.

The best ISAT($\mathcal{H}$) algorithms yet presented accomplish $O(n^2, dn)$-time where n is the number of interval variables and d is the number of constraints included in $\mathcal{H} \setminus \mathcal{C}$. The most efficient ISI($\mathcal{H}$) algorithms yet presented require $O(n^4, dn^3)$-time (see [6] and [3]). In this paper, we argue that the main shortcoming of most traditional approaches is their usage of time points as time primitives while reasoning. We present a graph oriented approach using a representation we call instantiation intervals that accomplishes $O(n^2)$-time performance for ISAT($\mathcal{H}$) and ISI($\mathcal{H}$).

2 Instantiation Intervals

Consider a time point variable A that is qualitatively constrained by other time point variables. In general, the time period in which A can be consistently instantiated is most adequately describable by an interval. In the following, we will call this time point representation an instantiation interval. Since ISAT and ISI contain no quantitative information, the instantiation interval's endpoints (II-endpoints) have to be qualitatively expressed in relation to other II-endpoints (see [5] for a similar approach). The first step of our reasoning algorithm is the transformation of all interval constraints into constraints relating pairs of interval endpoints. This is in accordance with most other traditional approaches (see [3] for an example). Many resulting constraints can be expressed within time point algebra $\mathcal{TP}$ [10] comprising the three basic relations $\{<, =, >\}$.[2] Afterwards, we mimick the construction of all possible instantiations in qualitative terms.

---

1. This research has been supported by the Bundesminister fuer Bildung, Wissenschaft, Forschung und Technologie (BMBF) under the grant 01 IN 509 D 0, INDIA - Intelligente Diagnose in der Anwendung.
2. In Section 4, we illustrate how constraints included in $\mathcal{H} \setminus \mathcal{P}$ are treated.

With the successive insertion of the given time point constraints' information, a graph based on [2] is built up.For example, Figure 1 depicts the transformation into time point relations and their qualitative instantiation of A{before, meets}B ∈ 𝒫.

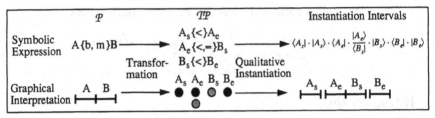

Figure 1. Transformation and qualitative instantiation of A{before, meets}B

With II-endpoints as time primitives, we can accomplish the expressiveness of 𝒯𝒫. Table 1 presents the necessary set of three relations (a, b are arbitrary II-endpoints, P denotes the time point the respective II-endpoint represents).

Table 1. II-endpoint relations and their semantics

Type	1	2	3
Relation	$a \cdot b$	$\dfrac{a}{b}$	$a \pm b$
Semantics	$a < b$	$a = b$	$P(a) \neq P(b)$

Time point constraints of the form $A \neq B$ cannot be expressed with II-endpoint relations of this form. We therefore define relation type 3 on the time point level.

With II-endpoints as reasoning time primitives, we can omit the computationally costly transformations of [2] to make implicit information explicit within the graph. Instead, we directly build up our corresponding graph we call II-graph with the successive insertion of the given time point constraints' information. Thus, we have to process every given time point constraint only once to obtain a solution for ISAT resulting in a reasoning algorithm of $O(n^2)$-time. Furthermore, instantiation intervals represent all possible instantiations of the given time points in a qualitative way. Therefore, the possible basic relations between pairs of time points, i.e., the solution for ISI, is given again without further transformations in $O(n^2)$-time.

3 Constructing Instantiation Interval Graphs

We subsequently present an II-graph construction example. Figure 2 shows a time point constraint net and its corresponding strongest implied relations.

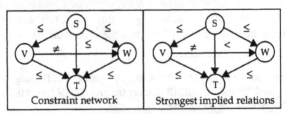

Figure 2. Constraint network example and its strongest implied relations

In Table 2, we show the stepwise construction of a semantically equivalent II-graph.

Table 2. Construction example of an II-graph

Constraint to instantiate	$S \leq W$	$S \leq V$	$S \leq T$
II-graph	$\langle S\| \cdot \dfrac{\|S\rangle}{\langle W\|} \cdot \|W\rangle$	$\langle S\| \cdot \dfrac{\|S\rangle}{\langle W\|\langle V\|} \cdot \begin{array}{l}\|W\rangle \\ \|V\rangle\end{array}$	$\langle S\| \cdot \dfrac{\|S\rangle}{\langle W\|\langle V\|\langle T\|} \cdot \begin{array}{l}\|W\rangle \\ \|V\rangle \\ \|T\rangle\end{array}$
Constraint to instantiate	$V \leq T$	$W \leq T$	$V \neq W$
II-graph	$\langle S\| \cdot \dfrac{\|S\rangle}{\langle W\|\langle V\|\frac{\langle T\|}{\|V\rangle}} \cdot \begin{array}{l}\|W\rangle \\ \|T\rangle\end{array}$	$\langle S\| \cdot \dfrac{\|S\rangle}{\langle W\|\langle V\|\frac{\langle T\|}{\|V\rangle\|W\rangle}} \cdot \|T\rangle$	$\langle S\| \cdot \dfrac{\|S\rangle}{\langle W\pm V\|\langle V\pm W\|\frac{\langle T\|}{\|V\pm W\rangle\|W\pm V\rangle}} \cdot \|T\rangle$

The first step, inserting $S \leq W$, is obvious due to the instantiation interval relations introduced in Section 2. The following two constraints are inserted in a similar way. The next constraint $V \leq T$ is more difficult to integrate because both constraint variables are already instantiated. Since $S = T$ is still a possible time point relation and the moving of $|V\rangle$ does not disable any valid time point relations, the type 2 relation of $\langle T|$ and $|V\rangle$ is realized without moving $\langle T|$. The constraint $W \leq T$ is integrated in the same way. The insertion of the last constraint $V \neq W$ is obvious due to the instantiation interval semantics.

Most interesting are the strongest implied relations between S and T. From our II-graph, we can conclude that S can be instantiated before T but not after T. Additionally, S and T cannot be instantiated at the same time point because V and W would also be instantiated at the same time point leading to a contradiction because of $V \neq W$.

4 Treating Disjunctive Information

Constraints in $\mathcal{H} \backslash \mathcal{P}$ contain disjunctive information that cannot be expressed within $\mathcal{TP}$. Fortunately, in $\mathcal{H}$, disjunctions only occur in a very restricted way. A constraint $C \in \mathcal{A}$ is included in $\mathcal{H} \backslash \mathcal{P}$ if its transformation into time point constraints results in a conjunction of time point constraints and a binary disjunctive constraint $A_1 < B_1 \vee A_2 < B_2$, where A_1, A_2, B_1, B_2 are Allen interval endpoints of C. The endpoint pairs A_1, B_1 and A_2, B_2 are not of the same Allen interval. Also, two of the transformed time point constraints have the form $A_1 \leq B_1$ and $A_2 \leq B_2$.

To cope with such a disjunctive constraint, we split both pairs of time point variables into four constraints $A_1 \leq B_{11}$, $A_1 < B_{12}$, $A_2 \leq B_{21}$, and $A_2 < B_{22}$. We have to store the disjunctive constraint as time order information parallel to our II-graph. The following computational effort additionally occurs within the graph:

- (*While building up the II-graph*) If the insertion of a further constraint results in $A_i < B_{i1}$, we have fulfilled the disjunctive constraint $A_1 < B_1 \vee A_2 < B_2$ which can be deleted. Furthermore, for both pairs of instantiation interval endpoints B_{11}, B_{12} and B_{21}, B_{22} , we respectively delete the II-endpoint that lesser constrains B.

- (*After building up the II-graph*) All disjunctive constraints not deleted after the insertion of the last time point constraint affect ISAT($\mathcal{H}$) if $\langle A| = |A\rangle = \langle B| = |B\rangle$

holds for the constructed II-graph with the disjunctive constraint $A_1 < B_1 \vee A_2 < B_2$. In this case, we have a contradiction.

For example, Figure 3 depicts the transformation process of $A\{s, o, fi\}B$ which is in $\mathcal{H} \setminus \mathcal{P}$.

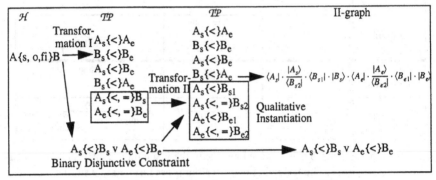

Figure 3. Transformation process of $A\{s, o, fi\}B$

Otherwise, constraints in $\mathcal{H} \setminus \mathcal{P}$ do not further differ in their treatment by the II-graph construction algorithm.

References

1. Allen, J. F.: Towards a general theory of action and time, in: *Artificial Intelligence 23*, 1984.
2. Gerevini, A., Schubert, L.: Efficient algorithms for qualitative reasoning about time, in: *Artificial Intelligence 74*, 1995.
3. Jonsson, P., Backstrom, C.: A Linear-Programming Approach to Temporal Reasoning, in: *Proceedings AAAI-'96*, 1996.
4. Kautz, H. A., Ladkin, P. B.: Integrating metric and qualitative temporal reasoning, in: *Proceedings AAAI-'91*, 1991.
5. Kockskaemper, S.: Modeling and Prediction of Dynamic Behavior for Model-based Diagnosis, in: *Proceedings IEA/AIE-95*, 1995.
6. Koubarakis, M.: Tractable Disjunctions of Linear Constraints, in: *CP-'96*, 1996.
7. Ladkin, P. B., Maddux, R.: On binary constraint networks, Technical Report, Kestrel Institute, 1988.
8. Nebel, B., Buerckert, H.-J.: Reasoning about temporal relations: a maximal tractable subclass of Allen's interval algebra, in: *Journal of the ACM 42*, 1995.
9. Van Beek, P., Cohen, R.: Exact and approximate reasoning about temporal relations, in: *Cmputational Intelligence 6 (1990)*, 1990.
10. Vilain, M. B., Kautz, H.: Constraint propagation algorithms for temporal reasoning, in: *Proceedings AAAI-'86*, 1986.

Agents in Traffic Modelling - From Reactive to Social Behaviour

Ana Lúcia C. Bazzan[1], Joachim Wahle[2], and Franziska Klügl[3]

[1] Inst. of Informatics, UFRGS, C.P. 15064, 91501-970 P. Alegre, RS, Brazil
bazzan@inf.ufrgs.br
[2] Physics of Transport and Traffic, Univ. of Duisburg, 47048 Duisburg, Germany
wahle@traffic.uni-duisburg.de
[3] Dept. of Art. Int., Univ. of Würzburg, Am Hubland, 97074 Würzburg, Germany
kluegl@informatik.uni-wuerzburg.de

Abstract. In modern societies the demand for mobility is increasing daily. One challenge to researchers dealing with transportation is to find efficient ways to model and predict traffic flow, even if the behaviour of people in traffic is not a trivial problem. The social nature of traffic (e.g. coordinated decisions) seems to be a key question, not well explored. We aim at creating a model of drivers as social agents, thus allowing their behaviour to be predicted and considered in the simulation. This may, on its turn, improve the accuracy of the existing Advanced Travel Information Systems (ATIS).

1 Introduction

Daily traffic jams reflect the fact that the capacities of the road network are satisfied or even exceeded. Thus, the modelling and prediction of traffic flow is one of science's future challenges. To be effective, such models have to make assumptions about the travel demand, and hence about travel choices and traffic behaviour. As obvious as it is, not so much attention has been paid to the social properties of traffic systems, in spite of their inherent social nature. However, the interdependence of actions associated with dynamic route guidance systems for example, lead to an increasing frequency of coordinated decisions. The use of such systems has the potential to change the nature of private car travelling in a yet unknown way. One typical scenario is the broadcasting of traffic messages to commuters. It is known that they have an impact on driver's behaviour, but currently drivers' reaction is neither registered nor considered in any forecast system.

The present work does not deal with the classical case of route choice simulation. There, the focus is on the *individual* driver without consideration of the *interaction* caused by coordinated decisions on the system as a whole. Hence, we depart from the classical view of route choice as an individual issue, and opt to study the social aspects of the problem.

Modelling of traffic scenarios with multi–agent systems (MAS) techniques is not new. However, the focus has been mainly in logistics regarding transportation scenarios, or coarse–grained level regarding traffic problems as e.g. traffic agents monitoring problem areas. The work proposed here focuses on a fine–grained level (traffic flow control). At this level, few works exist. For instance, Bazzan (1997) discusses a mechanism for the coordination of traffic signal. However, this paper deals mainly with the tactical level.

Artificial intelligence (AI) and, in particular, MAS techniques open the possibility to model the strategical level (as for instance the behaviour of drivers) in a more realistic way, at a level closer to the deliberative and social one. In the present work we focus on the use of mental states like beliefs and intentions.

2 The Tactical Layer

There are mainly two approaches to the modelling of traffic: the macroscopic and the microscopic. In the former, it is not possible to individualise classes of behaviours. In microscopic approaches, each individual can be described as detailed as desired, thus permitting the model of drivers' behaviours. To meet computational constraints, one basic idea of traffic flow modelling is to describe its dynamics as simple as possible. In this spirit, cellular automaton (CA) models were introduced by Nagel and Schreckenberg (1992) to describe the vehicular motion. This is implemented by means of four rules in the CA: collision–free acceleration, interaction,randomisation, and movement. A typical application is the on–line simulation of traffic in downtown Duisburg (http://www.traffic.uni-duisburg.de/OLSIM/).

The Nagel–Schreckenberg CA can be directly interpreted as a multi–agent system with reactive agents. This was done using the multi–agent simulation environment SeSAm (Shell for Simulated Multi–Agent Systems), described in Klügl and Puppe (1998). In several simulation experiments we were able to show that the multi–agent model of the Nagel–Schreckenberg cellular automaton reproduces the original model's behaviour with sufficient accuracy.

3 Social Agents and the Strategical Layer

As explained above, the use of microscopic traffic simulators allows travel and/or route choices to be considered. However, it is important to notice that such choices seem not to be influenced by the same attributes as is rational optimisation. The decision–making process in human beings uses not only logical elements, but also involves some emotional components that are typically non-logical. As a result, behaviour can be also explained by other approaches, which additionally consider emotion, intentions, beliefs, motives, cultural and social constraints, impulsive actions, and even simply willingness to try.

Dynamic route guidance systems will soon be available for a huge number of the road users. The influence of these systems on the actual traffic state cannot be modelled with the methods described above since they assume rational agents. Understanding travellers' route choice behaviour is an important consideration for the development and effectiveness of such systems. This is done by

BELIEFS	DESIRES
BEL (usual_route (R))	DES (min_time)
BEL (roadwork (R))	DES (on_time)
BEL (roadwork (R)) ⇒ BEL (congested (R))	DES (¬jam)
BEL (alt_route (A))	DES (few_lights)
BEL (congested (R)) ⇒ BEL (choose (A))	DES (via_highway)
BEL (alt_route (A)) ⇒ BEL (many_lights (A))	DES (¬stop)
BEL (broadcast (R, 'jam')) ⇒ BEL (congested (R))	DES (¬roadwork (R) ∧ usual_route (R))
BEL (¬broadcast (R) , 'any') ⇒ BEL(¬congested (R))	DES (choose (R) ∧ usual_route (R))
BEL (leave_later) ⇒ BEL (¬on_time)	DES (leave_later)

Table 1. Partial Knowledge Base for Agent Ag_1.

combining traffic messages generated by the CA–based simulation tool with a BDI formalism. The latter is based on the formalism of Rao and Georgeff (1991), i.e. based on the modalities for belief, goal, and intention.

To illustrate its use, we discuss a well–known scenario: the day–to–day travel choice of commuters. For simplicity, we assume that there are two possible routes, namely R and A, connecting the places of interest. Route R is shorter than alternative A but a heavy roadwork is announced for R. In this scenario, there is no optimal solution to the problem. If a significant number of commuters follow the recommendation and use alternative A, route R might be still faster. On the other hand, many drivers think the same way and stay with their typical choice.

To implement such a scenario using the BDI formalism, a typical agent has a knowledge base (KB) like that shown in Table 1. Others have similar KB's. The beliefs set is represented by formulae describing the world. Desires are all possible states that the agent can achieve. Notice that they can be conflicting, like DES (*on_time*) and DES (*leave_later*), or nearly unachievable as e.g. DES (¬*stop*). Goals are desires that are consistent with the beliefs, not conflicting, and believed to be achievable. Therefore, the set of goals is not necessarily a singleton. A similar relationship exists between plans and intentions. Hence, an agent can have many plans, each to achieve a given state, but only plans believed to be achievable will form intentions. Besides, intentions must be mutually consistent. Table 1 shows part of an agent KB. For the sake of simplicity, the identification of the agents is omitted from the logical declarations. This states that the agent Ag_1 believes that R is its usual route in this commuting scenario. The sixth line of the beliefs column states that if it is believed that A is an alternative route (to R), then it is believed that the agent will have to drive along a road with many traffic lights. Ag_1 has a set of desires, not all consistent with the beliefs. As it is believed that there is a roadwork on R, the usual route, R is believed to be congested, and an alternative route A should be chosen. These beliefs are definitely not consistent with the six last desires. As for DES(*min_time*) and DES(*on_time*), these are consistent as long as no broadcast over route A means that A is not congested. Hence Ag_1 can still be on time and the journey will take the minimum time *for that route*.

4   Conclusions

This paper discusses the need to change the modelling paradigm of a driver in an intelligent transportation system. No traffic forecast system is currently able to represent drivers as more than rational decision–makers who merely perceive small parts of their environment and react according to pre–established rules. Hence, this work extends the existing systems first by modelling a driver as a social agent based on multi–agent systems techniques, and second by generating a feedback to the simulation tool from such a model.

We have started with an existing microscopic traffic simulation tool, the CA–based Nagel–Schreckenberg model, and its interpretation as a multi–agent system. However, such sub–cognitive multi–agent implementation is valid mainly at a tactical level. In order to tackle the strategical one, we have developed a more deliberative model of agents, able to deal with not completely rational decision–making. To achieve this, we have used mental states like emotions, preferences, intentions, etc. Such mental states play a key role especially in a commuting–like scenario, since the actions tend to be repeated and the knowledge of the driver accumulates with time. Another important characteristic of this scenario, to which the BDI formalism fits very well, is its social nature. The individual decision has no optimal solution. If a significant number of commuters follow the route recommendation broadcasted, there is no guarantee that the recommended route will be a better choice.

In short, we have present two possible layers of a multi–agent system designed to simulate traffic flow and to model drivers. While the former can be tackled by a tactical level (where sub–cognition is enough to make drivers act), in the latter it is essential to embed not only cognition but also more sophisticated forms of decision–making involving the mental states mentioned above. The next challenge of the work is to integrate both tools and environments.

Acknowledgements
We would like to thank Prof. M. Schreckenberg and Prof. F. Puppe. This research is sponsored partially by the agencies DLR (German Nat. Aerospace Research Center) and CNPq (Brazilian Nat. Council for Research and Technology) to the project SOCIAT.

References

Bazzan, A.L.C.: An Evolutionary Game–Theoretic Approach for Coordination of Traffic Signal Agents. PhD Thesis (1997), University of Karlsruhe

Klügl, F., Puppe, F.: The Multi–Agent Simulation Environment SeSAm. In: Proc. of the workshop SiWis (1998), Paderborn

Nagel, K., Schreckenberg, M.: A cellular automaton model for freeway traffic. J. Phys. I France 2 (1992) 2221

Rao, A.S., Georgeff, M. P.: Modelling Rational Agents within a BDI architecture. In: Proc. of the Int. Conf. on Knowledge Representation and Reasoning (1991) 473-484, San Mateo, CA, Morgan Kaufmann

Time-Effect Relations of Medical Interventions in a Clinical Information System

Michael Imhoff[1], Marcus Bauer[2] and Ursula Gather[2]

[1] Surgical Department, Community Hospital Dortmund,
Beurhausstraße 40, D-44137 Dortmund
[2] Department of Statistics, Dortmund University,
Vogelpothweg 87, D-44225 Dortmund

Abstract. For the implementation of time-critical decision support algorithms in a clinical information system (CIS) a precise relation between medical interventions and effects needs to be established. We evaluated for selected drugs and infusions the relation in time between charted dose and effect on on-line hemodynamic variables. The time of the intervention was compared with the onset of the change of the hemodynamic variables as determined by new time series methods. The average time difference between intervention and calculated hemodynamic effect was 13.23 min (0 - 29) which did not differ significantly between different interventions. The marked lag between intervention and effect and the great variance of this lag pose an important problem for time-critical decision support. Even after optimizing data acquisition important factors will remain unaccounted for. Therefore, decision support systems may need extensive testing with real-world data before they are released into clinical practice. (*supported by the Deutsche Forschungsgemeinschaft, Sonderforschungsbereich 475 "Complexity Reduction in Multivariate Data Structures"*)

1 Introduction

Clinical Information Systems (CIS) can today provide the health care professional with the complete Electronic Patient Record (EPR) at the point of care. This data may include vital signs (e.g. heart rate, blood pressure), fluid intake and output, medications as well as entire clinical pathways. These CIS are complex database systems with sometimes more than 2,000 variables for each patient [3–5]. Over the last couple of years knowledge-based systems have been developed that use CIS data to support medical decision making [9]. Especially in operation-critical decision support such as in intensive care medicine a strong influence of the timing of interventions can be assumed. Therefore, in the implementation of time-critical decision support algorithms in a CIS a precise relation between medical intervention (time, dose) and effect needs to be established. For selected drugs and infusions we evaluated the relation in time between charted dose and effect on hemodynamic variables.

2 Methods

On a 16-bed surgical ICU all medication data was charted with a CIS, allowing the user 1 minute time resolution for all data. The system configuration was comprised of the CIS (Emtek Continuum 2000, Version 4.1M3), a Decision Support System (Sybase SQL server 4.9.2), and Statistical Software (SPSS version 6.1, SAS version 6.12) runnung on Sun Sparc under Sun Solaris 1.1.2 and 2.5.

On-line monitoring data was acquired from 148 consecutive critically ill patients (53 female, 95 male, mean age 64.1 years) with extended hemodynamic monitoring requiring pulmonary artery catheters in one minute intervals from the CIS. At total of 11,339 hours of monitoring of heart rate (HR), arterial mean pressure (MAP), and pulmonary artery mean pressure (MPAP) were analyzed.

Effects of interventions were defined as certain patterns of change in the time series of HR, MAP, and MPAP. Those were identified based on second order autoregressive (AR) time series models [6]. Each univariate time series was split into 90 minutes intervals starting with the first observation. The first 60 minutes of each interval were used as the estimation period of the respective AR-model, while the last 30 minutes of each interval served as the prediction period. This 90 minute window was moved over the entire time series in 30 minute increments. For each estimation period a second order autoregressive model was fitted and applied to the prediction period. The actual measurements were compared to the 95% confidence intervals (CI) for the prediction period. Values outside the CI were classified as an outlier, if less than 5 consecutive observations (= minutes) were outside the CI, and as a level change by 5 or more consecutive observations outside the CI. Only prediction periods that showed just one pattern were included. Only level changes of more than 5% were included in the analysis, because even with the 95% CI the pattern recognition discovered also some clinically non-relevant level changes. The time of the onset of a level change was defined by the first observation of this level change outside the CI. Level changes following this definition constituted what we call an effect here on the hemodynamic variables.

The time of the intervention was compared with the onset of the identified effect on the hemodynamic variables as determined by the time series analysis described above. Only time differences of less than 30 minutes were included, because of the 30 minute overlap between the time series window.

An intervention was defined by a change in the dose rate of dobutamine, adrenaline, noradrenaline, nitroglycerin, or by a change of the fluid balance by more than 500 cc in less than 10 minutes. Only effects that could be pharmacologically attributed to the respective intervention were included in the final analysis. The time of the intervention as it was charted in the CIS was compared with the onset of the identified change of the hemodynamic variables.

3 Results

¿From a total of 80,752 time series analyses, 12,599 included catecholamine or fluid interventions, of which 2.608 intervention-effect pairs met the inclusion criteria for further analysis. The average time difference between intervention as charted and detected hemodynamic effect was 13.23 minutes (0 - 29 min). This time lag did not differ significantly between catecholamines, vasodilators, and rapid infusions. The 90% percentiles for most intervention-effect combinations ranged from 0 to over 25 minutes. Only the median time lag between increase of vasopressors and increase of blood pressure was lower. Changes of fluid balance showed an especially wide variation in their time lag to the associated effect. The largest number of observations involved rapid fluid changes. As expected, a wide variation of time lags can be found here. Interestingly, the time lags for fluid interventions did not differ markedly from those of most drug interventions. As the half-lifes of the investigated drugs are longer than their action times and removal of fluids takes more time than administration, it could be expected that a decrease of an intervention would show a longer time lag than the respective increase. Except for adrenaline and noradrenaline this could not be confirmed. In summary, the time lags between medical interventions as charted and hemodynamic effects showed a wide variation. Although there was a tendency that the time lag between catecholamine dosage changes and pressure changes was shorter on the average, there were no relevant differences between any intervention-effect pair. These observations cannot be sufficiently explained by the pharmacological and physiologic properties of the drugs and infusions studied.

4 Discussion

CIS have improved precision and volume of bedside documentation in high acuity areas as shown in several studies [2, 4]. Investigations in dose documentation with CIS show a 99% accuracy in drug documentation [8]. Therefore, documentation with a CIS can be considered highly accurate for non time-critical items. No study known to the authors addresses the issue of time-critical documentation in the ICU. This may be surprising as recent publications emphasize the significance of time-oriented data for clinical decision support [10]. It is pointed out that correct documentation of temporal patterns is essential for decision support in critical care [7].

Our study investigates strong temporal relationships in hemodynamic therapy of the critically ill. It can be expected that changes of catecholamine drips and rapid fluid challenges have an immediate effect within a few minutes on hemodynamic variables, such as heart rate or blood pressure [1].

Surprisingly, our study finds an average latency between the documented change of a drug and the hemodynamic effect of about 13 minutes. Although the average time between increase of vasopressors (adrenaline, noradrenaline) and changes in arterial pressure was shorter at 7.5 minutes, the most striking finding remains the wide variation with a range of over 20 minutes for all interventions. This wide variation may be attributable to several factors: (a) Inaccurate time entries for dose changes or IV fluids by the user. (b) Gradual response of the patient, or rather his/her cardiovascular system, to the intervention which may take time to be recognized by a pattern recognition algorithm that is based on thresholds. (c) Interindividual differences in the reactivity of patients towards a certain drug. (d) Technical issues of drug and fluid application.

Thus several factors that introduce significant variance in the documentation of temporal patterns may be beyond the user's influence. This time variance can significantly affect the performance of time-oriented decision support algorithms [7, 10]. Some of the time variation may be reduced by improvements in data acquisition and processing: (a) Automatic data transfer from bedside devices wherever possible. (b) Training of users and standardization of charting procedures. (c) Optimization of algorithms for the detection of patterns and change points in time series data. (d) Adaptation of decision support algorithms to account for the variance in temporal patterns.

In summary, the relation in time between interventions and effects shows a large variance which poses a major problem for the implementation of time-critical decision support algorithms. While some of the variation may be reduced by interfacing IV devices with CIS and educating users in more precise documentation, important technical, physiological and pharmacological factors remain unaccounted for. This emphasizes the necessity to test decision support systems extensively with real-world data before they are released into clinical practice. Here further research is needed in time-oriented clinical decision support.

References

1. van Aken, H., Möllhoff, T., Lawin, P.: Pharmakologische Unterstützung des Herz-Kreislauf-Systems. In: Lawin, P. (ed): Praxis der Intensivbehandlung. Thieme, Stuttgart New York (1994) 551-566
2. Hammond, J., Johnson, HM., Varas, R., Ward, C.G.: Qualitative comparison of paper flowsheets vs a computer-based clinical information system. Chest **99** (1991) 155-157
3. Imhoff, M.: 3 years clinical use of the Siemens Emtek System 2000: Efforts and Benefits. Clinical Intensive Care **7 Suppl.** (1996) 43-44
4. Imhoff, M.: Clinical Data Acquisition: What and how? Journal für Anästhesie und Intensivmedizin **5** (1998) 85-86
5. Imhoff, M., Bauer, M.: Time series analysis in critical care monitoring. New Horiz. **4** (1996) 519-531
6. Imhoff, M., Bauer, M., Gather, U., Löhlein, D.: Statistical pattern detection in univariate time series of intensive care on-line monitoring data. Intensive Care Med. **24** (1998) 1305-1314
7. Kahn, M.G., Marrs, K.A.: Creating temporal abstractions in three clinical information systems. Proc. Annu. Symp. Comput. Appl. Med. Care (1995) 392-396
8. Lubarsky, D.A., Sanderson, I.C., Gilbert, W.C., King, K.P., Ginsberg, B., Dear, G.L., Coleman, R.L., Pafford, T.D., Reves, J.G.: Using an anesthesia information management system as a cost containment tool. Description and validation. Anesthesiology **186** (1997) 1161-1169
9. Morris, A.: Algorithm-based decision making. In: Tobin, M.J. (ed.): Principles and Practice of Intensive Care Monitoring. McGraw-Hill, New York (1998) 1355-1381
10. Shahar, Y., Combi, C.: Timing is everything. Time-oriented clinical information systems. West. J. Med. **168** (1998) 105-113

Author Index

Lecture Notes in Artificial Intelligence (LNAI)

Lecture Notes in Computer Science